W0051052

SURFACE PROPERTIES OF SEMICONDUCTORS AND DYNAMICS OF IONIC CRYSTALS

POVERKHNOSTNYE SVOISTVA POLUPROVODNIKOV I DINAMIKA IONNYKH KRISTALLOV

ПОВЕРХНОСТНЫЕ СВОЙСТВА ПОЛУПРОВОДНИКОВ И ДИНАМИКА ИОННЫХ КРИСТАЛЛОВ

The Lebedev Physics Institute Series

Editor: Academician D. V. Skobel'tsyn

Director, P. N. Lebedev Physics Institute, Academy of Sciences of the USSR

Volume 25	Optical Methods of Investigating Solid Bodies
Volume 26	Cosmic Rays
Volume 27	Research in Molecular Spectroscopy
Volume 28	Radio Telescopes
Volume 29	Quantum Field Theory and Hydrodynamics
Volume 30	Physical Optics
Volume 31	Quantum Electronics in Lasers and Masers
Volume 32	Plasma Physics
Volume 33	Studies of Nuclear Reactions
Volume 34	Photomesonic and Photonuclear Processes
Volume 35	Electronic and Vibrational Spectra of Molecules
Volume 36	Photodisintegration of Nuclei in the Giant Resonance Region
Volume 37	Electrical and Optical Properties of Semiconductors
Volume 38	Wideband Cruciform Radio Telescope Research
Volume 39	Optical Studies in Liquids and Solids
Volume 40	Experimental Physics: Methods and Apparatus
Volume 41	The Nucleon Compton Effect at Low and Medium Energies
Volume 42	Electronics in Experimental Physics
Volume 43	Nonlinear Optics
Volume 44	Nuclear Physics and Interaction of Particles with Matter
Volume 45	Programming and Computer Techniques in Experimental Physics
Volume 47	Radio Astronomy: Instruments and Observations
Volume 48	Surface Properties of Semiconductors and Dynamics of Ionic Crystals
Volume 49	Quantum Electronics and Paramagnetic Resonance

In preparation

Volume 46	Cosmic Rays and Nuclear Interactions at High Temperatures
Volume 50	Electroluminescence
Volume 51	Physics of Atomic Collisions
Volume 52	Quantum Electronics in Lasers and Masers, Part 2

Proceedings (Trudy) of the P. N. Lebedev Physics Institute

Volume 48

SURFACE PROPERTIES OF SEMICONDUCTORS AND DYNAMICS OF IONIC CRYSTALS

Edited by
Academician D. V. Skobel'tsyn
Director, P. N. Lebedev Physics Institute
Academy of Sciences of the USSR, Moscow

Translated from Russian

Translation edited by Albin Tybulewicz

CONSULTANTS BUREAU
NEW YORK–LONDON
1971

The Russian text was published by Nauka Press in Moscow in 1969 for the Academy of Sciences of the USSR as Volume 48 of the Proceedings (Trudy) of the P. N. Lebedev Physics Institute. The present translation is published under an agreement with Mezhdunarodnaya Kniga, the Soviet book export agency.

Library of Congress Catalog Card Number 77-136983
SBN 306-10854-2

ISBN 978-1-4684-1580-3 ISBN 978-1-4684-1578-0 (eBook)
DOI 10.1007/978-1-4684-1578-0

A Division of Plenum Publishing Corporation
227 West 17th Street, New York, N. Y. 10011

United Kingdom edition published by Consultants Bureau, London
A Division of Plenum Publishing Company, Ltd.
Davis House (4th Floor), 8 Scrubs Lane, Harlesden, NW10 6SE, England

CONTENTS

Contents

INVESTIGATION OF THE NATURE OF THE DOMINANT RECOMBINATION CENTERS ON THE REAL SURFACE OF GERMANIUM *

Yu. F. Novototskii-Vlasov

A method combining the large-signal sinusoidal field effect and the steady-state photoconductivity was developed for the investigation of the parameters of fast surface states and of surface recombination centers in a wide range of values of the surface potential. This method can be used to investigate the kinetics of changes in these parameters. An experimental investigation was made of the changes in the surface properties of germanium due to exposure to oxygen, ozone, water vapor, and several other liquids, as well as the changes due to heating in vacuum and in various gaseous media. This investigation was used as the basis of a model of surface recombination centers, which was in good agreement with all the results obtained by the present author and with the published data. This model was employed to develop a method for obtaining low and stable values of the surface recombination velocity on germanium.

Introduction

Surface recombination plays an important and sometimes decisive role in the operation of many semiconductor devices. If the geometrical dimensions of a device are less than the diffusion length of the excess carriers, every change in the surface recombination velocity will give rise to a change in the device characteristics and to instability. Investigations of surface recombination have become especially important in recent years in connection with the development of microelectronics. The effect of the surface on the characteristics of microminiature elements is much greater than the influence on conventional semiconductor devices. The stability of the characteristics is essential to the improvement of the reliability of the operation of microminiature elements which are used in exceedingly large numbers in complex electronic circuits.

Investigations of the nature of surface recombination centers should provide technologists and device designers with an opportunity to control the concentration of these centers and thus obtain stable and low values of the surface recombination velocity.

*Based on a thesis submitted for the degree of Candidate of Physico-Mathematical Sciences, defended in May 1964, at the P. N. Lebedev Physics Institute, Academy of Sciences of the USSR. Scientific supervisor A. V. Rzhanov, Corresponding Member of the Academy of Sciences of the USSR.

Studies of the surface recombination processes are not only of practical importance. The surface is a very convenient medium for the investigation of the influence of various impurities and defects on the properties of semiconductors. Simple methods involving changes in the ambient atmosphere or the field effect can be used to alter the position of the surface Fermi level within a wide range of values. In the bulk of a semiconductor such changes in the Fermi level are impossible at a constant temperature and this makes it necessary to investigate batches of many samples.

Moreover, the concentration of recombination impurities such as copper, gold, iron, and nickel cannot be altered in the bulk of a semiconductor during measurements. This complicates studies of these recombination impurities to the extent that considerable progress has been made only in the study of the mechanism of recombination at radiation defects because such defects can be controlled during measurements.

The processes of adsorption and desorption allow us to investigate surface recombination at various impurities by removing some impurities and depositing others and thus investigating the effect of a wide range of different impurities on the same sample. A warning must be added that this ease of deposition of impurities on the surface requires a very careful control of the ambient medium and all stages of the surface treatment.

Investigations are complicated considerably by the presence of a "background" of recombination centers of unknown origin which are introduced during chemical treatments. Therefore, many investigators have prepared atomically clean surfaces and have investigated the influence of the adsorption of various impurities on the properties of such a surface. At first, the purpose of these investigations has been to model the real surface by starting from a clean surface. Investigations of the real surface have been regarded as studies of contaminated conditions. However, it is now clear that clean and real surfaces are independent objects and that investigations of real surfaces are of fundamental interest. On the other hand, although the real surface of a semiconductor carries a considerable concentration of recombination centers of unknown origin, the parameters of these centers are usually fully reproducible and independent of the etching method. This shows that there is nothing accidental in the appearance of these centers. The purpose of the investigation reported in the present paper was to determine the nature of these surface recombination centers.

The author is deeply grateful to the Head of the Laboratory for the Physics of Semiconductors, B. M. Vul, for his lively interest in this investigation and for the valuable critical comments which strongly shaped this investigation. Thanks are also due to A. V. Rzhanov, who supervised this work.

The author is much indebted to his colleagues I. G. Neizvestnyi, S. V. Pokrovskii, and T. I. Galkin for their help in this investigation and the atmosphere of creative discussion which benefited this investigation. The author offers his sincere thanks to Senior Laboratory Technician G. A. Balandina for her great help in the experiments and her painstaking analysis of the results, as well as to all the staff of the Laboratory for the Physics of Semiconductors for their continuous friendly help which enabled the author to complete his investigation.

CHAPTER I

Electron Processes on the Surface of a Semiconductor

1. Surface States

The special role played by the surface of a semiconductor is due to the presence of surface states which are located in the forbidden band. The electrons in these states are localized near the surface and cannot move along the crystal. The surface states can be divided into two kinds, according to their origin:

a) intrinsic states due to the termination of the lattice of an ideal crystal, which include the Tamm [1] and Shockley [2] states;

b) states due to the presence of structure defects or impurities on the surface, for example, the states due to the presence of an oxide film, adsorbed gas molecules, atoms, and ions.

The possibility of the existence of surface states due to the termination of the periodic lattice of a crystal was demonstrated first by Tamm [1]. He considered the one-dimensional Kronig–Penney model of a crystal in which the periodic field was represented by an infinite series of square wells. Tamm took into account the influence of the surface by assuming that the series of square wells was bounded on one side by an infinitely high potential wall. Intrinsic surface states of a different kind were considered by Shockley [2], who also used a one-dimensional model. Shockley showed that surface states could exist even in the absence of separate potential wells on the surface, i.e., even in the case of overlap of the energy bands corresponding to different atomic levels. In the Tamm and Shockley models, each surface atom has two energy levels so that the density of intrinsic surface states is about 10^{15} cm^{-2}.

There is as yet no direct experimental proof of the existence of intrinsic surface states on such materials as germanium and silicon although these states are frequently used in explanations of experimental results [3, 4]. The most convincing proof of the existence of these states is provided by investigations of the properties of atomically clean surfaces.

Such surfaces can be obtained, for example, by bombarding the surface with argon ions [5] or by cleaving a crystal in very high vacuum [6]. The density of states on such surfaces, estimated from measurements of the field effect [7, 8], is fairly high (it may reach $10^{13}-10^{14}$ cm^{-2}) but it is definitely lower than the density of surface atoms. These states may be associated with structure defects generated during the treatments used to obtain an atomically clean surface. The discrepancies between the electrical properties of surfaces prepared by different methods provide an indirect confirmation of this conclusion. For example, a degenerate p-type layer is formed on a surface prepared by ion bombardment [9], whereas an n-type layer is found on a surface obtained by cleaving in vacuum [6].

The theory and experimental data may differ also because surface atoms on atomically clean germanium and silicon are displaced from their positions in the lattice. This effect is due to the mutual saturation of the free valence bonds of surface atoms [5].

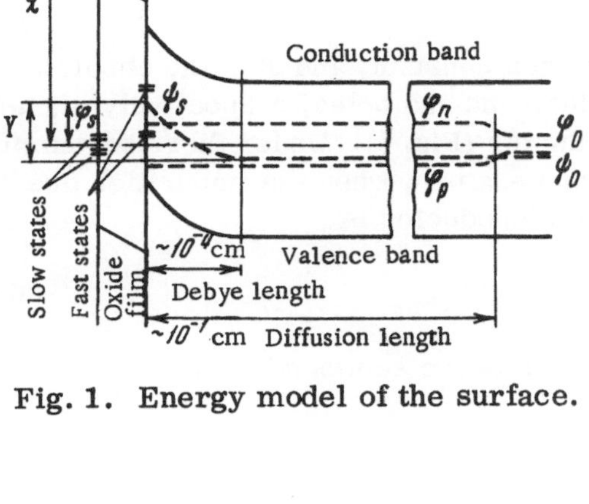

Fig. 1. Energy model of the surface.

Under normal conditions, there is always an oxide film on the surface of such semiconductors as germanium and silicon. In such cases, we have, in addition to surface states at the crystal-oxide interface, additional surface states on the outer boundary of the oxide film (Fig. 1). The former states are called internal or "fast"; the latter are called external or "slow." This terminology is used because the residence times of electrons at these two types of state differ very appreciably, being less than 10^{-7} sec for the fast states and over 10^{-2} sec for the slow states.

In principle, the fast states may be intrinsic. An oxide film can be regarded as a semiconductor with a wide forbidden band. Pratt [10] solved the problem of an interface between two crystals and found that, as in the case of the boundary between a crystal and vacuum [1, 2], interfacial states may, under some conditions, exist in the common forbidden band of both crystals and that the density of these states is equal to the density of atoms at the interface. However, numerous experimental data [4, 11-20] demonstrate that the density of the fast states on the surface of germanium is 10^{11}-10^{12} cm^{-2}. Thus, the nature of the fast states cannot be explained by a change in the periodic structure at the boundary of a crystal, and is due to imperfections at the semiconductor-oxide interface [21], due to impurity atoms remaining at this interface after etching [22], or due to adsorbed atoms [23].

The slow states are associated with the presence of an oxide film on the surface of a semiconductor. These states are not found on atomically clean surfaces [6]. The slow states are due to the presence of adsorbed atoms, molecules, or ions on the surface of the oxide film [24-27]. The energy positions and densities of the slow states vary within a wide range when the ambient medium is altered. Usually, their density exceeds 10^{13} cm^{-2} [22, 28-30].

2. Surface Conductivity

The concept of surface states was used successfully by Bardeen [31] to explain the experimental observation that the contact potential was independent of the volume concentration of a dopant [32] and that the rectifying characteristics of point-contact diodes were independent of the work function of a metal [33]. The presence of surface states gives rise to a space-charge layer at the surface and the charge in this layer is equal in magnitude and opposite in sign to the charge in the surface states. In metals, such a layer is concentrated in a thickness of the order of 10^{-8} cm because of the high density of the electron gas. The density of free carriers in semiconductors is considerably lower than in metals and, consequently, the space-charge layer extends fairly deeply (10^{-5}-10^{-4} cm).

The distribution of the electrostatic potential in a surface space-charge layer is given by Poisson's equation

$$\nabla^2\psi = -\frac{4\pi}{\varepsilon}\rho, \tag{1}$$

where ψ is the electrostatic potential; ε is the permittivity of the semiconductor; ρ is the charge density. The solution of this problem is given by Kingston and Neustadter [34] and by Garrett and Brattain [35]. We shall follow the more widely used variant of Garrett and Brattain [35].

The densities of holes and electrons in a nondegenerate semiconductor can be represented as follows:

$$p = n_i \exp[\beta(\varphi_p - \psi)],$$
$$n = n_i \exp[\beta(\psi - \varphi_n)]. \tag{2}$$

Here, $\beta = q/kT$ (q is the electronic charge, k is the Boltzmann constant, and T is the absolute temperature); φ_p and φ_n are the quasi-Fermi levels of electrons and holes, respectively; n_i is the density of electrons and holes in an intrinsic semiconductor (Fig. 1). Under thermodynamic equilibrium conditions, we have $\varphi_p = \varphi_n = \varphi_0$. At room temperature, when the impurities are completely ionized, the charge density at any point in a semiconductor is

$$\rho = q[(p - p_0) - (n - n_0)], \tag{3}$$

where n_0 and p_0 are the electron and hole densities in the bulk of the semiconductor.

It is assumed that φ_p and φ_n are constant throughout the space-charge layer. This assumption is correct if:

a) the diffusion length of the minority carriers in a semiconductor is large compared with the thickness of the space-charge layer;

b) the difference between the electrostatic potentials of the surface and the interior does not exceed 15 kT/q [36].

An additional restriction is due to the inapplicability of Eq. (2) in the space-charge region. The error committed by the substitution of the average electrostatic potential in the Boltzmann equation, i.e., by ignoring statistical fluctuations of this potential, increases with increasing deviation of the average electrostatic potential from its value in the bulk of the semiconductor. The expressions in Eq. (2) are valid when

$$\frac{q^2}{\varepsilon L'} \ll kT, \tag{4}$$

where $L' = \left(\frac{\varepsilon}{2\pi N q \beta}\right)^{1/2}$ and N is the highest local density of electrons or holes. For germanium at 300°K we can write Eq. (4) in the form

$$N \ll 1.6 \cdot 10^5 n_i \tag{5}$$

or, using the limit of validity of Eq. (2) in the form

$$N = 1.6 \cdot 10^4 n_i, \tag{6}$$

we can find the values of the electrostatic potential above which the expressions in Eq. (2) cease to be valid:

$$(\varphi_p - \psi) = (\psi - \varphi_n) = 9.7 \quad kT/q. \tag{7}$$

Substituting the two expressions from Eq. (2) into Eq. (3), we obtain

$$\rho = q n_i \{\exp[\beta(\psi_0 - \varphi_n)] - \exp[\beta(\varphi_p - \psi_0)] + \exp[\beta(\varphi_p - \psi)] - \exp[\beta(\psi - \varphi_n)]\}, \tag{8}$$

where ψ_0 is the value of the electrostatic potential in the interior of the semiconductor. We shall assume that ψ and ρ vary only along a direction perpendicular to the semiconductor surface and we shall take this direction to be the x axis. Then, integrating Eq. (1), we obtain

$$\left(\frac{d\psi}{dx}\right)^2 = -\frac{8\pi}{\varepsilon}\int_{\psi_0}^{\psi} \rho d\psi \tag{9}$$

and, substituting Eq. (8) into Eq. (9), we find that

$$\frac{d\psi}{dx} = \frac{2}{\beta L} F(y, \lambda, P, N), \tag{10}$$

where

$$L = \left(\frac{\varepsilon}{2\pi q n_i \beta}\right)^{1/2}, \qquad \lambda = \frac{p_0}{n_i} = \frac{n_i}{n_0},$$

$$P = \beta(\varphi_p - \varphi_0), \qquad N = \beta(\varphi_n - \varphi_0), \qquad y = \beta(\psi - \psi_0),$$
$$F(y, \lambda, P, N) = \mp [\lambda e^P (e^{-y} - 1) + \lambda^{-1} e^{-N} (e^y - 1) + (\lambda - \lambda^{-1}) y]^{1/2},$$

the negative sign of the root being taken for y > 0 and the positive root for y < 0.

Using Eq. (10), we can derive expressions for the excess densities of holes and electrons Γ_p and Γ_n. By the excess density we mean the difference between the total number of carriers in a real system and their number in the absence of a space-charge layer. Using Y and ψ_s to denote the values of y and ψ on the surface, we can write

$$\Gamma_p = \int_0^\infty (p - p^*)\, dx = -\int_0^Y (p - p^*) \left(\frac{dx}{dy}\right) dy = -{}^1/_2 n_i L \lambda e^P \int_0^Y \frac{e^{-y} - 1}{F(y, \lambda, P, N)}\, dy, \tag{11}$$
$$\Gamma_n = \int_0^\infty (n - n^*)\, dx = -{}^1/_2 n_i L \lambda^{-1} e^{-N} \int_0^Y \frac{e^y - 1}{F(y, \lambda, P, N)}\, dy,$$

where n^* and p^* are the steady-state values of the electron and hole densities outside the space-charge region.

The total space charge in the surface layer, due to the presence of excess electrons and holes, is given by:

$$q(\Gamma_n - \Gamma_p) = -q n_i L F(y, \lambda, P, N). \tag{12}$$

The change in the surface conductivity due to the presence of the space-charge layer is given by

$$\Delta G = q\mu_p (\Gamma_p + b\Gamma_n), \tag{13}$$

where μ_p is the mobility of holes and b is the ratio of the electron and hole mobilities. Substituting Eq. (11) into Eq. (13), we obtain

$$\Delta G = {}^1/_2 q \mu_p n_i L \lambda^{-1/2} g, \tag{14}$$

where

$$g = \lambda^{1/2} \int_Y^0 \frac{\lambda(e^{-y} - 1) + b\lambda^{-1}(e^y - 1)}{F(y, \lambda)}\, dy. \tag{15}$$

The values of g for an n-type semiconductor are given in Fig. 2. When the surface potential is higher than the volume potential (Y > 0), the surface conductivity is positive and increases with increasing value of this potential. This is due to an increase in the electron density near the surface (accumulation layer). When the surface potential is a little less than the volume potential, the surface conductivity becomes negative because the electron density in the surface layer is low (depletion layer). When the surface potential is reduced still further, the conductivity increases because the density of holes in the surface region rises (inversion layer). The surface conductivity minimum is observed at the following value of the surface potential:

$$Y = \ln\left(\frac{\lambda^2}{b}\right). \tag{16}$$

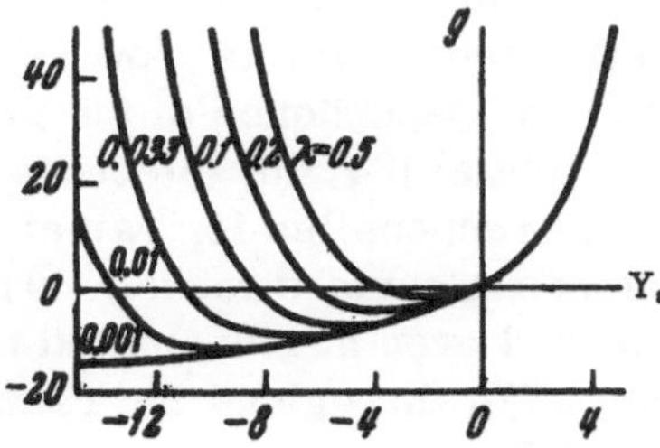

Fig. 2. Integral of the surface conductivity as a function of the surface potential for various values of the parameter λ.

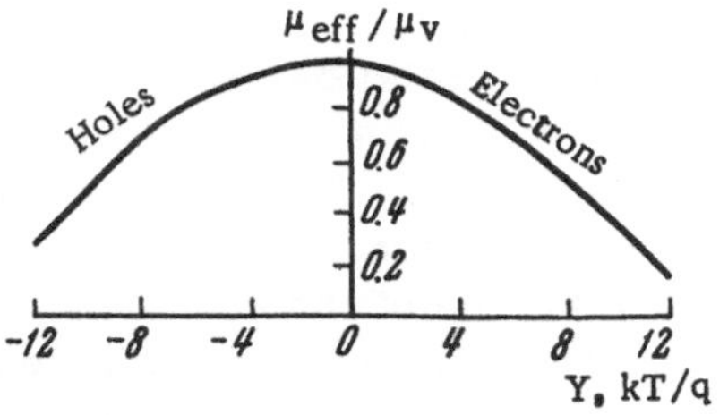

Fig. 3. Ratio of the effective and volume mobilities as a function of the surface potential (according to Schrieffer [37]).

Garrett and Brattain [35] ignored the changes in the carrier mobility which occur in the space-charge layer due to the additional scattering by the surface. The surface mobility was calculated first by Schrieffer [37] (Fig. 3), who showed – by analogy with metals – that the scattering of carriers by the surface of a semiconductor is completely diffuse. The experimental data on the field effect [18, 38] are basically in agreement with Schrieffer's calculations. However, some investigators have pointed out that at high values of the surface potential the formulas given by Schrieffer [37] yield values of the mobility which are too low [39, 40]. These discrepancies have been attributed to the incompletely diffuse nature of the scattering of carriers by the surface. However, further refinements of Schrieffer's calculations [41-43] have demonstrated that the discrepancies can be explained by the contribution of the light holes to the mobility. Good agreement between the experimental results and the theoretical calculations given in [41] confirm that carriers are scattered diffusely by the surface [44].

The surface potential can be altered by changing the gaseous environment. This alters the distribution and density of the slow states which govern the position of the Fermi level at the surface [31]. Brattain and Bardeen investigated the contact potential [28] and showed that the successive introduction of ozone, dry and wet oxygen, as well as wet nitrogen into the atmosphere surrounding a sample of germanium could be used to obtain a wide range of changes in the contact potential difference between the germanium and a standard platinum electrode. Further experiments [21] showed that the surface potential in a gas cycle of this type varied within the range 0.3-0.4 V. Exposure to an ozone atmosphere produced a p-type surface layer. Chlorine and H_2O_2 vapor had a similar effect. Water, alcohol, and ammonia vapors had the opposite effect because they produced an n-type surface layer.

The Brattain–Bardeen cycle method can be used conveniently to study surface states. These states can be investigated by measuring the field effect. The field effect represents the generation of an induced charge in a semiconductor when an electric field is applied normally to its surface. The application of the field produces initially a flux of carriers (mainly majority carriers) which flows through the contacts into or from a sample, depending on the sign of the charge induced in the sample (the contacts are assumed to be ohmic). This process takes place in a time interval comparable with the dielectric relaxation time (10^{-12} sec). After this stage, the whole charge appears simply as a change in the carrier density in the space-charge layer. Then, the capture or liberation processes establish an equilibrium between the majority-carrier band and the fast states in a time interval of the order of 10^{-8} sec. In the case of the formation of an inversion layer, the relaxation process also includes the recombination or generation of electrons and holes, which takes place at the same fast surface states or at volume recombination centers. In this case, the conductivity relaxation time is comparable with the effective minority-carrier lifetime in the sample.

The fast relaxation process, involving the induced charge, is followed by the interaction of the slow states (slow conductivity relaxation). A characteristic feature of this slow relaxation is its nonexponential nature. This can be explained either by the dependence of the probability of electron transitions to the slow states on the surface potential [24, 45] (electron transition model), or by the inhomogeneity of the surface, which is responsible for variation of the relaxation time from one region to another [11, 46] (heterogeneous surface model). The nonexponential nature of the slow conductivity relaxation is observed even at low applied transverse fields [11], when the surface potential is constant to within kT/q throughout the relaxation process, and it evidently supports the inhomogeneous surface model.

A new interpretation of the slow relaxation on germanium has appeared recently [47]: the experimental observations are accounted for by assuming a uniform distribution of slow states across the oxide film thickness. However, this explanation is incompatible with the strong dependence of the slow relaxation time on the oxide film thickness [22, 48, 49], which tends to support the view that the slow states are located on the outer side of the oxide film. An increase in the oxide film thickness is attributed in [48, 49] to dehydration of the surface, and some of the published data indicate that such dehydration increases the time constant of the slow relaxation [11, 30, 50, 51]. This is interpreted in [50, 51] as the transformation of fast states into slow, and conversely.

The field-effect investigations have yielded information on the distribution of charge in the fast states. If the surface conductivity is modulated by a transverse electric field, a surface conductivity minimum [Eq. (16)] can be used to calculate the charge in the fast surface states by comparing the experimentally determined dependence of the surface conductivity on the induced charge with the theoretical dependence [52]. The induced charge Q_n is split into contributions to the mobile charges in the space-charge region Q_{sc} and to the static charges in the surface states Q_{ss}. If the minima of the theoretical and experimental curves are made to coincide along the ordinate, the difference between the abscissas of these curves, at a given value of δG, is equal to Q_{ss} considered as a function of the surface potential Y. Thus, using this method (described by Brown in [52]), we can deduce the dependence $Q_{ss}(Y)$ from measurements of the field effect.

The density of the carriers in the fast states is given by the Fermi distribution function

$$Q_{ss} = -qN_t/\left[1 + \exp\left(\frac{E_t - E_i - q\varphi_s}{kT}\right)\right] \tag{17}$$

if there is only one group of states, or by the sum of such distributions if there are several groups of states. It is assumed in Eq. (17) that the surface states are not spin-degenerate. Double degeneracy is assumed in [22, 53]. The number of fast states is much lower than the number of surface atoms. Hence, it is concluded that the fast states are localized and, because of the electrostatic interaction, the second electron in each state differs in energy from the first electron. Therefore, each state can be occupied only by one electron so that the distribution function is:

for donor states

$$f = 1/\left[1 + {}^{1}/{}_{2}\exp\left(\frac{E_t - E_i - q\varphi_s}{kT}\right)\right],$$

and for acceptor states

$$f = 1/\left[1 + 2\exp\left(\frac{E_t - E_i - q\varphi_s}{kT}\right)\right].$$

However, changes due to the inclusion of the factor $1/2$ or 2 in Eq. (17) are not very important because the differences due to this factor lie within the limits of the experimental error and the actual form of the distribution function cannot be deduced from measurements of the field effect [17].

3. Surface Recombination

The concept of surface recombination was introduced by Shockley [54]. He assumed that, at the boundary of a sample, the difference between the number of recombining minority carriers and the number of generated carriers is directly proportional to their density, and that the coefficient of proportionality is s. Thus, the surface recombination velocity can be defined as

$$s = \frac{u}{\Delta p}, \tag{18}$$

where u is the recombination rate given by the number of electron-hole pairs per unit surface area per unit time; Δp (or Δn) is the excess minority-carrier density in the plane which forms the boundary of the space-charge layer, where $\psi = \psi_0$ (Fig. 1). Equation (18) allows us to replace an analysis of the recombination at centers located on a real surface by an analysis of the process of "leakage" of excess-carrier pairs through this imaginary plane at the boundary of the space-charge layer. This representation describes satisfactorily the case of the uniform generation of carriers across a sample. In the surface generation case, the effective surface recombination velocity given by Eq. (18) may not be equal to the real recombination velocity on the surface of a sample. Bir [55] formulated the criterion of validity of the concept of the effective surface recombination velocity, which is independent of the method of generation of excess carriers. This criterion is satisfied by germanium at all values of the surface potential which can be realized experimentally [56]. Thus, in the case of germanium, we are justified in using the surface recombination velocity defined by Eq. (18).

In the case when the electron and hole distribution functions are nondegenerate and the surface recombination occurs at a local level E_t in the forbidden band, an analysis of the recombination statistics can be used to obtain an expression for the recombination rate, which is similar to that deduced by Shockley and Read [57]:

$$u = \frac{N_t c_p c_n (p_s n_s - n_i^2)}{c_p (p_s + p_{s1}) + c_n (n_s + n_{s1})}, \tag{19}$$

where c_n and c_p are the probabilities of capture by a single center per unit time, given by products of the relevant effective capture cross sections and the thermal velocity of electrons or holes; n_s and p_s are the total densities of free carriers on the surface; n_{s1} and p_{s1} are the surface densities of carriers in the case when the Fermi level passes through the energy level of the recombination centers.

If the volume and surface carriers are in mutual equilibrium or (which is equivalent) if the quasi-Fermi levels are constant in the space-charge region, we always find that in the case of germanium [36]

$$p_s n_s = np \equiv (n_0 + \Delta n)(p_0 + \Delta p), \tag{20}$$

where n_0, p_0 are the equilibrium values and n, p are the total values of the carrier density in the interior.

If there are no trapping centers in the interior or on the surface, which is true of germanium at room temperature, we find that

$$\Delta p = \Delta n \tag{21}$$

and Eq. (19) can be rewritten in the form

$$u = \frac{N_t c_p c_n (n_0 + p_0 + \Delta p)\Delta p}{c_p (p_{s0} + \Delta p_s + p_{s1}) + c_n (n_{s0} + \Delta n_s + n_{s1})}\,. \tag{22}$$

At low injection levels, the quantities Δp, Δp_s, and Δn_s can be neglected compared with n_0, p_0, p_{s0}, n_{s0}, p_{s1}, and n_{s1} and then

$$u = \frac{N_t c_p c_n (n_0 + p_0)\Delta p}{c_p (p_{s0} + p_{s1}) + c_n (n_{s0} + n_{s1})}\,. \tag{23}$$

Using Eq. (18), we obtain

$$s = \frac{N_t c_p c_n (n_0 + p_0)}{c_p (p_{s0} + p_{s1}) + c_n (n_{s0} + n_{s1})} \tag{24}$$

or, introducing the relevant energy levels, we find that

$$s = \frac{N_t c_p c_n (n_0 + p_0)}{n_i \left[c_n \exp\left(\frac{E_t - E_i}{kT}\right) + c_p \exp\left(\frac{E_i - E_t}{kT}\right) + c_n \exp\frac{q\varphi_s}{kT} + c_p \exp\left(-\frac{q\varphi_s}{kT}\right)\right]}\,. \tag{25}$$

Stevenson and Keyes [58] assume that $c_p = c_n$; in this case, Eq. (25) describes a curve which is symmetrical with respect to the point $\varphi_s = 0$. However, this assumption does not correspond to the true situation [59] and, therefore, we shall consider a more general case when $c_p \neq c_n$. It is convenient to introduce an energy level given by the following expression [60–62]:

$$q\varphi_0 = \frac{kT}{2}\ln\frac{c_p}{c_n}\,, \tag{26}$$

so that Eq. (25) can be rewritten to obtain finally

$$s = \frac{N_t (c_p c_n)^{1/2} \frac{n_0 + p_0}{2n_i}}{\mathrm{ch}\left(\frac{E_t - E_i - q\varphi_0}{kT}\right) + \mathrm{ch}\,\frac{q}{kT}(\varphi_s - \varphi_0)}\,. \tag{27}$$

The $s(\varphi_s)$ curve is symmetrical relative to a maximum which occurs at $\varphi_s = \varphi_0$ (Fig. 4). Thus, the position of an experimentally determined $s(\varphi_s)$ curve can be used to determine directly the capture cross-section ratio c_p/c_n. The width of the $s(\varphi_s)$ curve is determined by the

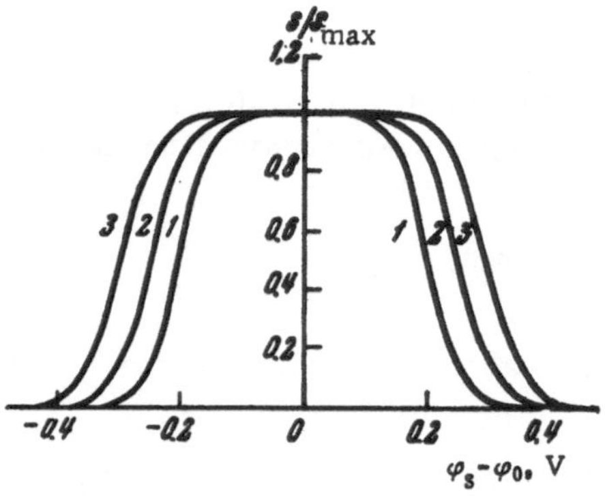

Fig. 4. Surface recombination velocity as a function of the surface potential. The parameter is the energy position of the recombination centers ($E_t - E_i$): 1) 0.21 eV; 2) 0.26 eV; 3) 0.31 eV.

energy position of the level so that, in the case when $\mathrm{ch}\left(\frac{E_t - E_i - q\varphi_0}{kT}\right) \gg 1$, we can approximately write that at the point $s = 0.5\, s_{max}$ we have

$$E_t - E_i - q\varphi_0 = \pm q\,(\varphi_s - \varphi_0), \tag{28}$$

i.e., the $s(\varphi_s)$ curve at a given temperature corresponds to two possible values of $(E_t - E_i)$. The amplitude of the $s(\varphi_s)$ curve is given by the product $N_t\,(c_p c_n)^{1/2}$. Thus, the $s(\varphi_s)$ curve yields three equations for finding four parameters of a recombination center (N_t, E_t, c_p, c_n) which define fully such a center.

The missing fourth equation is found by analysis of the $Q_{ss}(Y)$ curve. Simultaneous measurements of the field effect and of the surface recombination velocity have indicated [63-65] that the fast states manifested in the field effect are identical with the centers which contribute to the surface recombination velocity. Determination of the slope of the $Q_{ss}(Y)$ curve at the point corresponding to the energy position of the recombination centers can be used to determine N_t. We shall consider in Chap. IV how far such analysis of the experimental dependence $Q_{ss}(Y)$ and $s(\varphi_s)$ is justified.

Rzhanov [66] developed his own recombination statistics and used them, as well as the carrier lifetime at excited recombination levels, to deduce a new relationship for the surface recombination velocity [67]:

$$s = N_t\,(c_p\,c_n)^{1/2}\,\frac{n_0 + p_0}{2n_i} \Big/ \left\{\mathrm{ch}\,\frac{q(\varphi_s - \varphi_0)}{kT} + \mathrm{ch}\,\frac{E_t - E_i - q\varphi_0}{kT} + M\left[\mathrm{ch}\,\frac{q(\varphi_0 - \eta)}{kT} + \mathrm{ch}\,\frac{E_t - E_i - q\eta}{kT}\right]\right\}. \tag{29}$$

The new quantities, compared with Eq. (27), are

$$M \equiv \left(\frac{c_p c_n}{r_p r_n}\,p_1^{**}\,n_1^{*}\right)^{1/2}, \tag{30}$$

$$\frac{q\eta}{kT} = {}^1\!/_2 \ln\left(\frac{r_p}{r_n}\,\frac{n_1^{*}}{p_1^{**}}\right), \tag{31}$$

where r_p and r_n are the probabilities of carrier transitions from the excited levels to the ground state of a center;

$$n_1^{*} = n_i \exp\left(\frac{E_t^{*} - E_i}{kT}\right), \qquad p_1^{**} = n_i \exp\left(\frac{E_i - E_t^{**}}{kT}\right)$$

are quantities which represent the energy positions of an excited level capable of capturing electrons ($E_t^{*} - E_i$) and of an excited level capable of capturing holes $E_i - E_t^{**}$).

Thus, the new relationship represents a recombination center by eight independent parameters. Since the experimentally determined $s(\varphi_s)$ curve gives only three conditions for the determination of these parameters, Eq. (29) cannot be used in any practical analysis (it has five arbitrary parameters). Therefore, we shall employ Eq. (27) in our analysis of the experimentally determined $s(\varphi_s)$ curves.

Equation (27) is valid for recombination centers of one type only. If several recombination centers participate in surface recombination and each makes a comparable contribution to the surface recombination velocity, we cannot analyze the total curve into elementary curves of the type given by Eq. (27) unless some additional data are available [68].

The experimental data on the dependence of the surface recombination velocity on the surface potential of germanium are interpreted in different ways. Some investigators explain their results by the presence of recombination centers of one type, whereas others use similar experimental dependences $s(\varphi_s)$ to draw the precisely opposite conclusion, i.e., that there are several centers. This contradiction is due to the fact that practically none of the obtained experimental dependences $s(\varphi_s)$ agrees exactly with a theoretically obtained elementary curve of the type given by Eq. (27).

The first results on the dependences $s(\varphi_s)$ were reported in [28]. It was found that the surface recombination velocity was practically unaffected by changes in the surface potential within a wide range (0.3-0.4 V). A similar weak dependence $s(\varphi_s)$ has been observed in many measurements in gaseous media [60, 69, 70]. Usually the surface recombination velocity measured in such media is either nearly constant in a wide range of values of the surface potential or it increases monotonically with increase in this potential. No maximum has been found in the $s(\varphi_s)$ curves. It is not possible to describe the experimental dependences by means of elementary curves of the type given in Eq. (27).

However, the dependence $s(\varphi_s)$ obtained in vacuum has a maximum. Although this dependence still cannot be described by a single elementary curve, the difference between two experimental curves obtained in a normal atmosphere and in vacuum gives a dependence typical of only one group of recombination centers [70]. The same result is obtained when the ambient atmosphere is altered [60]. This has led many experimentalists to the conclusion that surface recombination on germanium is primarily due to centers of one type [13, 14, 60, 70-74]. An exactly opposite interpretation of the results is given in [20, 65, 75, 76] and in some other papers. The main argument used is that the observed dependence of the surface recombination velocity on the surface potential is less strong than that expected theoretically. It is suggested in [65] that the discrepancy between the experimental and theoretical $s(\varphi_s)$ curves is due to the "patchiness" of the surface. This explanation is supported in [62] and is confirmed indirectly by the nonexponential relaxation of the slow states, which is usually attributed to the inhomogeneous distribution of the surface potential on germanium. Additional data supporting this view are given in Chap. IV.

In spite of the disagreement on whether centers of one type dominate recombination on the surface of germanium, all the investigators are in agreement that the recombination centers are of the acceptor kind. The position of the point s_{max} along the surface potential axis indicates that in practically all the experiments we find that $(c_p/c_n) > 1$ and that the differences between the positions of s_{max} lie within the limits of the experimental error.

The parameters of the $s(\varphi_s)$ curve may vary with the crystallographic orientation of the surface of a sample [76]. For example, in the case of the (111) orientation, the half-width of the recombination curve is 8 kT/q at 300°K, whereas the half-width of the $s(\varphi_s)$ curve in the case of the (100) orientation is only 5-6 kT/q. Since the positions of the maxima of the recombination curves are identical in these two cases, it follows from Eq. (27) that the energy positions of the recombination levels are different for the same values of the ratio of the effective capture cross sections. In the case of the (110) orientation, the recombination curves have several special features: they are strongly asymmetrical, the position of the maximum is shifted, and the temperature dependence of this position differs considerably from the corresponding dependence for the (111) orientation.

The problem of the influence of various chemical etchants on the surface parameters is important in connection with the nature of the surface recombination centers. In these investigations the results obtained by Morrison [77] stand out from all the others. Morrison found that very small amounts of impurities in an etchant or the washing water had a considerable influence on the resultant surface characteristics. Moreover, he found that etching in CP-4 yielded surface characteristics that were very poorly reproducible. The results of Morrison differ essentially from those of other investigators. This difference may be due to the unsatisfactory methods employed by Morrison who did not determine directly the dependence $s(\varphi_s)$ but deduced indirectly the parameters of the recombination centers. Margoninski [17] reported that the reproducibility of the surface parameters obtained after successive treatments in CP-4 was very good and that the same surface treatment always gave the same position of the surface states. Moreover, Margoninski reported that the influence of various chemical etchants on the surface states was surprisingly weak and was usually within the limits of experimental error. This conclusion about the weak dependence of the surface-state parameters on the preliminary chemical treatment of the germanium surface was confirmed by the results of many other investigators [13, 60, 61, 64, 65]. In any case, it was found that variations in the parameters due to changes in the type of etchant or in the crystallographic orientation of the investigated surface were considerably smaller than changes in the same parameters due to the influence of the ambient medium.

Brattain and Bardeen [28] reported that the surface recombination velocity was independent of the ambient medium and they concluded that the concentration of the recombination centers was independent of the medium. However, Rzhanov, Novototskii-Vlasov, and Neizvestnyi [60] and other investigators [63, 70, 72] found that the concentration of the fast centers, including the surface recombination centers, was strongly affected by changes in the ambient medium. The density of the fast surface states on the germanium surface increases severalfold when this surface interacted with ozone [60, 63, 78]. Storage of a sample in vacuum or, better, heating in vacuum, had a similar effect [14, 70, 71, 76, 79].

The data on the nature of the surface recombination centers are very contradictory, with the exception of a series of investigations carried out in our laboratory [13, 14, 18, 20, 50, 76, 80-82]. Thus, it is suggested in [83] that the surface recombination centers are regions covered by an oxide of the GeO_2 type. Other workers [74] associate the fast states with the adsorption of oxygen at the germanium–oxide interface. Boddy and Brattain [84] attribute the fast states to the presence of impurities in the etchant.

Since the nature of the etchant has very little effect on the parameters of the fast states, including the surface recombination velocity, the nature of the recombination centers on real germanium surfaces should be related to those properties which are common to all etchants. The common components of all etchants are water and oxygen and, therefore, investigations of the nature of the surface recombination centers, carried out in our laboratory, were concentrated on the influence of oxygen and water on the surface characteristics of germanium. These investigations will be discussed more fully later in this paper.

CHAPTER II

Investigation Methods

In order to investigate the parameters of surface states and, in particular, the parameters of surface recombination centers, it is necessary to investigate the surface recombination velocity and the charge captured by the surface states as a function of the surface potential in the widest range of values of this potential.

In principle, there are two methods for altering the surface potential: a) the gas cycle method; b) the field effect method.

In the gas cycle method, the surface charge and, consequently, the surface potential are altered by molecules and ions adsorbed on the surface. All the gas cycle techniques are based on the Brattain–Bardeen cycle [28]: wet nitrogen, wet oxygen, dry oxygen, and ozone. Such cyclic variation of the ambient medium allows us to alter the surface potential within wide limits (10-15 kT/q), which are sufficient for the acquisition of the necessary data on the parameters of the capture and recombination centers. However, the ambient medium itself alters the fast-state parameters [60, 61, 70, 77] and this makes the gas cycle method unsuitable for investigations of this type.

1. Field Effect Method

The field-effect method involves the application of an electric field to a capacitor in which one of the electrodes is a semiconductor. The applied transverse electric field induces a charge which is localized near the surface of the semiconductor and which alters the position of the Fermi level in the surface layer.

The field measurements are carried out using direct, sinusoidal, or pulsed voltages. The dc field effect is usually unsuitable for investigating fast states because the external field is screened by the charge located in the slow states and the screening effect is particularly strong in a wet atmosphere. The dc effect is used only to investigate some special cases such as a clean surface in very high vacuum [15] or samples with thick oxide films [48, 49] when the slow states are either absent or have very long relaxation times.

The pulsed field effect is used fairly widely [3, 4, 30, 73, 85-89]. However, the pulsed field-effect method is suitable primarily for measurements in fairly high vacuum ($p \leq 10^{-4}$ mm Hg). This method can be used only with difficulty in gaseous media because the relaxation time of the slow states in such media becomes shorter than that in vacuum. The pulsed field method is completely unsuitable for investigations under conditions of high humidity and the use of pulsed fields can even distort the phenomena under investigation. The "accumulation effect" [60, 85], observed in pulsed fields, results in the charging of the slow states so that the surface potential measured during an interval between two consecutive pulses differs (and the difference increases with the humidity) from the initial potential before the application of a pulsed field. This difference grows with increasing amplitude of the applied pulses. Consequently, an increase in the pulsed field amplitude produces no corresponding change in the surface conductivity. The accumulation effect is observed in high-humidity media even when the off-duty factor is of the order of 1000. Under these conditions, a false maximum can be obtained in the surface conductivity curve [60].

Moreover, the pulsed field effect, like the dc effect, yields isolated points on the surface conductivity curve and, therefore, all the experimental conditions have to be kept rigorously constant for the fairly long time which is required to record a complete conductivity curve. For this reason, the dc and pulsed field effects are unsuitable for investigating fast processes.

The ac field effect is free from the limitations imposed by the slow states with relatively short relaxation times. This method can be used in all cases when the pulsed and dc field effects can be employed but it has the advantage that it can be used also under high humidity conditions. The accumulation effect still occurs [13, 62, 70]. However, the slow states charged during one half-period of the sinusoidal voltage do not discharge completely during the next half-period and, consequently, the "zero" conductivity point (the conductivity in the absence of a transverse electric field) is shifted from the position at which the instantaneous transverse field is equal to zero. However, the whole field-effect curve shifts with the "zero" point and there is no distortion in the dependence of the surface conductivity on the induced charge, which would be observed in the case of the accumulation of charge in pulsed fields. The conductivity in the absence of a transverse field and the initial surface potential can be determined easily by simultaneous measurements of the field effect and of the effective lifetime of carriers.

In contrast to the dc and pulsed field effects, the ac effect allows us to investigate the kinetics of the processes taking place on the surface of a semiconductor.

2. Measurement of the Surface Recombination Velocity

The method for measuring the surface recombination velocity should be compatible with the ac field effect, i.e., it should enable us to measure the value of the surface recombination velocity at any value of the surface potential established by the application of a transverse electric field. The requirement of compatibility with the field effect eliminates many well-known methods for measuring the effective carrier lifetime and limits the applicability of other methods.

The method of Navon, Bray, and Fan [90] is inapplicable because high (up to tens of volts) pulse voltages have to be applied to a sample. Since the sample also acts as the capacitor electrode in the field effect, the application of a strong longitudinal pulsed field would induce a charge near the surface. Moreover, this charge would be induced nonuniformly along the sample. In this method, the longitudinal field pulses are of the same order of magnitude as the amplitude of the transverse field in the field effect and, therefore, the longitudinal pulses much distort the field-effect curve as well as the lifetime measurements.

The method of Many [91, 92] differs from the Navon–Bray–Fan method in the application of lower pulse voltages (which are of the order of a few volts). However, even in this case, measurements near zero transverse voltages ($U_t = 0$) are as unreliable as in the Navon–Bray–Fan method. The other difficulty encountered in the use of the Many method is this: if the surface recombination velocity depends strongly on the applied transverse electric field, the effective lifetime changes considerably during a measuring pulse and this gives rise to a non-exponential decay of the excess conductivity. The only solution is to measure the effective lifetime at points where the transverse field reaches its maximum and dU_t/dt has its minimum value [70]. This procedure is very inconvenient. The application of a measuring pulse at other moments of time [13] distorts the results, particularly at the points $U_t = 0$ where dU_t/dt has its maximum value. This explains the strong dependence of the surface recombination velocity on the amplitude of the transverse ac field reported in [13]. Subsequent investigations, carried out using the steady-state photoconductivity method, have shown that there is no such dependence. Figure 5 shows the results of measurements in which the amplitude of the transverse electric field was varied by a factor of 6. We can see that the points obtained in different fields coincide within the experimental error.

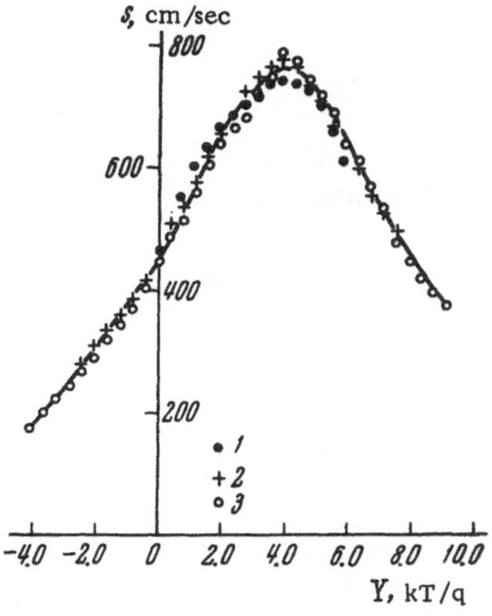

Fig. 5. Dependence of the surface recombination velocity on the surface potential for different amplitudes of the transverse field. Sample No. 14, p-type, after heating in vacuum at 350°K. U_{eff} (V): 1) 22; 2) 56; 3) 120.

Finally, the bridge method of Many and the field effect have incompatible requirements in respect of the contacts with a sample: in the field effect we need ohmic contacts, whereas in the Many method a single injecting contact is required. It is very difficult to find a compromise solution between these requirements, particularly in investigations of the temperature dependence [93]. Attempts to bypass this difficulty by the use of three contacts (one injecting and two ohmic [13]) have proved unsatisfactory because the symmetry of the injecting pulse is altered and the decay of the excess conductivity becomes non-exponential.

The photoconductivity decay method is free of the numerous disadvantages of the Many method but it is still necessary to measure the effective lifetime at the points corresponding to maximum values of the transverse field.

The photogalvanomagnetic effect [94] is unsuitable because the frequency requirements of this effect and of the sinusoidal ac field effect are incompatible.

The galvanomagnetic effect [95, 96], which seems to be ideal for direct measurements of the surface recombination velocity, is unsuitable for investigations in a wide range of conditions because of the following special limitations:

a) this effect is strong in intrinsic semiconductors but weak in extrinsic materials even if their resistivities are high;

b) this method requires different surface treatments of the two large faces of a sample and the recombination velocity on one of those faces should not vary with time, which presents a considerable difficulty [97].

Finally, we must stress especially that the Navon–Bray–Fan, Many, and photoconductivity decay methods yield isolated points on a curve and are unsuitable for investigation of the kinetics of the processes occurring on the surface. For example, in the case of the most widely used of these methods – the Many technique [13, 17, 70, 98, 99] – about one hour is required for recording the dependence of the surface recombination velocity on the surface potential and during this hour the characteristics of the surface must be kept constant.

The steady-state photoconductivity method is free of all the disadvantages characterizing the methods discussed in this section. The special advantage of this method is the possibility of investigating the kinetics of the processes occurring on the surface.

3. Steady-State Photoconductivity Method

Following van Roosbroeck [100], we obtain the following expression for the photoconductivity, which applies in the case of the surface generation of carriers (the absorption coefficient is assumed to tend to infinite):

$$G_{\text{pc}} = q(\mu_n + \mu_p)\tau_0 R \frac{S_2(\operatorname{ch} h/L - 1) + \operatorname{sh} h/L}{(1 - S_1 S_2)\operatorname{sh} h/L + (S_1 + S_2)\operatorname{ch} h/L}, \tag{32}$$

where τ_0 is the volume lifetime; R is the rate of generation of carrier pairs; h is the thickness of a sample; $L = (D\tau_0)^{1/2}$;

$$S_1 = \frac{s_1 L}{D}, \quad S_2 = \frac{s_2 L}{D};$$

D is the diffusion coefficient; s_1 and s_2 are the surface recombination velocities on the illuminated and dark faces, respectively.

Let us consider the case $L \geq 4h$, which corresponds to $L \geq 0.16$ cm for $h \leq 0.04$ cm. Then,

$$\begin{aligned} \operatorname{sh}\frac{h}{L} &= \frac{h}{L} \quad \text{(within 1\%)}, \\ \operatorname{ch}\frac{h}{L} &= 1 \quad (3\%), \\ \operatorname{ch}\frac{h}{L} - 1 &= \frac{1}{2}\left(\frac{h}{L}\right)^2 \quad (0.5\%). \end{aligned} \tag{33}$$

Using Eq. (33), we can rewrite Eq. (32) in the form

$$\frac{1}{G_{\text{pc}}} = \frac{2}{K} \frac{s_1 s_2 h + (s_1 + s_2) D + hD^2/L^2}{h(hs_2 + 2D)}, \tag{34}$$

where $K = q(\mu_n + \mu_p)R$. Equation (34) is identical with Eq. (32) to within 5% for L = 4h; at higher values of L these equations agree even better.

We shall now consider some special cases.

A. $s_1 \neq s_2$, $s_2 \ll 2D/h$. In this case, we have

$$\frac{1}{G_{pc}} = \frac{2}{K}\,\frac{s_1 s_2 h + (s_1 + s_2)D + hD^2/L^2}{2hD} = \frac{1}{K}\left(\frac{s_1 s_2}{D} + \frac{s_1 + s_2}{h} + \frac{1}{\tau_0}\right). \tag{35}$$

The quadratic term can be neglected because

$$\frac{s_1}{D} s_2 \ll \frac{s_1}{D}\,\frac{2D}{h} = \frac{2s_1}{h},$$

and Eq. (35) transforms to

$$\frac{1}{G_{pc}} = \frac{1}{K}\left(\frac{s_1 + s_2}{h} + \frac{1}{\tau_0}\right), \tag{36}$$

i.e., for any value of the surface recombination velocity on the illuminated face, the photoconductivity is proportional to the usual effective carrier lifetime:

$$\tau_{eff} = \left(\frac{s_1 + s_2}{h} + \frac{1}{\tau_0}\right)^{-1} \tag{37}$$

B. $s_1 \neq s_2$, $s_2 \gg 2D/h$. We now obtain

$$\frac{1}{G_{pc}} = \frac{2}{K}\,\frac{s_1 s_2 h + (s_1 + s_2)D + hD^2/L^2}{h^2 s_2} = \frac{2}{K}\left(\frac{s_1}{h} + \frac{D}{h^2} + \frac{s_1 D}{s_2 h^2} + \frac{D^2}{L^2 s_2 h}\right). \tag{38}$$

The third and fourth terms in Eq. (38) are small compared with the first two terms

$$\frac{s_1}{h}\,\frac{D}{s_2 h} \ll \frac{s_1}{h}\,\frac{h}{2D}\,\frac{D}{h} = \frac{s_1}{2h}, \quad \frac{D^2}{L^2 h s_2} \ll \frac{D^2}{h^2 h s_2} \ll \frac{D s_2 h/2}{h^2 h s_2} = \frac{D}{2h^2},$$

and, therefore, Eq. (38) can be reduced to

$$\frac{1}{G_{pc}} = \frac{1}{K}\left(\frac{2s_1}{h} + \frac{2D}{h^2}\right). \tag{39}$$

We can see that s_2 does not occur in the above formula and the photoconductivity is governed solely by the surface recombination velocity on the illuminated face and by carrier diffusion, i.e., the rate of flow of excess carriers to the dark face is limited by diffusion.

C. $s_1 = s_2 = s$. We now find that

$$\frac{1}{G_{pc}} = \frac{2}{K}\,\frac{s^2 h + 2sD + hD^2/L^2}{h(hs + 2D)} = \frac{2}{K}\,\frac{s(hs + 2D) + hD^2/L^2}{h(hs + 2D)} = \frac{2}{K}\left[\frac{s}{h} + \frac{D^2}{L^2(hs + 2D)}\right] = \frac{1}{K}\left(\frac{2s}{h} + \frac{1}{\tau_0}\,\frac{1}{1 + hs/2D}\right). \tag{40}$$

In this case, for any value of the surface recombination velocity, the photoconductivity is very nearly proportional to the effective lifetime given by Eq. (37) except that the second term now has the factor $1/(1 + hs/2D)$. Fortunately, this factor is of no importance. When s is small, this factor is close to unity; and when s is large, the second term in Eq. (40) is small compared with the first and its lower value (because of the above-mentioned factor) has no effect on the quantity in parentheses. Therefore, in the case $s_1 = s_2 = s$ we can write

$$G_{pc} = K\tau_{eff} = K\left(\frac{2s}{h} + \frac{1}{\tau_0}\right)^{-1} \tag{41}$$

Equation (41) is identical with Eq. (40) to within 1% for values of s ranging from zero to infinity, provided the condition $L \geq 4h$ is satisfied.

Thus, when all the faces of a sample are treated in the same way (this case is encountered most frequently), the steady-state photoconductivity is proportional to the usual effective lifetime for any value of the surface recombination velocity. The reason for this difference between the steady-state photoconductivity method and the other methods is due to the fact that excess carriers are generated near surface recombination centers and no time is needed for the transport of carriers by diffusion from the point of generation to the point of recombination.

On the other hand, the requirement $s_1 = s_2$, which is a necessary condition for the effective exploitation of the advantages of the steady-state photoconductivity method, means that the application of a transverse field should produce the same changes in the surface recombination velocity on both the large faces of a sample, i.e., the conditions should be such as to ensure that the field effect is two-sided. The two-sided field effect can be combined with the steady-state photoconductivity requirements by the use of transparent field-effect electrodes.

Cases A and B correspond to the method using the steady-state photoconductivity, employed by Garrett and Brattain [64, 65], in combined measurements of the field effect and of the surface recombination velocity when the field-modulated surface is in darkness and light is incident on the free surface. In this arrangement, all the advantages of the steady-state photoconductivity method are lost because the range of the measured surface recombination velocities is limited by the condition $s \ll 2D/h$, as in all other methods of measuring the effective lifetime. Only after carriers have been generated on the field-modulated surface does the steady-state photoconductivity method enable us to measure any surface recombination velocity. Moreover, if we use the one-sided field effect we must postulate that the surface recombination velocity under the field electrode is equal to the velocity on the free face of a sample. This assumption is acceptable in the case of an air gap in a capacitor and only in the absence of the accumulation effect. If a mica spacer is used and this spacer is in direct contact with the sample, the assumption just stated is not justified and is frequently responsible for considerable errors.

Lyashenko, Snitko, and Sytenko [99] calculated the recombination velocity on a surface covered by an electrode used in the one-sided field effect. They assumed that $s_1 = s_2$ in the absence of a field. They found that $s_2 = 0$ when $Y = +6$ kT/q. Since the curve representing the carrier lifetime, given in [99], showed no tendency to saturation, extrapolation indicated that $s_2 = -80$ cm/sec for $Y = +8$ kT/q. Actually, the surface recombination velocity on the side not covered by the electrode was considerably less.

4. Combined Measurement Method

In this method, we were able to induce large charges on the surface of germanium by ensuring that the field-effect capacitor had a large capacitance. This was done by placing mica sheets, about 10μ thick, between the sample and the field electrodes. Effective sinusoidal voltages of 100-200 V could induce a charge up to 10^{-7} C per unit area of the sample and this made it possible to cover a range of surface potential extending over 12-15 kT/q. The transverse electric field had a frequency of 20-100 cps in order to ensure that, at every moment, the fast surface states were in equilibrium with the allowed energy bands and the slow states were unable to screen the transverse field.

The large-amplitude sinusoidal field effect made it possible to observe, on the screen of an oscillograph, the field-effect curve with a conductivity minimum. Using the Brown

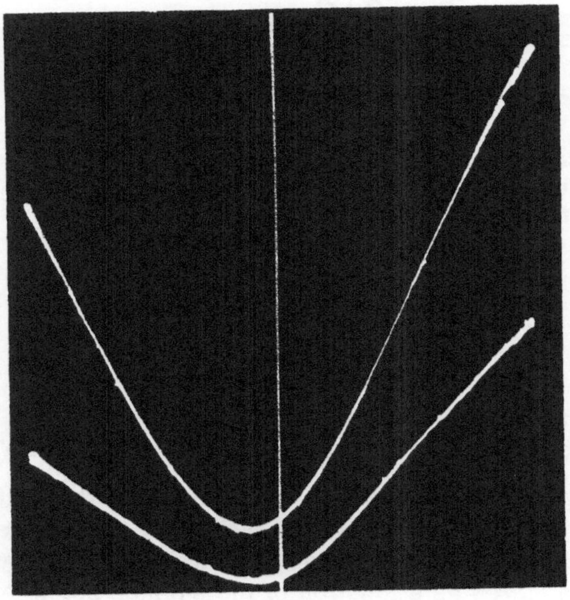

Fig. 6. Oscillograms of changes in the conductivity of a sample due to the application of a transverse electric field in darkness (lower trace) and during illumination (upper trace).

method [52] and the theoretical dependence of the conductivity on the surface potential [35], which was calculated taking into account the change in the carrier mobility near the surface [41], we were able to calibrate the experimentally obtained curve in units of the surface potential. We were also able to calculate the dependence of the charge, captured by the fast surface states, on the surface potential: $Q_{ss}(Y)$.

A sample was illuminated with light modulated at a frequency which was not a multiple of the transverse field frequency. This made it possible to observe simultaneously two field-effect curves on the screen of an oscillograph: the curve in darkness and the curve during illumination (Fig. 6). The duration of the light pulses was about 3 msec, which ensured that the photoconductivity was of the steady-state type. The vertical gap between the two curves corresponded, at a given value of the transverse field, to the photoconductivity in that field [64, 101, 102]. The intensity of illumination was selected so that the value of the electric surface potential during illumination practically coincided with the potential in darkness. The photoconductivity at any value of the field could then be referred to the surface potential found from the field-effect curve in darkness.

The use of this method in the case of large signals [20, 101] was not strictly justified. In fact, when the surface potential during illumination is Y_2 and in darkness is Y_1, the steady-state photoconductivity in the case of an arbitrary signal can be represented in the form

$$G_{pc}(Y_1, Y_2) = K \frac{1}{\Delta Y} \int_{Y_1}^{Y_2} \tau_{eff}(Y)\, dY,$$

where

$$\Delta Y = Y_1 - Y_2.$$

In the case of small signals ($\Delta Y \ll 1$), we can use the approximation

$$Y_1 = Y_2, \quad \tau_{eff}(Y_1) = \tau_{eff}(Y_2).$$

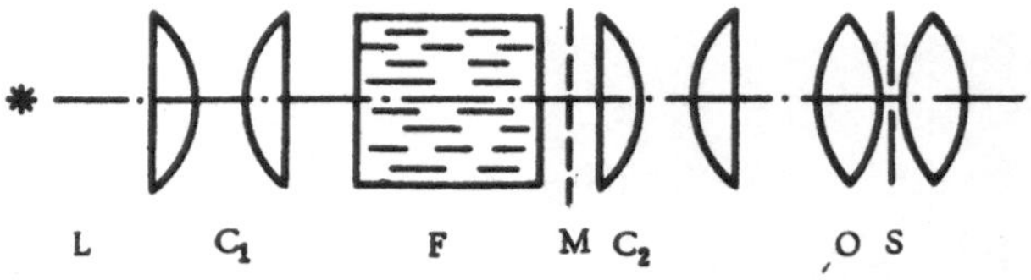

Fig. 7. Optical system: L, STs-65 lamp; C_1, first condenser; F, water filter; M, modulator; C_2, second condenser; O, objective lens; S, optical stop.

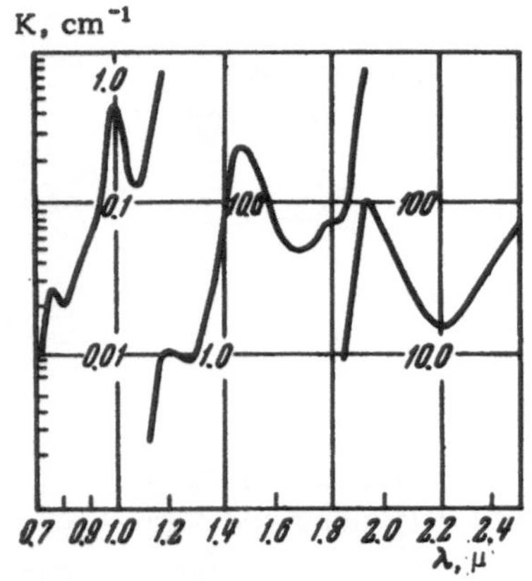

Fig. 8. Absorption spectrum of water.

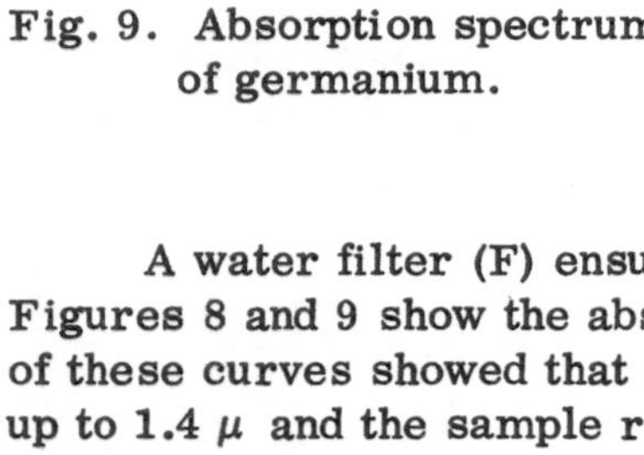

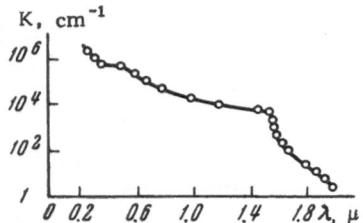

Fig. 9. Absorption spectrum of germanium.

Then,

$$G_{pc}(Y_1, Y_2) \approx G_{pc}(Y_1) = K\tau_{eff}(Y_1).$$

At higher values of ΔY this operation is incorrect [20, 101] and G_{pc} (and, consequently, s) cannot be associated with Y_1 or Y_2.

Using the large-signal ac field effect together with the steady-state photoconductivity, we were able to find the surface recombination velocity and the charge captured by the fast surface states as a function of the surface potential. By photographing the oscillograms we were able to reduce the time required to record the curves representing the state of a sample to the film exposure time, which amounted to 0.1 sec for a sensitive film. This not only eliminated completely the possibility of distortion of the results by slow variation of the spectrum of surface recombination and capture centers (due to adsorption and desorption) but also enabled us to investigate (in some cases) the kinetics of the surface processes.

5. Optical System

Figure 7 shows schematically the optical system employed. The source of light was an STs-65 lamp with a basket-shaped filament, which illuminated a sample uniformly. This lamp (L) had a dc supply in order to avoid modulation of its intensity. The first condenser lens (C_1), with a focal length of 35 mm, directed a flux of light sufficient to produce an easily measurable modulation of the conductivity of an unetched sample or a sample whose surface was bombarded with ions.

A water filter (F) ensured the surface generation of carriers in a germanium sample. Figures 8 and 9 show the absorption spectra of water [103] and germanium [104]. A comparison of these curves showed that a layer of water 5 cm thick absorbed completely infrared radiation up to 1.4 μ and the sample received only that part of the spectrum in which the absorption coefficient of germanium exceeded 10^4 cm^{-1}.

A modulator (chopper), denoted by M in Fig. 7, was located at the focus produced by the first condenser. The linear dimensions of the modulator blades were selected so that light pulses had rise fronts of 5-8% duration compared with the duration of the pulse itself. Consequently, we obtained two different curves (in darkness and during illumination) on the screen of an oscillograph (Fig. 6).

A second condenser (C_2) focused the image of the STs-65 lamp filament in the plane of a stop (S) of an objective (O) to ensure that the use of this stop did not alter the dimensions of the illuminated surface and the uniformity of illumination. A field stop, which ensured that only the middle of the sample was illuminated, was placed in the second condenser. This was done in order to avoid illumination of the contacts, which would have produced a stray photo-emf. Moreover, the illuminated part of the sample should be far from the contacts in order to prevent the extraction of excess carriers (generated by illumination) through the end contacts.

Obviously the longer was the distance between the edge of the illuminated area and the end of a sample, the higher were the driving fields which could be applied to a sample without appreciable extraction of the excess carriers. On the other hand, a reduction in the dimensions of the illuminated strip reduced the signal. We were unable to increase the signal by increasing the intensity of the incident light because the maximum intensity was limited by the requirement of a small ($\Delta Y \leq 0.2$ kT/q) difference between the surface potential during illumination and the surface potential in darkness. A compromise between these two requirements suggested that the illuminated area should be 1 cm wide in the case of a sample 2 cm long.

Table 1 lists the errors in the measurement of the lifetime due to the extraction of the excess carriers by the driving field (these values are given for different driving fields and different lifetimes). The calculations were carried out for a p-type sample at 300°K using the formulas given in [105]. The requirements were not so stringent for n-type samples.

The objective with the stop was used to obtain, in the plane of the sample, an image of the field diaphragm in the second condenser and for smooth variation of the intensity of the incident light. Since, during each experiment, the surface recombination velocity changed by at least one order of magnitude, the intensity of the incident light had to be altered after each treatment in order to ensure, on the one hand, that the change in the surface potential due to illumination was small and, on the other, that the photoconductivity signal was sufficiently large for reliable measurements. Changes in the intensity of the incident light were deduced from the photoconductivity of the investigated sample. If, during the variation of the stop, the effective lifetime in a sample remained constant, the change in the photoconductivity was directly proportional to any change in the intensity of the incident light.

The light flux from the STs-65 lamp was kept constant during a given experiment by stabilizing the voltage across the lamp using a special circuit. The voltage across the lamp

TABLE 1. Fraction of Excess Carriers Extracted by the Field Through a Contact

τ_{eff}, μsec	E, V/cm									
	0.2	0.3	0.4	0.5	0.6	0.7	0.8	0.9	1.0	2.0
30	0.0	0.0	0.0	0.0	0.0	0.0	0.0	0.0	0.0	0.3
50	0.0	0.0	0.0	0.0	0.0	0.1	0.1	0.1	0.2	1.5
75	0.0	0.1	0.1	0.1	0.2	0.3	0.5	0.6	0.8	4.7
100	0.1	0.1	0.2	0.4	0.6	0.8	1.1	1.5	1.9	9.0
150	0.3	0.5	0.8	1.3	1.9	2.6	3.4	4.3	5.4	18.5
200	0.6	1.1	1.7	2.6	3.7	4.9	6.3	7.9	9.7	27.0
250	1.1	1.9	3.0	4.4	6.1	7.9	10.1	12.3	14.6	35.1
300	1.6	2.8	4.4	6.4	8.7	11.1	13.8	16.3	19.0	41.1
400	2.9	5.0	7.7	11.4	14.0	17.6	20.9	24.0	27.4	50.2
500	4.1	7.6	11.1	15.3	19.6	23.6	27.7	31.5	35.2	58.2

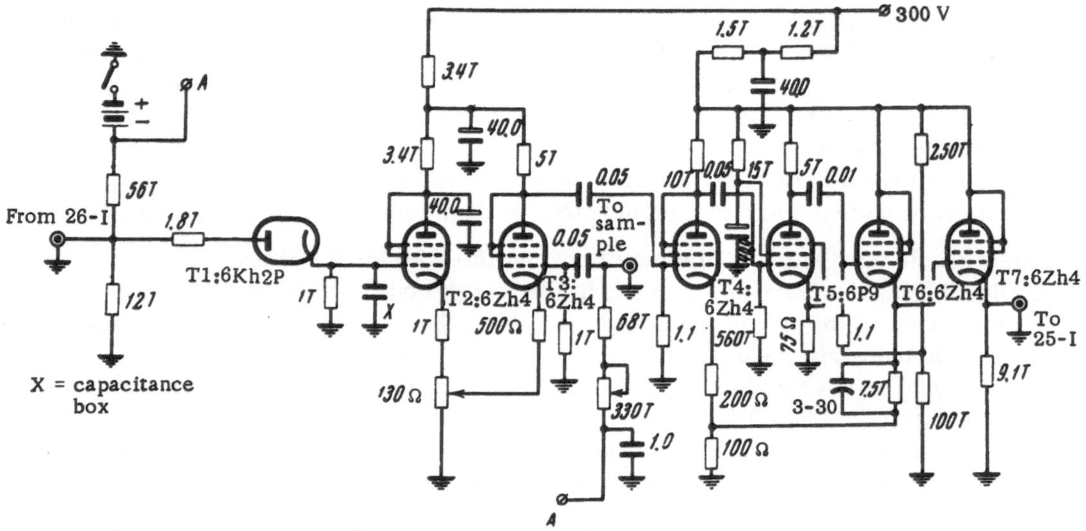

Fig. 10. Basic circuit of the compensated amplifier for measuring the lifetime by the photoconductivity decay method.

was compared with a reference voltage. The unbalance signal was applied to a phase-sensitive amplifier of the UÉ-110 type whose output was connected to an RD-09 motor coupled mechanically to a rheostat in the supply circuit of the STs-65 lamp. The direction of rotation of the motor was determined by the sign of the unbalance signal. This arrangement maintained the voltage across the lamp to within 0.4%, corresponding to a possible oscillation of the intensity within 1%.

The photoconductivity signal was calibrated by measuring the lifetime in the absence of a transverse electric field using the method of the photoconductivity decay. This was done in order to determine the coefficient of proportionality K in Eq. (41).

A pulse-discharge lamp, type IST-10, produced a light pulse with a decay time of about 3 μsec. The pulse repetition frequency was varied within the 2-25 cps range (the repetition frequency was low in order to increase the service life of the lamp). The signal generated by a sample was applied to the grid of an electron tube T3 (Fig. 10). An exponentially varying voltage, produced by a calibrated RC circuit in the cathode of the diode T1, was applied to the cathode of the tube T3 in synchronism with the signal from the sample. A positive pulse from a 26-I generator charged the capacitance of this RC circuit. At the end of the pulse, the diode T1 disconnected the RC circuit from the 26-I pulse generator and the capacitance was discharged through a calibrated resistor of 100 Ω. A capacitance box X was used to alter the time constant of the RC circuit from 0.1 to 100 μsec. The potentiometer in the cathode of the tube T2 was used to regulate the amplitude of the calibrating exponential signal. The difference between the signal from the sample and the calibrating exponential signal was taken from the anode of the tube T3 through a negative-feedback amplifier of the vertical plates of a 25-I oscillograph. The amplitude and time constant of the exponential function generated by the RC circuit were varied until a zero signal was obtained on the oscillograph screen. The accuracy of the measurement of the carrier lifetime was 5%.

6. Measuring Circuit

The block diagram of the apparatus used in our experiments is shown in Fig. 11. A transverse field was applied to a sample by a ZG-10 oscillator and the output voltage of this

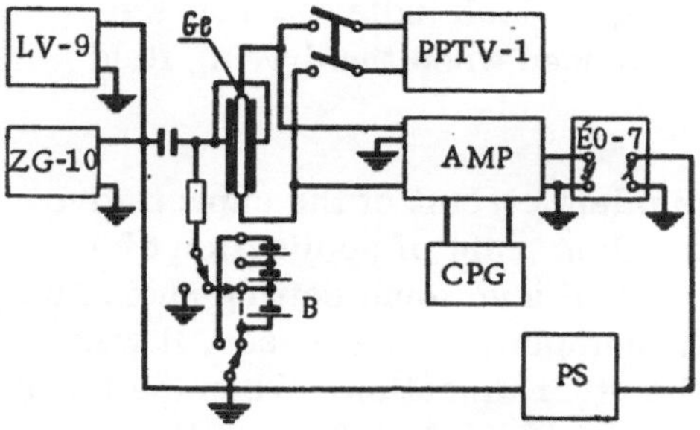

Fig. 11. Block diagram of the measuring apparatus: AMP, amplifier; CPG, calibrating pulse generator; PS, phase shifter.

oscillator was measured with a vacuum-tube voltmeter LV-9. The range of investigated surface potentials was extended using a battery B, whose voltage shifted the whole field-effect curve to the left or right. The current was supplied by a constant-current generator circuit (Fig. 12). The resistances connected in series with the sample were selected to ensure that $R_1/R_s \geq 50$ (here, R_1 is the series resistance and R_s the resistance of the sample). A variable resistance R_2 was used to vary the current through the sample. The driving field along the sample was measured with a PPTV-1 potentiometer.

The application of a sinusoidal voltage to the capacitor enclosing the sample produced a stray signal due to the capacitative component of the current. The signal was compensated in the input circuit of the amplifier by means of variable resistances R_3 and R_4, which formed a bridge circuit with the distributed capacitance of the sample-electrode system. These resistances were used to balance out the stray capacitative signal until it became the same in amplitude and phase at both ends of the sample. Such balancing was usually carried out in the absence of a driving field. However, in some cases, hysteresis of the field-effect curves was observed after the application of a driving field to a sample subjected to this balancing procedure. This phenomenon occurred frequently in a wet ambient medium. We were able to distinguish two types of hysteresis loop: a) loops at the ends of the field-effect curve; b) loops near the zero value of the field. The loops obtained in the first case indicated that the conditions were not quasistationary, i.e., the conditions did not ensure that the fast states at any moment were in equilibrium with the energy bands. These loops could be eliminated only by altering the transverse field frequency. The loops of the second kind were eliminated by an

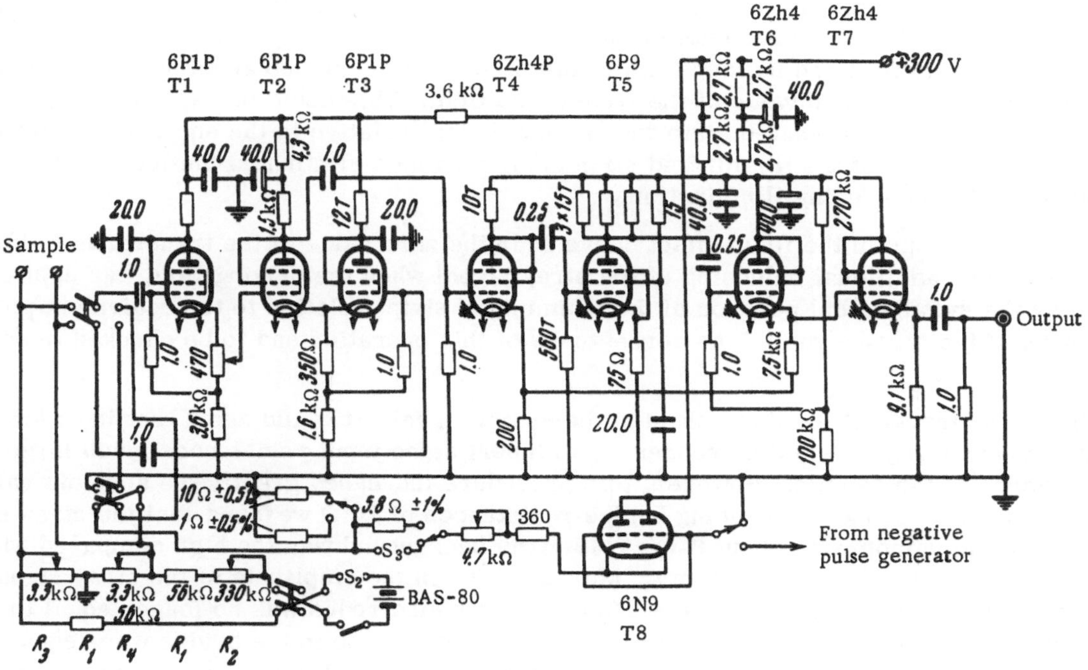

Fig. 12. Basic circuit of the amplifier for measuring the field effect.

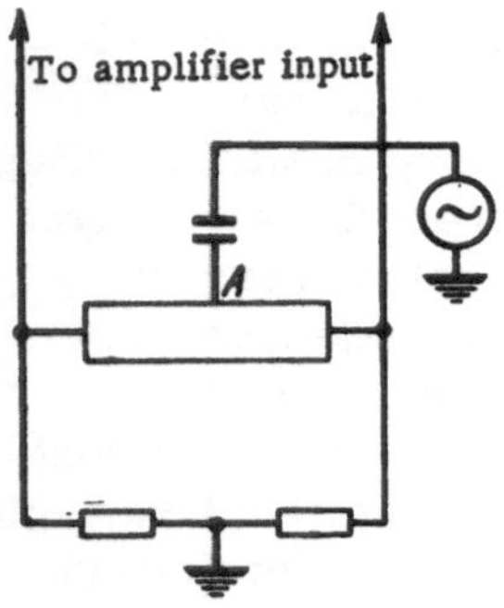

Fig. 13. Equivalent circuit for the compensation of the capacitative current.

additional balancing of the input bridge, which indicated that the stray capacitive signal became unbalanced when the driving field was switched on.

Let us now consider the equivalent circuit of the capacitative signal shown in Fig. 13. The equivalent point of application of a field to a sample (A) in the ideal case of a uniform setting should be located exactly in the middle of the sample. In real cases, it was always shifted slightly in one or the other direction. When the transverse field frequency was 50 cps, the shift of the phase of the capacitative current relative to the voltage generated by the ZG-10 oscillator was about 90°, i.e., the capacitative current had a maximum at the moment when the sinusoidal voltage passed through zero. Since the contacts were not perfectly ohmic, the flow of current through a sample altered its conductivity near the contacts. Consequently, one-half of the sample had different resistance from the other and the bridge became unbalanced. An increase in the effective carrier lifetime increased the injection through the contacts and, therefore, this unbalance increased in wet media when the lifetime was short. Moreover, because of an increase in the losses in the capacitor containing the sample, the stray signal increased in its absolute amplitude. The departure from uniformity of the resistance of a sample, which gave rise to loops in the field-effect curves, was very small. When the effective voltage produced by the oscillator was 150 V and the frequency was 50 cps, a stray capacitative signal of 0.35 V appeared at the ends of a sample. The separation between the two branches of the loop was about 0.1 mV. Such an unbalance could be due to a change in the resistance of one half of the sample due to injection representing a few hundredths of a percent, which was a reasonable value. This was confirmed by recording the dependences of the loop amplitude on the frequency and the amplitude of the transverse field and on the amplitude of the driving field.

The unbalance effect in the stray signal was observed most clearly when a sample was illuminated during the measurements of the field effect. It was frequently found that the field-effect curve in darkness had no loops, while the field-effect curve obtained during illumination split into two near the zero value of the transverse field. This could be explained quite easily by assuming that the illuminated strip was asymmetrical relative to the equivalent point A. It was found that a shift of the illuminated strip along a sample eliminated the loops observed in the field-effect curve during illumination.

Thus, the loops in the field-effect curve near the zero value of the transverse field were eliminated by an additional balancing of the stray signal when the driving field was applied, as well as by the symmetrical position of the illuminated strip relative to the equivalent point of application of the stray signal. The correctness of this operation had to be checked in each case.

The resistances at the input shunted the useful signal. It would seem that in order to increase the sensitivity it would be necessary to insert some input resistances much larger than the resistance of the sample. However, this procedure increased greatly the absolute value of the stray signal. For example, using bridge resistances of $10^5\,\Omega$ we found that the stray signal at the end of the sample was about 8 V. This stray background was too high compared with a useful signal, which was usually 10 mV. Moreover, when the amplitude of the stray signal was large, the effective voltage between the sample and its electrodes was no longer equal to the voltage supplied by the ZG-10 oscillator. When the resistances in the bridge were equal to the resistance of the sample, the useful signal decreased only by a factor of two compared with that in the absence of a shunt, whereas the stray signal decreased proportionally to the decrease in the shunting resistance.

In view of the conditions used to balance out the stray signal, neither of the sample ends could be grounded and, therefore, we had to use an amplifier with a symmetrical input. The signal from the sample has to be applied to the grid and the cathode of the tube T2 (Fig. 12). In order not to shunt the sample by the low cathode resistance of this tube, we used a cathode follower T3. Since the transfer constant of the follower was less than unity, the signal applied to the grid of the "subtracting" tube T2 was also passed through a cathode follower (T1) with a potentiometer at the output in order to achieve exact balance. The stray signals reaching the grid and cathode of the "subtracting" tube had the same phase and produced no signal at the anode of this tube. The useful signal, generated because of the modulation of the conductivity of the sample by the transverse field and the light, reached the grid and cathode of the tube T2 in antiphase and was added at the anode.

The signal from the anode of the "subtracting" tube was applied to an amplifier with strong negative feedback and a gain of about 100. The noise and background level, measured at the input, was about 20 μV. The signal taken from the amplifier was applied to the vertical plates of an ÉO-7 oscillograph and the sinusoidal voltage produced by the ZG-10 oscillator was applied to the horizontal plates. Since the phase of the signal reaching the vertical plates could be different from the phase of the voltage produced by the oscillator, a phase shifter was included in the scanning voltage circuit (Fig. 11).

The amplitude of the useful signal was calibrated by applying pulses of known amplitude to the amplifier input. These pulses were produced by interrupting (using an electronic circuit) a direct current flowing along a calibrated resistor. This current was measured with an accurate milliammeter of the 0.5 grade, which could measure currents up to 10 mA. The resistors used to produce the calibration signal were 1 and 10 Ω, to within 0.5%. Consequently, the total error in the determination of the calibration pulse amplitude was 1%.

The current flowing through the calibration resistor was varied from 1 to 10 mA by a variable resistance in the cathode circuit of a tube T8 (Fig. 12). Thus, the amplitude of the calibration pulses could be varied from 1 to 100 mV, which covered completely the range of possible values of the useful signal.

The tube T8 was blocked by applying negative pulses of 10-15 V amplitude to the grid of this tube. These pulses were produced by a negative-pulse generator with a grounded positive terminal of the anode supply circuit. When the tube T8 was operated as a pulse generator, the milliammeter in the cathode circuit was replaced by a constant resistance equal to the resistance of the ammeter.

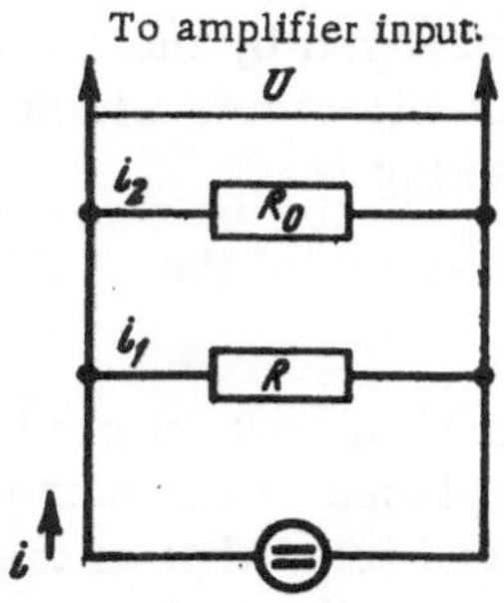

Fig. 14. Equivalent circuit of the input bridge of the amplifier.

We shall now deduce a formula which relates the signal from a sample to the change in its conductance. Figure 14 shows the equivalent circuit of the input bridge. The emf of the equivalent generator is $I_2\delta R_0$. The useful signal is taken from the bridge resistance R (because the bridge resistance is equal to the resistance of the sample, we can ignore the shunting effect of R_1). The voltage at the amplifier input is

$$\delta U = I_2 \delta R_0 \frac{R}{R_0 + R}.$$

Since

$$I_2 = \frac{U}{R_0}, \qquad U = I \frac{R_0 R}{R_0 + R},$$

we obtain

$$\delta U = \frac{U^2}{I}\frac{\delta R_0}{R_0^2},$$

or, changing to the conductance of the sample, we find that

$$\delta U = -\frac{U^2}{I}\delta G.$$

In the case of a field applied to two sides of the sample, the conductivity is

$$\delta G_{\square} = -\frac{l}{2d}\frac{I}{U^2}\delta U, \tag{42}$$

where l is the length and d is the width of the sample.

7. Thermostatting of the Sample

All the measurements were carried out at 300°K. The temperature was measured by means of a germanium sample cut from an intrinsic crystal. The temperature coefficient of this sample was about 3% at room temperature. The response time of the system was improved by using an intrinsic crystal of 10 × 1 × 1 mm dimensions. This crystal was not etched in order to diminish as fully as possible the influence of the ambient medium on the crystal's resistance.

The temperature dependence of the resistance of the measuring crystal was determined before the main experiments. This was done by measuring the resistance with a dc bridge. A regulating potentiometer of the ÉPV-01 type was connected in the output of the bridge and a relay of this potentiometer closed or opened the heating circuit when the resistance of the measuring crystal departed from its set value.

The resistance heater was wound on the cartridge which enclosed the sample. The length of the heater was 5 times its diameter to avoid the establishment of temperature gradients along the sample, which was located at the center of the cartridge. The temperature-measuring crystal was pressed against the holder in which the sample under investigation was enclosed. This system maintained the temperature at 300°K to within 0.1°K.

8. Heating of the Sample by a Current

In the first variant of the apparatus, a sample was heated externally. The heating circuit was similar to the thermostatting circuit: the temperature-sensitive pickup was a piece of germanium cut from intrinsic material; the resistance of this pickup was measured by means of a bridge. The unbalance signal of the bridge was applied to an ÉPV-01 potentiometer which switched the heater current on and off. This variant had a number of disadvantages.

First, we found that the system had a very slow response so that oscillations of the temperature during its regulation amounted to about 10°K.

Secondly, a radial (across the enclosing cartridge) temperature gradient was established by the external heater. Since the temperature-sensitive pickup could not be placed at the same point as the sample under investigation, the temperature of the sample differed from that of the pickup and the difference increased with rising temperature. For example, when the pickup was in contact with the outer wall of the sample holder, the temperature of the sample was 475°K and that of the pickup 500°K, i.e., the temperature difference amounted to 25°K. A check of the temperature of the sample being investigated could be carried out in this range because we used samples of fairly high resistivity (20–30 Ω · cm) which became intrinsic above 350°K so that their temperature dependence could be easily calculated. We found that by measuring the resistance of the sample during its heating we could find its temperature quite accurately. However, an additional circuit for measuring the resistance of the investigated sample would complicate the apparatus. Moreover, the temperature gradient could change during heating and this would require constant attention which would nullify all the advantages of the automatic temperature control.

Thirdly, when a sample was heated externally, the whole system was heated as well: the walls of the vacuum cartridge and the metal leads to the cartridge were also heated. The sample was the coldest part of the system and the impurities evolved from other parts of the system were deposited on the sample. Consequently, the surface of the sample became contaminated by accidental impurities during the heating cycle.

Finally, the contacts with the sample had to be able to withstand the high temperatures employed in external heating. Technically, it was difficult to prepare contacts which could withstand high temperatures and, at the same time, satisfy the requirements of the field-effect method. The known high-temperature alloys did not form ohmic contacts with germanium. On the other hand, contacts of electrolytically deposited copper, plated with tin, made excellent ohmic contacts with high-resistivity samples of p- and n-type germanium but, unfortunately, such contacts could be used only at temperatures below 500°K.

All these disadvantages were eliminated in a system in which the sample was heated directly by the current flowing through it. In this system, some set resistance of the sample was maintained automatically. As mentioned earlier, the samples employed in our investigation became intrinsic above 350°K and the temperature dependence of the conductivity of intrinsic germanium was described very accurately by the empirical formula of Morin and Maita [106]. Table 2 lists the values of the resistivity of intrinsic germanium calculated by means of this formula in the range 350–800°K. Knowing the geometrical dimensions of a sample, we were able to calculate its resistance at any temperature above 350°K.

The system used in heating a sample by the current flowing through it (Fig. 15) consisted of a bridge circuit in which one of the arms was the sample being heated and the other was a resistance box capable of carrying high currents. The other two arms of the bridge were two equal large resistances.

TABLE 2. Calculated Temperature Dependence of the Resistivity of Intrinsic Germanium

T, °K	ρ, Ω·cm	T, °K	ρ, Ω·cm	T, °K	ρ, Ω·cm	T, °K	ρ, Ω·cm
350	6.25	470	0.247	580	0.0419	700	0.0116
360	4.37	480	0.203	590	0.0370	710	0.0106
370	3.13	490	0.168	600	0.0327	720	0.00975
380	2.29	500	0.141	610	0.0290	730	0.00896
390	1.69	510	0.118	620	0.0258	740	0.00828
400	1.28	520	0.101	630	0.0230	750	0.00765
410	0.974	530	0.0853	640	0.0207	760	0.00705
420	0.753	540	0.0736	650	0.0186	770	0.00653
430	0.591	550	0.0634	660	0.0169	780	0.00605
440	0.469	560	0.0551	670	0.0153	790	0.00560
450	0.375	570	0.0479	680	0.0139	800	0.00525
460	0.303			690	0.0127		

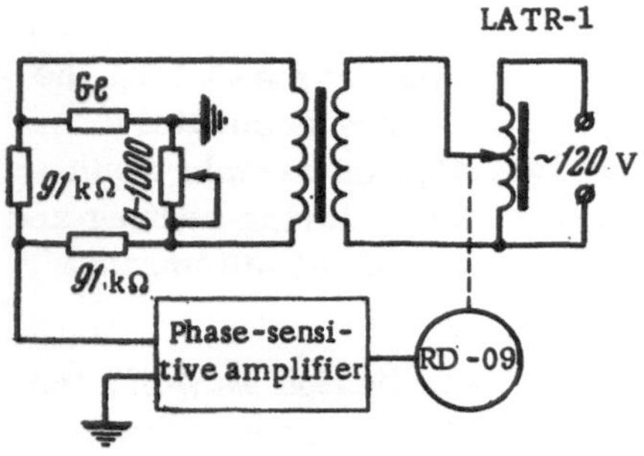

Fig. 15. Block diagram of the apparatus used for heating a sample by a current.

The bridge was supplied with an alternating current. The unbalance signal was applied to the input of a phase-sensitive amplifier with an RD-09 reversible motor at its output. Depending on the sign of the unbalance signal, the motor rotated in one direction or the other. The axle of this motor was coupled mechanically to the handle of an LATR-1 autotransformer which supplied the bridge. When the resistance of the germanium sample was higher than the resistance of the box, the RD-09 motor rotated the autotransformer handle to increase its output voltage. This increased the current through the sample while the resistance of the sample decreased because of the additional heating. The reverse process occurred when the resistance of the germanium sample was lower than the resistance of the box, i.e., the system checked the resistance of the sample and kept it equal to the resistance of the box.

This heating system had a practically instantaneous response because the response was governed by the inertia of the RD-09 motor. In this way, the temperature of the sample was kept constant to within ±1°K.

The second disadvantage of external heating – the difference between the temperature of the sample and that at the pickup – was automatically eliminated because the new system controlled the temperature in the heated sample itself. Special experiments were carried out to check the accuracy of the resistance of a sample calculated from the formulas of Morin and Maita [106] and from the geometrical dimensions of the sample. In these experiments small pieces of pure tin and lead (about 1 mm in size) were placed along a sample and the melting points of these pieces were noted. These experiments were carried out in vacuum in order to avoid the possibility of additional cooling of the tin and lead by convection. These pieces were cut so that they had acute angles and the moment of melting could be observed easily because the pieces shrank into circular drops. The results of these experiments were in full agreement with the calculations: the tin and the lead melted when the resistance of the sample corresponded to 505°K in the case of tin and 610°K in the case of lead. Moreover, similar pieces melted simultaneously at different points along the sample. Thus, this heating system allowed us to maintain a given temperature accurately and uniformly.

Obviously, the third disadvantage of external heating was also eliminated by the new heating method. We heated only the sample and (slightly) the holder but the sample was the hottest part of the system. The rest of the system was practically at room temperature. Consequently, the sample was not contaminated during its heating.

Finally, we found that if a sample was heated by the current flowing through it the usual contacts of electrolytically deposited copper, with tin-soldered current leads, could be used up to 750-800°K. (Higher temperatures were not investigated because the bulk properties of germanium changed greatly at these temperatures.) Liquid tin thus provided good electrical and mechanical contacts with the sample. Evidently the surface tension of liquid tin was sufficiently high to support the whole weight of the germanium sample. The fact that the tin was actually liquid and did not form any alloy was proved by shaking the cartridge. When this was done, the sample became detached from its current leads and dropped away.

It would seem that the same behavior of tin should be observed also in the case of external heating. However, there was an important difference between the two heating methods. The heating by an internal current was controlled by the state of the contact: if the contact failed at the melting point of tin (505°K), the current ceased to flow instantaneously and contact was reestablished. Such "self-regulation" resulted in a reliable liquid contact. In the case of external heating, a sample received heat even when the contact failed and this destroyed the contact completely.

The bridge circuit made it possible to heat internally a sample under conditions which could not be achieved in an external heater. This method could be used successfully to heat germanium in liquids to temperatures considerably higher than the boiling points of these liquids [80, 81]. Moreover, the system could be used to heat a sample placed in an enclosure kept at liquid nitrogen temperature [107].

9. Preparation of the Sample

Our measurements were carried out on samples of p- and n-type germanium of 20-40 Ω cm resistivity. We used high-resistivity crystals in order to increase the accuracy of the measurements and to satisfy easily the inequality $L > 4h$, which was necessary in the steady-state photoconductivity method.

To eliminate the influence of the scatter in the surface parameters (which could be due to the initial treatment), we standardized carefully all the stages of preparation of a sample. In our earlier investigations [14] we found that the parameters of a surface depended on its crystallographic orientation and, therefore, all the measurements were carried out on samples with their large faces oriented along the (111) planes.

After having been cut, the samples were ground or polished until dimensions of 20 × 5 × 0.4 mm were achieved. These dimensions were set by the requirements of the field-effect method and by the method used to measure the lifetime. In the field-effect method, the surface conductance should represent an appreciable fraction of the total conductance. The thinner the sample, the higher was the sensitivity of the field-effect method. On the other hand, the thickness of a sample could not be reduced indefinitely for two reasons: 1) thin samples fractured easily; 2) contacts could be soldered only to samples of certain minimum thickness (0.4 mm). On the other hand, this thickness satisfied the condition of validity of the steady-state photoconductivity method: $L > 4h$. This condition implied that the diffusion length should always exceed 1.6 mm, which was always satisfied by the mass-produced high-resistivity germanium that was used in this investigation.

The length of a sample was selected to avoid appreciable extraction of excess carriers through the end contacts. Table 3 lists the drift lengths of carriers for different effective

TABLE. 3. Drift Length (cm) of Minority Carriers in p-Type Germanium for Various Values of the Driving Field and Effective Lifetime

$\mathcal{E}$, V/cm	τ_{eff}, μsec				
	100	200	300	400	500
0	0.095	0.134	0.164	0.189	0.212
0.1	0.115	0.175	0.227	0.274	0.320
0.2	0.137	0.225	0.304	0.380	0.456
0.3	0.164	0.281	0.392	0.505	0.610
0.5	0.222	0.400	0.588	0.769	0.909
1.0	0.384	0.715	1.00	1.43	1.67

lifetimes and driving fields (these values were calculated using the formulas given in [105]). The drift length was defined as the distance in which the excess carrier density decreased by a factor e. To ensure that the error in the measurement of the lifetime, due to carrier extraction, does not exceed 3%, the distance from the edge of the illuminated area to the end of a sample should not be less than two drift lengths. The values given in Table 3 indicate that this distance should not be less than 4-5 mm and, therefore, we made our samples 20 mm long.

Carrier recombination occurred on all faces of a sample. Consequently, we used the formula

$$\frac{1}{\tau_{\text{eff}}} = \frac{1}{\tau_0} + \frac{2s}{h} + \frac{2s_d}{d} + \frac{2s_l}{l}, \tag{43}$$

where d and l are, respectively, the width and length of the sample; s_d and s_l are the surface recombination velocities on the lateral and end faces of the sample. Since the length of the sample was selected so that excess carriers did not reach its ends, the last term in Eq. (43) was practically equal to zero so that Eq. (43) could be reduced to

$$\frac{1}{\tau_{\text{eff}}} = \frac{1}{\tau_0} + \frac{2s}{h} + \frac{2s_d}{d}. \tag{44}$$

In the $s = s_d$ case, the lateral faces made a contribution to the effective lifetime, which was proportional to the ratio h/d. Since the value of s_d could not be measured directly, its contribution could not be taken into account in the determination of the surface recombination velocity on one of the large faces. Therefore, the error could be reduced solely by selecting the dimensions of the sample so that $d \gg h$.

The experiments were carried out on samples which were ground or polished before etching. Comparison of the results obtained after these two treatments forced us to drop the polishing treatment. We expected the polished samples to have advantages over the ground crystals in the value of the capacitance in the field effect. However, we found that, in spite of the fact that the etchants employed (for example, H_2O_2 + HF) did not alter the surface relief, the average values of the specific capacitances of the field-effect capacitors with 10-μ mica spacers were similar for the ground and polished samples (280 and 290 pF/cm^2, respectively). The effective lifetime on a polished surface was usually twice as long as that on a ground sample. This was because of the rougher surface relief of the ground samples which made the effective surface area of these samples about twice as large as the geometrical area. This difference between the effective lifetimes forced us to use ground samples. This was done because only the large faces could be polished and the error in the determination of the surface recombination velocity on these faces, due to carrier recombination on the ground lateral faces, was twice as large because of the longer lifetime.

Moreover, in contrast to the ground faces, the state of the surface of the polished samples depended very critically on their cleanness during the polishing process and on the duration of this process. Frequently this meant that the properties of the two large faces were not identical and the steady-state photoconductivity method could not be used in the two-sided field effect.

The samples were ground in two stages. Initially, the samples were ground down to dimensions of 20 × 5 × 0.8 mm and copper contacts were deposited on them so that we could measure the distribution of the resistivity along the length of each sample. After these measurements had been made, we selected only those samples which exhibited a resistivity scatter not exceeding 3%. Samples with a greater inhomogeneity of the resistivity along the length were unsuitable for accurate measurements for two reasons. First, the field-effect curve and the position of its minimum depended on the resistivity of a material. If a sample was inhomogeneous in respect of its resistivity, the field-effect curve was different for regions with different resistivities and, consequently, the experimentally determined field-effect curve was the average result with strong distortions in the region of the conductivity minimum. Secondly, the illumination of an inhomogeneous sample produced a photo-emf which interfered with measurements of the photoconductivity.

After the measurement had been made of the distribution of the resistivity along the sample, we passed on to the next grinding stage, carried out using a fine emery paper, in which we reduced the size to $20 \times 5 \times 0.4$ mm. The ends of the sample had ~0.2 mm wide bevels on both sides. This was done to suppress the influence of a transverse field on the end contacts because this could give rise to uncontrolled effects. Moreover, when only the middle parts of the end surface were covered by the contacts it was easier to ensure that the two large faces were etched away more uniformly along the length of the sample.

After a sample had been ground, we deposited electrolytic copper contacts on the ends of the sample. The composition of the electrolyte used in this process was (g/l): 37 $CuSO_4$, 65 Na_2SO_3, 46 KCN. Such an electrolyte produced mechanically strong coatings. The excess copper was dissolved in nitric acid. In this way, only the ends of the sample were covered and they were then plated with pure tin. If the tin was mixed with rosin, only the copper strip at the end was coated. Current leads were then soldered to the contacts. These leads were in the form of tinned copper wires of 0.15-0.16 mm diameter. Using double leads for each end face, we were able to pass up to 10 A through the sample (the current required to heat a standard-geometry sample to 800°K was 5-7 A).

As mentioned earlier, electrolytically deposited copper produced noninjecting contacts with high-resistivity p- and n-type germanium. The ohmic nature of the contacts was checked by investigating the changes in the shape of the field-effect curve when the direction of the current through a sample was reversed and the driving field was varied. No changes in the field-effect curve were found in driving fields up to 2 V/cm. Since the measurements were carried out in considerably lower driving fields, the distortion of the field-effect curves due to injection through the contacts was fully avoided.

Chemical etching of a sample should produce a flat surface in order to ensure a uniform distribution of the charge induced in the field effect. We investigated in detail three different etchants (CP-4, H_2O_2 + HF [108], and H_2O_2 + NaOH) and found that CP-4 did not produce a flat surface. The surface obtained after treatment in CP-4 was undulating. Moreover, whenever CP-4 or H_2O_2 + HF was used, the contacts had to be protected during etching and this required a watchmaker's skill: on the one hand, the coating had reliably to protect the contacts and leads and, on the other hand, the working surface of the sample had to be free of this coating. Moreover, the subsequent removal of the coating in organic solvents could contaminate the surface.

For these reasons, we used mainly an etchant consisting of 230 ml of 30% H_2O_2 and 2 ml of 10% NaOH. A dish filled with the etchant was placed on a thermostatted water bath and a sample was immersed in the solution kept at 80.8°C. To ensure uniform etching, the sample was rotated in the dish. The temperature of the etchant was maintained to within ±0.1°C throughout the etching treatment (6 min). After it had been etched, each sample was washed in boiling deionized water (500 cm^3) and dried with distilled ethyl alcohol and ethyl ether. The sample was then kept in a closed weighing bottle for 2-3 h to stabilize the oxide film before it was placed in the holder. The surface recombination velocity after such treatment was 10-20 cm/sec.

The sample holder was made of optically polished Pyrex plates of $45 \times 30 \times 5$ mm dimensions. A transparent electrode of tin dioxide of 25×10 mm dimensions was prepared by

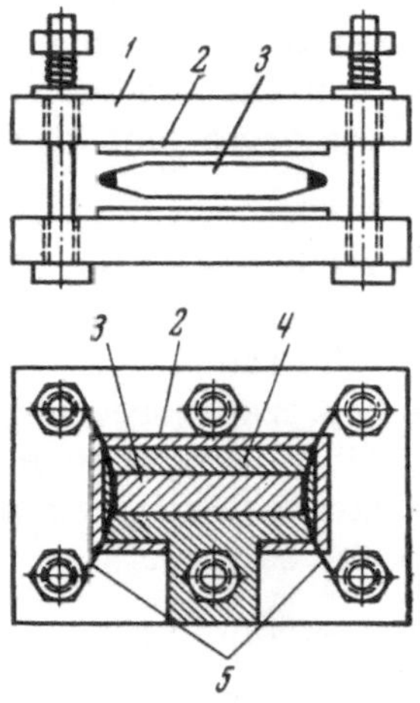

Fig. 16. Sample holder in the assembled state: 1) glass plates; 2) mica; 3) sample; 4) layer of tin dioxide; 5) current leads.

the evaporation of tin chloride on one surface of the Pyrex plate. The quality of the tin dioxide film was deduced from its color. These films transmitted 85-90% of the incident light and their resistance was 100-200 Ω, which was sufficiently low to avoid an appreciable voltage drop across the film when a capacitative current of frequencies up to the megacycle range passed along the film.

Mica spacers 10 μ thick, were placed between the electrodes and the sample. The mica spacers had to be carefully cleaned because they were in direct contact with the sample. This was done by washing in potassium bichromate followed by rinsing for 1 h in running water. The spacers were then washed in distilled water and in hot deionized water, and dried in ethyl alcohol and ethyl ether.

Since the contact between a mica spacer and the sample could affect the results, we carried out special experiments to determine the influence of mica. We determined the field effect in the case of an air gap between a sample and its electrode. The obtained dependences of the surface recombination velocity on the surface potential were identical in shape and amplitude with the dependences obtained using mica spacers (these experiments were carried out by S. V. Pokrovskaya). The results were in agreement with those reported in [85, 98].

In the two-sided field effect the two capacitors should yield identical curves if their capacitances were identical. To ensure that this was so, the two mica spacers were cut from the same plate, which guaranteed the same thickness. The difference between the capacitances of the two sides in the field effect did not exceed 3%. The area of the mica spacers was 30 × 20 mm, which was sufficient to cover the electrodes completely and to eliminate the possibility of a short circuit between the current leads and the field-effect electrodes. The assembled holder is shown in Fig. 16.

It was necessary to ensure that the surface of the sample and the electrode were parallel to each other. This was checked by examination of the equal-thickness interference fringes which could be observed in the case of transparent electrodes.

10. Vacuum and Gas Admission System

The vacuum and gas admission system (Fig. 17) enabled us to investigate a sample in vacuum, in dry and wet gases, and in vapors of various liquids. Three working cartridges made it possible to carry out parallel measurements on samples subjected to different environ-

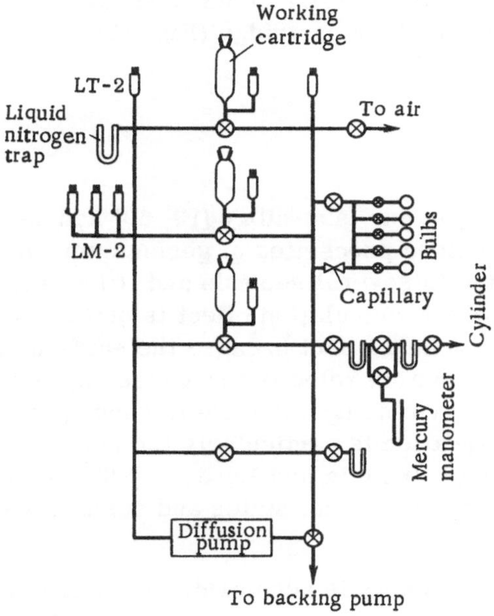

Fig. 17. Vacuum-gas system.

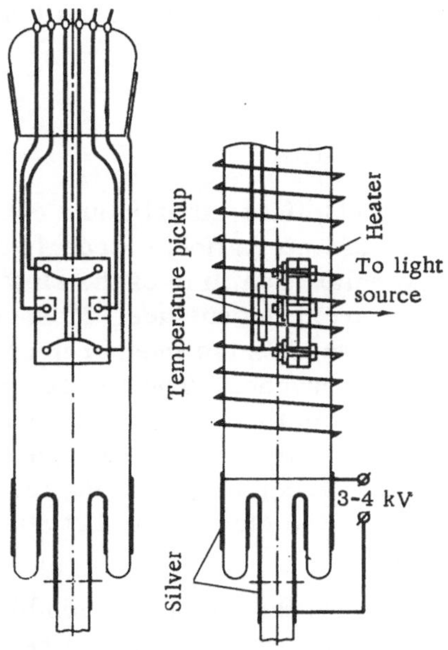

Fig. 18. Working cartridge.

ments. Each of these cartridges could be connected to a backing pump, a diffusion pump, and to systems for the admission of gases or vapors.

A high vacuum was produced by an SDN-1 diffusion pump fitted with a liquid-nitrogen trap. The working cartridge could be evacuated to 10^{-6} mm Hg. Each cartridge was fitted with a thermocouple vacuum gauge of the LT-2 type. The LT-2 gauges were disconnected from the cartridges during heating of the films before measurements. This precaution was necessary because special experiments indicated that gases driven off from the filaments of these gauges could contaminate the surface of a sample and alter its surface recombination velocity by 20-30%, giving rise to spurious results.

We avoided passing a gas through the working cartridge because complete drying could not be achieved under these conditions. When a cartridge had to be flushed out with a gas, this was done by alternate admission of a gas and evacuation of the cartridge. The gas system employed enabled us to dry a gas twice by means of traps filled with liquid nitrogen.

Water could be removed completely from a gas by freezing the water vapor twice in these traps. This was checked by observations of changes in the surface potential of a sample which was very sensitive to the presence of even the smallest traces of water. A third trap was used to absorb water evolved from the surface of the sample during heating.

The vapors of selected liquids were stored in special bulbs. They were admitted into the working cartridge via a valve or a capillary. If a valve was used, the saturated vapor pressure was established almost instantaneously in the cartridge, whereas in the case of a capillary the vapor pressure could be varied continuously by controlling the rate of flow through the capillary and observing the kinetics of changes in the surface properties of a sample. Each of the bulbs could be pumped out to a high vacuum. We were also able to evacuate a bulb containing a liquid by means of a backing pump: this was done in vacuum-cleaning of a liquid.

Thus, dry and wet gases, water vapor, and vapors of other liquids could be admitted into the working cartridge in any sequence. Moreover, a backing vacuum or a high vacuum could be established in the cartridge. In addition, we were able to produce an ozone atmosphere around the sample by means of an ozonizer built into the working cartridge (Fig. 18).

CHAPTER III

Experimental Results

The first field-effect experiments conducted in our laboratory [13, 60] had demonstrated that the presence of water affects strongly the surface properties of germanium. In a wet atmosphere, the relaxation time of the slow states is several seconds and it increases when the water is removed from the surface. Moreover, the accumulation effect is quite strong and this effect sometimes prevents the measurement of the field effect because the surface conductivity minimum cannot be reached. The surface recombination velocity increases appreciably when water is removed from the surface of germanium by placing a sample in vacuum or, which is more effective, by heating in vacuum. This observation is particularly interesting because it has been previously accepted (following the work of Brattain and Bardeen [28]) that the ambient medium affects only the slow states but does not alter the fast states and surface recombination [12, 19, 64].

Since the rise of the surface recombination velocity is accelerated when a sample is heated in vacuum, a method for vacuum heating in stages has been developed [14]. In this method, a sample is heated in small steps up to 450-500°K and after each step the sample is cooled to 300°K to carry out the measurements. After the measurements have been made, the sample is heated to a higher temperature, and the cycle is repeated.

This multistep heating method is very effective because it allows us to alter the state of the surface within very wide limits.

1. Heating in Vacuum. Polar Molecules

Several cycles of heating to 500°K have demonstrated [14] that such heating increases strongly the surface recombination velocity (by a factor of 10-20) compared with the initial velocity of an etched sample placed in a vacuum system. It is worth mentioning particularly that the values of the maximum surface recombination velocity lie within the limits 2000-2500 cm/sec for all p-type samples heated to 500°K. This shows that the results are reproducible. The maximum value of the surface recombination velocity, observed after heating at 500°K, is independent of the initial treatment of the sample and the initial value of this velocity. Thus, for example, it is known that etching in CP-4 gives rise to higher surface recombination velocities than does etching in hydrogen peroxide. In the first case, the recombination velocity is 200-400 cm/sec; in the second case, it amounts to a few tens of centimeters per second. In spite of this large difference (which may amount to an order of magnitude) between the initial surface recombination velocities, heating to 500°K equalizes the velocities on the samples subjected to either of these chemical treatments.

Similar results have been obtained for samples stabilized by storing for several months in air after etching. The values of the surface recombination velocities for such samples are about 400 cm/sec before heating but after heating at 500°K they again lie in the 2000-2500 cm/sec range.

Figure 19 shows the dependences of the surface recombination velocity on the surface potential, plotted on a reduced scale, in which the maximum surface recombination velocity is taken as unity. The good agreement between the points obtained after various preliminary heat treatments indicates that heating alters only the concentration of some recombination centers, whereas other parameters, which influence the form of the curve and the position of its maximum (E_t, c_p, c_n), remain unchanged.

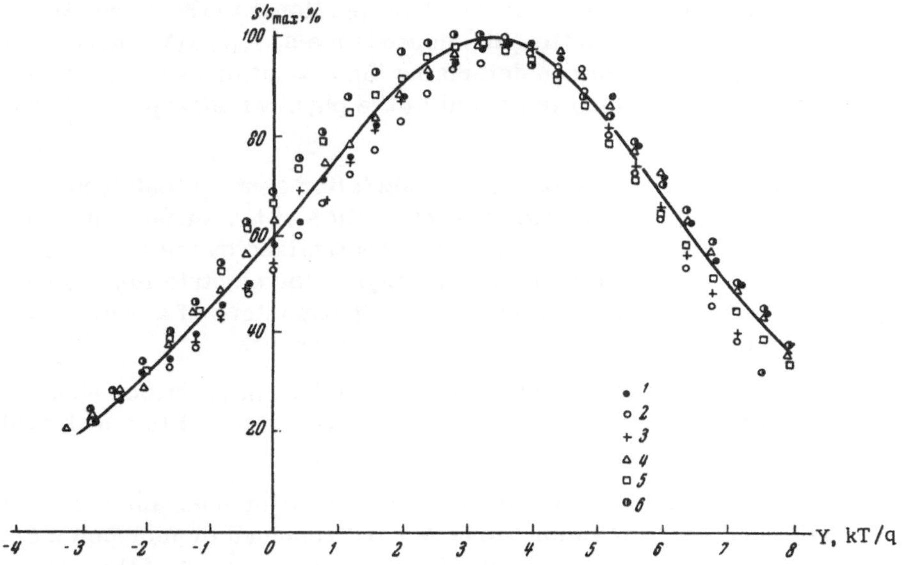

Fig. 19. Multistage heating in vacuum. Sample No. 13 (p-type). 1) No heating; after heating at (°K): 2) 350; 3) 400; 4) 450; 5) 475; 6) 500.

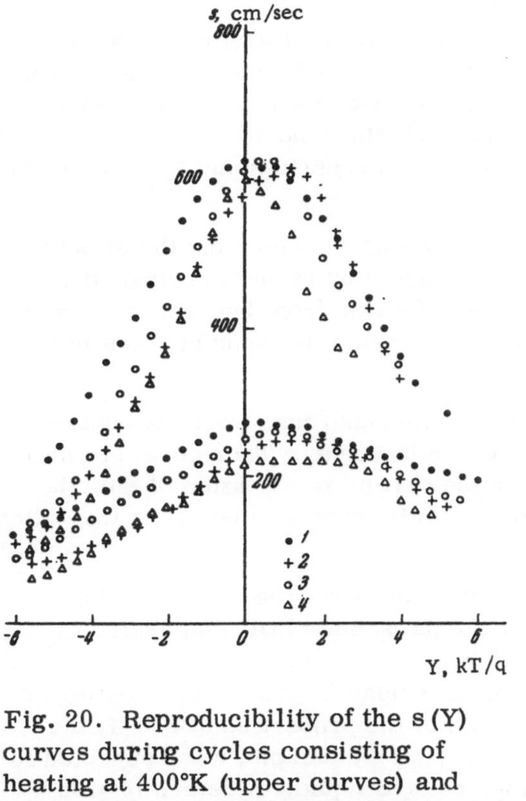

Fig. 20. Reproducibility of the s (Y) curves during cycles consisting of heating at 400°K (upper curves) and exposure to water vapor (lower curves). Sample No. 152 (n-type). 1) First cycle; 2) second cycle; 3) fourth cycle; 4) seventh cycle.

The most interesting effect of such heating is the reversibility of the surface characteristics in a cycle consisting of heating and the admission of water vapor. The original value of the surface recombination velocity is re-established when a sample heated in vacuum to a temperature not higher than 500°K comes into contact with water vapor.

Figure 20 shows the degree of reversibility in a heating-water vapor cycle. This figure gives the s (Y) dependences recorded during heating in vacuum to 400°K followed by the admission of water vapor. These measurements are carried out in vacuum. Water vapor is admitted at a pressure of 20 mm Hg into the cartridge containing a sample, the sample remains in contact with the water vapor for 30-40 min, and then the cartridge is evacuated until a pressure of 10^{-6} mm Hg is reached. The measurements are made after this treatment. It is found that a 30-40-min exposure to water vapor is sufficient to re-establish the initial surface properties.

These experiments demonstrate that the water molecules are desorbed from the surface during heating in vacuum and, therefore, the surface recombination velocity increases after desorption. Since the form of the s (Y) curve is the same for all heat treatments, the maximum surface recombination velocity after each heating must be proportional to the concentration of newly generated recombination centers. Plotting the dependence $s_{max}(1/T_h)$, where T_h is the temperature used during heat treatment, we can determine the activation energy. Estimates show [14] that this energy is about 0.2 eV, which is typical of the physical adsorption of water on many solids.

This value for the activation energy of recombination centers, obtained after heating in vacuum, as well as the reversibility of the characteristics of the surface in a heating-water vapor cycle suggest that recombination centers are neutralized by the water adsorbed physically on the germanium surface. Since water is a polar liquid, the electric field of the water molecule adsorbed physically on the surface may alter the parameters of a nearby recombination center so that it ceases to be active in recombination processes.

However, we cannot exclude the possibility of chemical interactions since water is capable of dissolving the oxide film and of forming germanium hydroxide. If this hydroxide is formed, the OH radical plays the main role.

In order to determine which of the possible neutralization mechanisms did actually take place, water molecules on the germanium surface were replaced by molecules of other substances having one of the properties of water [51]. The replacement of the water adsorbed on the germanium surface with alcohol molecules excluded the possibility of dissolving the oxide film on the germanium. Using polar substances which do not contain the OH radical, we could eliminate both "chemical" interactions. This left us only with the possibility of an interaction between electric dipoles and recombination centers.

When the germanium samples were placed in contact with vapors of various substances, special attention was paid to the elimination of traces of water from such vapors. Therefore, the liquids employed should have vapor pressures lower than the vapor pressure of water at room temperature. On this basis, the following liquids were selected: normal amyl alcohol, isoamyl alcohol, chlorobenzene, and nitrobenzene. Some of their properties and those of water are given in Table 4.

Benzene was used in a control experiment whose purpose was to determine the effectiveness of the method used for removing traces of water from a liquid by vacuum drying. It is evident from Table 4 that the conditions for removing traces of water from benzene were less favorable than for the other four liquids because the vapor pressure of benzene at room temperature was higher than the vapor pressure of water.

TABLE 4. Some Properties of Selected Substances

Substance	Dipole moment, 10^{-18} abs. esu	Vapor pressure at 300°K, mm Hg
n-Amyl alcohol	1.6	3
Isoamyl alcohol	1.8	4
Chlorobenzene	1.6	15
Nitrobenzene	3.9	0.3
Ethyl ether	1.2	550
Benzene	0	100
Water	1.8	30

The most unfavorable conditions were encountered in the vacuum drying of ethyl ether. Therefore, special attention was paid to the careful purification of the ether and to the drying of a bulb immediately before it was filled with the ether.

Vapors of various liquids were admitted to the working cartridge after the sample had been heated to 500°K (this was done in order to remove the physically adsorbed water from the surface). After such heating and after the measurements of the properties at 300°K, the working cartridge, carrying the sample, was connected to a bulb containing the selected liquid. A determination was then made of the changes in the surface parameters caused by the adsorption of molecules of this liquid on the surface of the sample.

The experiments carried out in the amyl alcohols showed that the vapors of normal and iso alcohols behaved similarly to water vapor and that they neutralized practically completely the recombination centers introduced by heating at 500°K. The s (Y) curves, obtained after the exposure of a sample to amyl alcohol vapor for 1 h, were practically identical with the initial s (Y) curve recorded before heating.

We mentioned earlier that these experiments did not eliminate completely the possibility of chemical neutralization because the OH radical in alcohol molecules could be responsible for the observed effects. Therefore, the main stress was laid on the investigation of the influence of chlorobenzene and nitrobenzene on the surface recombination because these two liquids have nothing chemically in common with water.

The results of these experiments showed that chlorobenzene neutralized completely those recombination centers which were introduced by heating in vacuum. On the other hand, although the value of s_{max} after treatment in a nitrobenzene vapor was considerably lower than s_{max} after heating to 500°K, it was still twice as high as s_{max} in the initial state. This result could be explained by the fact that, as shown in Table 4, the vapor pressure of nitrobenzene at room temperature was considerably lower than the vapor pressures of other liquids, whereas the exposure of a sample to vapors of all the investigated liquids was the same (about 1 h).

Therefore, we recorded the time dependences of the surface parameters during additional experiments carried out using nitrobenzene vapor. In contrast to other investigated polar liquids and water, nitrobenzene had practically no influence on the surface potential. Water vapor shifted the surface potential in the n-type direction. Therefore, it was usually difficult to reach the surface conductivity minimum in wet atmospheres and, consequently, it was difficult to carry out quantitative investigations. Moreover, the maximum of the surface recombination velocity was located near the surface conductivity minimum (in the case of p-type samples). All these factors made it difficult to investigate the kinetics of the process of neutralization of the surface recombination centers in wet atmospheres.

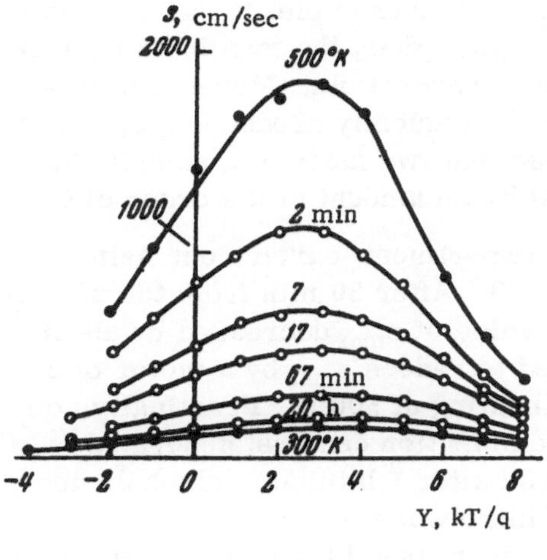

Fig. 21. Changes in the surface recombination velocity due to the adsorption of nitrobenzene vapor. The duration of exposure to nitrobenzene is given above each curve (p-type sample No. 70).

The surface potential of a sample in contact with nitrobenzene vapor was close to the surface potential of the same sample in vacuum. This enabled us to investigate the kinetics of neutralization of the recombination centers by nitrobenzene molecules. Since the vapor pressure of nitrobenzene was low, the neutralization process was fairly slow, which avoided the possibility of different changes on different parts of the surface.

The effect of nitrobenzene vapor on the surface recombination velocity of a sample heated initially to 500°K is shown in Fig. 21. After 20 h from the admission of nitrobenzene vapor practically all the recombination centers generated by heating became neutralized. As in the case of heating in steps, the successive curves recorded after different time intervals from the admission of nitrobenzene vapor were identical in shape, which showed that the recombination centers were all of the same type.

The experimentally obtained time dependence of the maximum surface recombination velocity could be represented by a simple power law

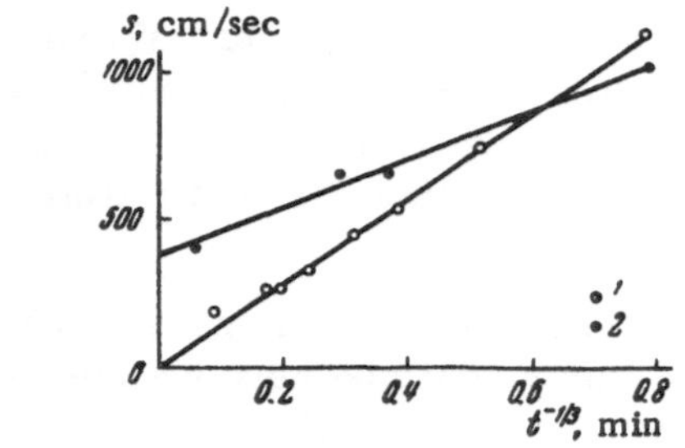

Fig. 22. Dependence of the maximum surface recombination velocity on the duration of treatment of two p-type samples in vapors of: 1) nitrobenzene (p-type sample No. 70); 2) ethyl ether (p-type sample No. 67).

$$s_{max} = a + bt^{-1/3},$$

where a and b are constants. Figure 22 shows the experimentally determined points (circles). They fit well a straight line which passes through the origin. This means that all the centers should be neutralized as $t \to \infty$. However, the point corresponding to t=20 h deviated considerably from the straight line. Obviously, some of the centers could not be reached by the nitrobenzene molecules.

A study of the effect of ether yielded results similar to those for other polar liquids. The surface recombination velocity s_{max} decreased but the shape of the dependence s (Y) was retained, i.e., only the concentration of the recombination centers changed but not the parameters of these centers. A different result was reported in [80, 81] for germanium samples heated and immersed in liquid ethyl ether: this treatment shifted markedly the position of the maximum in the s (Y) curve. We shall consider this discrepancy in the next chapter.

In our experiments, the ether vapor differed from the other liquids by producing only a partial neutralization of the recombination centers generated by heating at 500°K. This result could not be explained by incomplete adsorption because the vapor pressure of the ether was very high (Table 4) and the last set of measurements was carried out four days after the admission of the ether vapor. Moreover, the experimental points fitted well the straight line $s_{max} = a + bt^{-1/3}$. However, in contrast to nitrobenzene, this line did not pass through the origin but intersected the ordinate at s_{max} = 400 cm/sec. The partial neutralization of the recombination centers by the ether vapor could be due to some chemical reaction between ether and the surface of germanium, which would alter the oxide film structure. For example, it was observed that when a freshly etched sample was immersed for 1 h in liquid ether, the maximum surface recombination velocity increased severalfold. Moreover, prolonged immersion in liquid ether frequently produced a strongly microinhomogeneous surface so that two faces of a sample differed in their surface potential by an amount of the order of 0.1 V.

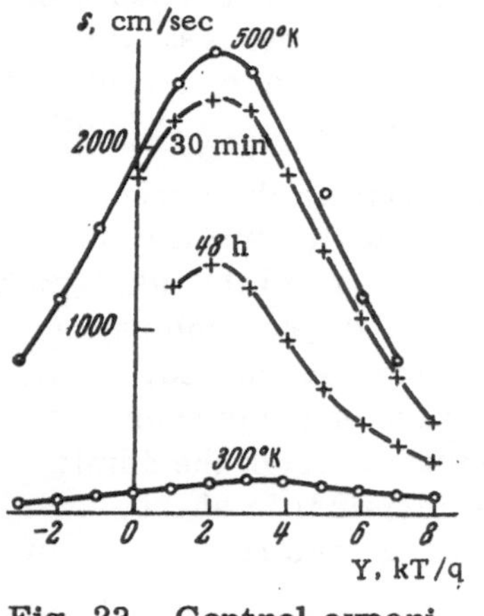

Fig. 23. Control experiment on exposure to benzene. Sample No. 48 (p-type).

The results of a control experiment, carried out using benzene, are presented in Fig. 23. After 30 min from the admission of the benzene vapor, the value of s_{max} decreased by about 10% but two days were required to reduce s_{max} by a factor of 2 compared with its value after heating at 500°K. Practically complete neutralization of the recombination centers, generated by heating in vacuum, was observed after 1 h in amyl alcohols and chlorobenzene. In the case of nitrobenzene and ethyl ether, the surface recombination velocity decreased by a factor of about 2 after 2 min from the admission of the vapor. Thus, a control experiment carried out using benzene indicated that traces of water, present on the walls of the working cartridge, as well as in the vapors of the investigated liquid, could not make an appreciable contribution to the effects observed on the surface of a sample in contact with the vapor of a polar liquid.

2. Effect of Ozone on a Freshly Etched Germanium Surface

The published results of the investigations into the effects of various gaseous environments on the properties of germanium surfaces have revealed something unexpected [60]: ozone not only strongly changes the surface conductivity of germanium, as reported by Brattain and Bardeen [28], but also increases the concentration of surface states, including recombination centers. This observation is not compatible with the generally accepted view [21, 28, 64, 65] that the environment affects not the fast surface states but only the slow ones.

Repeated investigations of the effect of ozone on freshly etched germanium [13] have demonstrated the incorrectness of the assumption that the system of fast states is not affected by gaseous media. Moreover, these investigations have provided a powerful method for investigating surface recombination. The strong effect of the adsorption and desorption of water and oxygen on the surface recombination velocity has made it possible to investigate the nature of surface recombination centers.

Detailed studies of the effect of ozone on surface recombination in germanium, carried out in our laboratory [13], have shown that the characteristics of the recombination centers generated by the exposure to ozone are completely identical with the characteristics of the centers which appear when water is desorbed from the surface of germanium by heating in vacuum. This indicates that the recombination centers on the surface of germanium are associated with oxygen atoms or ions distributed at random at the germanium–oxide interface.

The following observations also seem to confirm this hypothesis. The parameters of the surface recombination centers on samples subjected to various chemical treatments and investigated in different gaseous media are the same [13, 14, 61, 70, 72]. This means that the surface recombination centers are not associated with accidental impurities in an etchant [77] but with some regular defects located at the germanium–oxide interface and these defects appear invariably after the oxidation-reduction process which takes place during the etching of a sample. Since all etching solutions contain oxygen and water, it is reasonable to assume that the surface centers are associated with either of these molecules. We can also attribute recombination to the Tamm states on the surface but the low concentration of the recombination and capture centers is several orders of magnitude less than the expected value of the Tamm level density. Therefore, it seems that we have to reject this possibility.

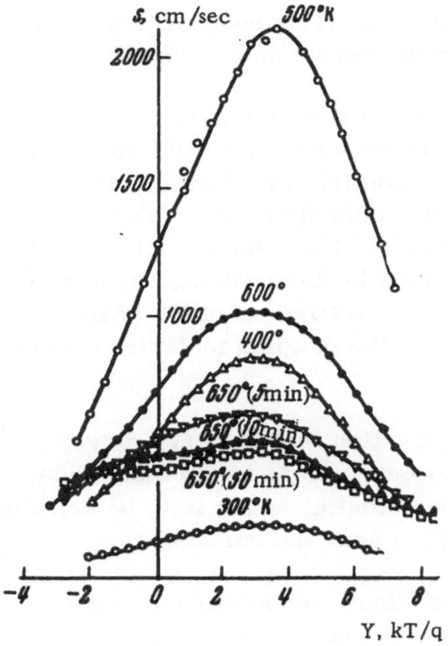

Fig. 24. Changes in the surface recombination velocity due to multistage heating in vacuum. Sample No. 3 (p-type).

3. Heating in Vacuum to 650°K

The development of a system for heating by a current [109] has made it possible to reach higher temperatures and thus obtain new data on the dependence of the concentration of surface recombination centers on the heating temperature. It has been found that heating at 500°K does not saturate the concentration of the recombination centers, as suggested in [14], but the concentration reaches a maximum and begins to decrease at still higher temperatures [18].

A family of s (Y) curves, recorded by heating in steps to 650°K in vacuum, is shown in Fig. 24. It is evident that, as in the case of heating to 500°K, the heating to 600°K

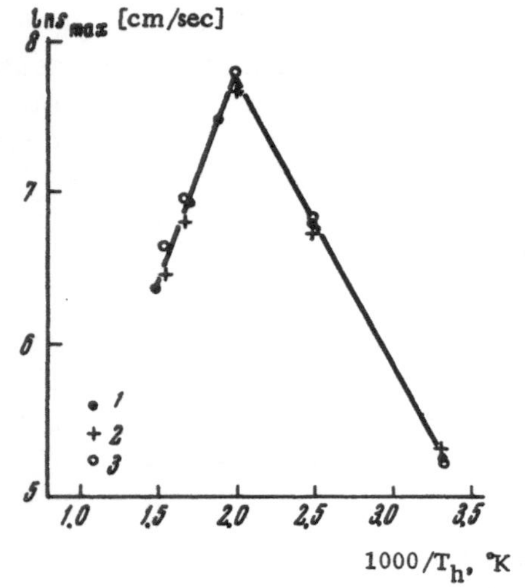

Fig. 25. Dependence of the maximum surface recombination velocity on the reciprocal of the heating temperature T_h: 1), 2) p-type samples Nos. 3 and 5 heated in vacuum; 3) n-type sample No. 118 heated in dry oxygen.

does not alter the form of the s (Y) characteristic. However, the characteristic recorded after heating to 650°K differs in shape from the other curves. The difference increases with increasing duration of heating at 650°K. The s (Y) curve obtained after heating at 650°K can be easily analyzed into a constant component and a curve similar to the s (Y) dependences obtained after heating to lower temperatures. This shows that the recombination centers affected by heating are all of the same nature. Consequently, the maximum values of the surface recombination velocity are proportional to the concentration of the recombination centers.

Figure 25 shows the dependence of ln s_{max} on the reciprocal of the heating temperature T_h for the case of heating in steps to 650°K in vacuum. We can see that heating above 500°K reduces greatly the concentration of the recombination centers so that after heating at 650°K this concentration is only 30% of its maximum value. Thus, new recombination centers are generated by heating in the 300-500°K range, whereas heating to temperatures of 500-650°K annihilates some of the centers.

Assuming that surface recombination is due to oxygen present at random positions on the germanium–oxide interface, we find that the experiments can be interpreted by postulating that oxygen is removed from the germanium surface in the 500-650°K range. On the other hand, the published data [110] suggest that these temperatures are far too low for the desorption of oxygen from the germanium surface.

The alternate mechanism may be the diffusion of oxygen atoms from the surface of germanium into the interior. In this case, the surface recombination velocity should decrease even if we assume that an atom of oxygen, diffusing from the surface into the interior, continues to act as a recombination center. This is not self-evident but experiments on the etching of samples heated in vacuum to 650°K seem to confirm this hypothesis. The effective lifetime in such samples is considerably shorter than in samples not subjected to this heat treatment and the method of successively etching away layers shows that the volume lifetime in the heat-treated samples is about 150 μsec, whereas the volume lifetime in the original material is 600-1000 μsec.

Further support of this hypothesis is provided by the observation that the s (Y) curve, obtained after heating to 650°K, can be analyzed into a bell-shaped curve close to the theoretical dependence and a constant component independent of the surface potential. This is to be expected for recombination centers located outside the space-charge region near the surface.

The hypothesis of a reduction in the concentration of the surface recombination centers due to the desorption of oxygen or its diffusion into the interior was checked by an experiment in which a sample of germanium was heated to 650°K in an oxygen atmosphere. The results, obtained for a large number of samples, showed that the curves for the samples heated in oxygen were completely identical with the curves obtained after heating in vacuum (Fig. 25). An oxygen atmosphere (40-200 mm Hg) had no effect on the shape or amplitude of these curves.

4. Effect of Ozone on a "Dehydrated" Surface

An increase in the surface recombination velocity, caused by the exposure to ozone, is the main experimental observation supporting the hypothesis that oxygen is responsible for the surface recombination in germanium. Therefore, it seemed interesting to investigate in more detail the influence of ozone on the surface recombination mechanism. There were still many unsolved problems associated with the influence of ozone on the surface of germanium. First, the investigations reported in [13] were carried out using low concentrations of ozone set by the limitations of the method employed. Consequently, the maximum surface recombination velocity obtained after treatment in ozone was only 150 cm/sec. It was naturally interesting to find the maximum concentration of recombination centers which could be achieved by the exposure to ozone.

Our experiments were carried out using ozone flowing continuously through the working enclosure. When the flow of ozone was stopped the surface recombination velocity began to decrease and eventually reached its initial value (this process was accelerated by the presence of water vapor). This indicated that the new centers were either unstable or that no new centers were produced. The point was this: a large number of surface recombination centers was present on a freshly etched sample but these centers were inactive since they were neutralized by physically adsorbed water. Consequently, if the ozone simply introduced new recombination centers, this should have had no effect on the surface recombination velocity during the ozone treatment. We should have observed an increase in the number of the recombination centers only after the removal of the "compensating" water (by heating to 500°K). However, the experimental results were precisely opposite: new centers appeared in an ozone atmosphere without a preliminary heating. The necessary condition for this effect was the removal of the compensating water.

Consequently, the new recombination centers introduced by ozone could become active if the ozone neutralized in some way the "compensating" water. However, this effect was the equivalent of pumping or heating, i.e., the necessary condition was also the sufficient condition for an increase in the surface recombination velocity and, therefore, the effects of ozone could be explained without assuming that it produced new recombination centers.

A careful analysis of the first experiments on ozone failed to prove the generation of new recombination centers. Therefore, additional experiments on the effect of ozone were planned so as to remove any ambiguity. As mentioned earlier, the highest value of the concentration of the recombination centers was obtained after heating the germanium to 500°K. This treatment removed practically all the physically adsorbed water responsible for the compensation of the surface recombination centers. The effect of oxygen after heating to 500°K should show whether ozone produced any additional recombination centers.

Since the results of heating germanium samples in oxygen were identical with the results of heating in vacuum, all these experiments were carried out in an atmosphere of carefully dried oxygen.

The samples were heated for 5 min at 500°K. The measurements were carried out after cooling to 300°K. Next, a built-in ozonizer was switched on and changes in the surface parameters were recorded.

The results of one of such experiments are presented in Fig. 26. We found that the exposure of dehydrated germanium to ozone did not increase the surface recombination velocity, as in the case of a freshly etched surface [13, 60]. On the contrary, ozone reduced this velocity by a factor of 2.5 in 20 min from the moment of switching on the ozonizer. The new value of the maximum surface recombination velocity (about 800 cm/sec) remained fairly stable

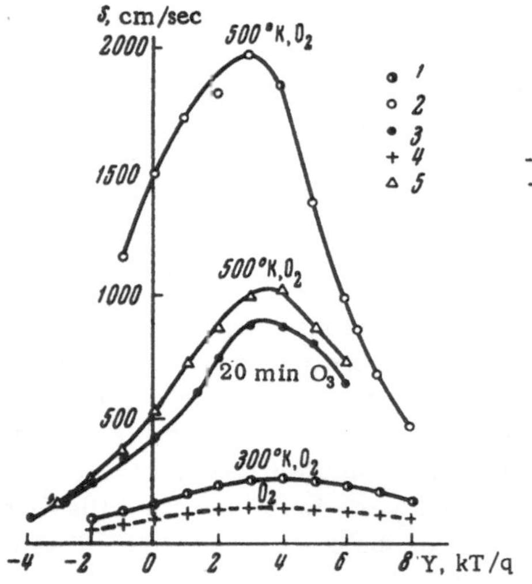

Fig. 26

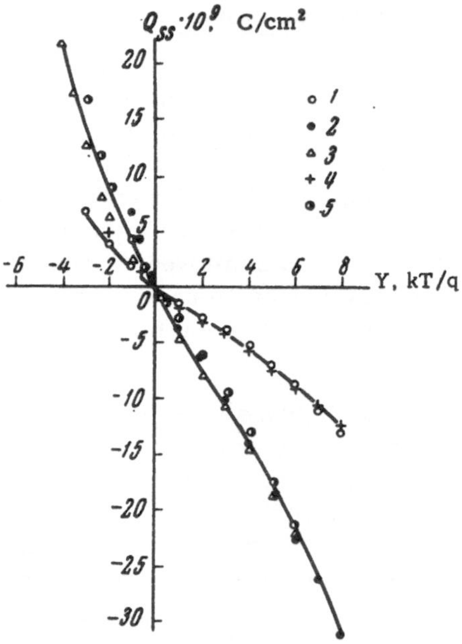

Fig. 27

Fig. 26. Effect of ozone on the surface recombination velocity. Sample No. 54 (p-type). 1), 2), 3) First cycle; 4), 5) after exposure to water vapor.

Fig. 27. Changes in the charge captured by the fast states. Sample No. 34 (p-type). 1) Before heating; 2) after heating at 500°K; 3) after 20-min exposure to ozone; 4) after exposure to water vapor; 5) after second heating at 500°K.

after the ozonizer was switched off and this indicated the absence of traces of water vapor from the working atmosphere. Since the shape of the s (Y) curve was not greatly affected by this treatment, we could conclude that the concentration of the recombination centers decreased by a factor of 2-2.5. These experiments were repeated on many other samples and the results showed convincingly that the ozone destroyed a considerable proportion of the recombination centers.

It is interesting to note that the exposure to ozone did not alter, within the limits of experimental error, the charge captured by the fast states Q_{ss} (Y) and, consequently, it did not alter the density of the fast states in the investigated range of the surface potential (Fig. 27). In this respect, the effect of ozone was similar to heating at 600-650°K [20]. When a sample was exposed to water vapor the surface recombination velocity decreased strongly and the maximum value of this velocity fell to about half the value for an unheated sample (Fig. 26). The subsequent heating at 500°K yielded a maximum value of the surface recombination velocity which was half that obtained after the first heating. This showed that the ozone destroyed irreversibly about half the recombination centers. The Q_{ss} (Y) curves repeated exactly the corresponding curves in the first cycle (Fig. 27).

Since the observed effect of ozone on dehydrated surfaces seemed to contradict the results reported in [13, 60], we carried out the following experiments: we studied the effect of ozone on a freshly etched surface before the establishment of an equilibrium value of the surface recombination velocity and this was followed by heating at 500°K. The results are presented in Fig. 28.

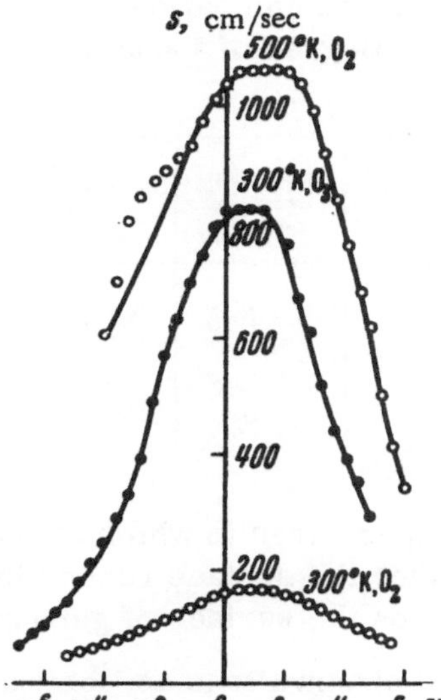

Fig. 28. Effect of ozone on the surface recombination velocity of a freshly etched n-type sample No. 112.

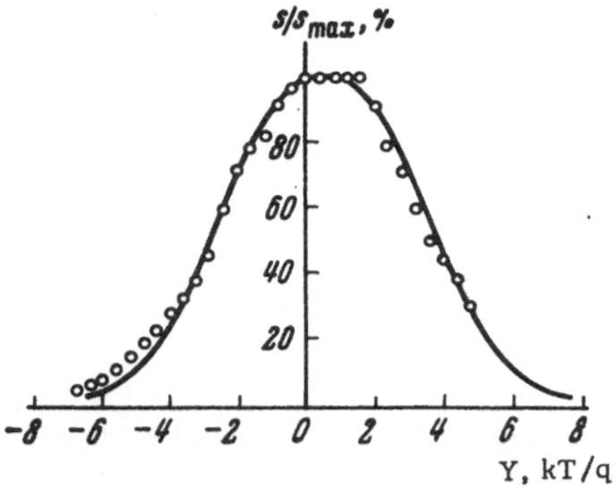

Fig. 29. Dependence s (Y) for a freshly etched sample (n-type sample No. 112) exposed to ozone. The experimental results are represented by points and the continuous curve is a theoretical dependence for a single level at $E_t - E_i = 0.11$ eV, $c_p/c_n = 11$.

As found in our previous investigations, the surface recombination velocity increased strongly in the ozone atmosphere (the surface recombination velocity rose more strongly than in the published investigations because the concentration of ozone in our study was higher). However, the subsequent annealing of a sample at 500°K had practically no effect on the maximum surface recombination velocity, which was found to be about half that observed for samples not subjected to preliminary ozonization.

It was interesting to note that the exposure of a freshly etched surface to ozone produced recombination centers of only one type. The experimental dependence s (Y) agreed well with a theoretical curve plotted using the parameters $E_t - E_i = 4.4$ kT, $c_p/c_n = 11$ (Fig. 29). Subsequent heating of the sample at 500°K did not alter the concentration of these centers. It is evident from Fig. 28 that the s (Y) curves obtained after the exposure of the sample to ozone and after its heating at 500°K differed only by a shift along the ordinate, i.e., the difference was simply a straight line s (Y) = const, which represented recombination centers of some different kind (we shall not discuss the nature of these centers).

5. Influence of Water Vapor after Heating above 500°K

Since ozone produced no new recombination centers but, in many respects, caused effects similar to those of heating [111], the question of the nature of the recombination centers on a real surface of germanium arose again. This question could not be answered by heating to 500°K because the surface properties obtained after such heating were governed completely by the presence of physically adsorbed water. Information on the nature of these centers could be provided by experiments involving heating to temperatures above 500°K, which could alter the concentration of the surface recombination centers.

Analysis of the data obtained after heating to temperatures of 500-650°K raised the question whether the coincidence of the range of temperatures in which the concentration of the

TABLE 5. Reduction in the Concentration of Recombination Centers after Heating at 650°K as a Function of Degree of Vacuum

Sample No. (p-type)	s_{max} before heating, cm/sec	s_{max} after heating at 500°K, cm/sec	s_{max} after heating at 650°K, cm/sec	$\frac{s_{max}\ 650°\ K}{s_{max}\ 500°\ K}$, %	Vacuum during measurements, mm/Hg	Vaccum during heating at 650° K, mm/Hg
3	200	2120	630	30	$2.1 \cdot 10^{-6}$	$2.1 \cdot 10^{-5}$
66	350	2110	750	35	$1.6 \cdot 10^{-6}$	$3 \cdot 10^{-5}$
56	86	1640	1240	76	$1.5 \cdot 10^{-5}$	$1.2 \cdot 10^{-4}$
116 (n-type)	120	1880	1970	104	$1.2 \cdot 10^{-5}$	$2.2 \cdot 10^{-4}$

recombination centers decreased and of the range of temperatures in which chemisorbed water monolayers were removed [112] was accidental, or whether the surface recombination centers were associated with the presence of chemisorbed water on the surface of germanium.

This question followed also from such experimental observations as, for example, the influence of the degree of vacuum on the ln $s_{max}(1/T_h)$ curves. The results presented in Sec. 3 in the present chapter were obtained by heating in a vacuum of about 10^{-6} mm Hg, which was maintained during the cooling of a sample and during the measurements on it. This was achieved by heating a sample at a rate sufficiently slow for pumping away the water evolved from the surface. In those cases when no precautions were taken and the temperature was raised very quickly, it was found that the vacuum in the cartridge fell to 10^{-5}-10^{-4} mm Hg during cooling. The results obtained under these conditions differed appreciably from those obtained by heating in higher vacuum. Table 5 gives the changes in s_{max} after heating at 650°K, compared with S_{max} after heating at 500°K, as a function of the degree of vacuum during the period of cooling and during the measurements. We can see that heating at 650°K reduces s_{max} by a factor of about 3 if the vacuum after heating is 10^{-6} mm Hg but if this vacuum is 10^{-5} mm Hg, the value of s_{max} is practically unaffected.

We determined the highest temperature after which the initial properties of the surface could be re-established by exposure to water vapor. After each heating and measurement, a sample was placed in an atmosphere of saturated water vapor. After pumping for 1 h, we carried out the next heating at a higher temperature. In addition to the main result (after heating in vacuum above 500°K, the subsequent exposure to water vapor did not re-establish completely the initial surface properties), we obtained an interesting secondary result: the maximum surface recombination velocity after a sample had been heated in the 500-650°K range did not decrease but increased quite strongly. This was evident from the following data obtained for a p-type sample No. 47 after a cycle consisting of heating in vacuum, exposure to saturated water vapor, and heating in vacuum:

Treatment (T, °K)...	300	400	450	500	550	600	650
s_{max}, cm/sec....	220	1590	1780	1980	2150	2800	4000

[Since the s (Y) curves were identical, only the values of s_{max} are given in the above table.] A distinguishing characteristic of the heating cycle was that it was carried out in a poor vacuum of about $(2-4) \cdot 10^{-4}$ mm Hg (this pressure was governed by the partial pressure of the water vapor in the vacuum system).

Thus, this experiment seemed to confirm the hypothesis that chemisorbed water was responsible for surface recombination. However, such a definite conclusion could not be drawn. This was because a sample was exposed to saturated water vapor for days before each heating and such exposure could give rise not only to adsorption but also to chemical processes on the

TABLE 6. Changes in s_{max} (cm per sec) after Heating in Vacuum and Exposure to Water Vapor

Treatment (T, °K)	Sample No. 28 (p-type)	Sample No. 31 (p-type)
300	250	250
400	980	900
500	1740	1750
600	1000	1150
H_2O (10^{-2} mm Hg)	1350	1220

surface of a sample. A sample exposed to saturated water vapor could become covered by a film of liquid water and such a film would be capable of dissolving the oxide layer.

Therefore, in subsequent experiments, water vapor was admitted through a capillary so that we were able to establish a water vapor pressure sufficiently low to avoid the formation of a film of liquid water. Since the purpose of these experiments was only to ensure the chemisorption of water, it was sufficient to expose a sample to water vapor at pressures of about 1 mm Hg. According to Law [112], only one water monolayer was chemisorbed at this pressure. The following cycle of operations was then carried out: heating in steps up to 600°K, the admission of a small amount of water, and observations of changes in the surface recombination velocity.

The results of these experiments (Table 6) showed that the admission of water vapor produced a small rise of the surface recombination velocity. After a sample had been heated at 600°K, the value of s_{max} decreased by a factor of 2 compared with the value after the sample had been heated at 500°K, whereas the rise of s_{max} caused by the admission of water vapor was only 20-30%.

This result could be easily understood bearing in mind that, according to Law [112], the desorption of chemisorbed water occurred not in the form of molecules but in accordance with the reaction

$$Ge + xH_2O = GeO_x + xH_2,$$

i.e., this reaction "healed" the sites at which water molecules were located in the germanium–oxide interface, and when the sample was exposed to a wet atmosphere the water molecules could not reoccupy their original sites in the interface. Prolonged exposure to saturated water vapor could give rise to etching at these sites, and could increase the surface recombination velocity (as described earlier).

Thus, the results of the experiments on the effect of water vapor on a sample subjected to heating up to 600-650°K confirmed, to some extent, the hypothesis that chemisorbed water was responsible for surface recombination on germanium but these results could not be regarded as a direct proof.

Such a proof was provided by experiments in which germanium samples were heated to high temperatures in vacuum.

6. Heating at High Temperatures

Preliminary studies of the influence of heating above 650°K on the surface recombination velocity were carried out without the use of the field-effect method. In this case, the measured value of s depended not only on the preceding heat treatment but also on changes in the surface potential resulting from such treatment. In the field-effect experiments, we compared s_{max} obtained after various treatments; in the experiments without the field effect we compared the surface recombination velocities at surface potentials which were set by the treatment itself. However, the field-effect data showed that, in the absence of a transverse electric field, the surface potential Y_0 did not change greatly after a sample had been heated to 650°K and, usually, we found that $s_0 = s(Y_0) \approx s_{max}$. Moreover, s_0 could differ at most by a factor of 2-3 from the

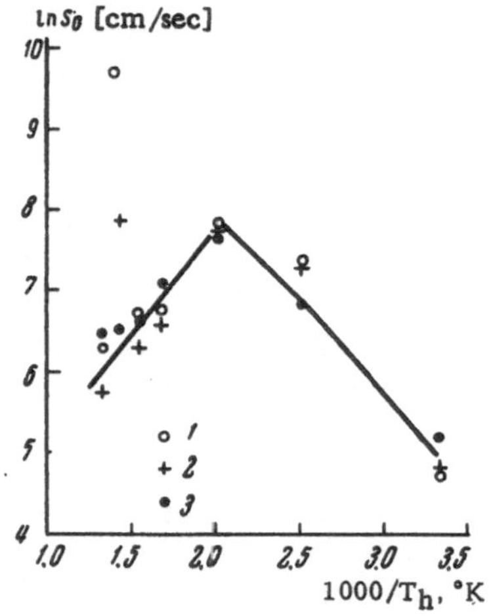

Fig. 30. Dependence of s_0 on the reciprocal of the temperature (T_h) of heating in 10^{-5}–10^{-4} mm Hg vacuum. 1) Sample No. 9 (p-type); 2) sample No. 116 (n-type); 3) sample No. 58 (p-type) heated in 10^{-6} mm Hg vacuum.

value of s_{max}, whereas various treatments altered s_{max} by a factor of 10–20. Consequently, we could ignore small changes in s_0.

The value of s_0 increased considerably after the heating at 700–720°K (samples Nos. 9 and 116 in Fig. 30). When the temperature was increased to 750°K, s_0 again fell to low values.

It was interesting to note that the experimental points obtained after the heating to 650°K were similar for all samples but after the heating at 700–720°K the values of s_0 differed strongly from sample to sample. After the heating at 750°K, the values of s_0 were again practically identical for all the samples.

The peak in the value of s_0, observed after the heating at 700–750°K, was several times larger than the value of $s_0 \approx s_{max}$ after heating at 500°K. This indicated that the high-temperature heat treatment produced a large number of active surface recombination centers which disappeared after treatment at 750°K.

It is evident from Fig. 30 that after a sample had been heated at 700°K the values of s_0 for different samples differed by a factor of 10. This scatter could not be attributed to the dependence s (Y). It was found that the degree of vacuum in the system during measurements and the value of s_0 were correlated: the poorer the vacuum, the higher was the value of s_0.

Thus, some component of the residual gas in the system had a very strong effect on the surface properties and increased greatly the number of surface recombination centers.

To identify the residual gas component responsible for the recombination we heated our samples in various gases and vapors and we also exposed the samples to gases and vapors after heating at 750°K. We found that the admission of water vapor after heating to 750°K increased the value of s_0 by a factor of 10–20.

We also carried out heating in high vacuum. In this experiment, the measurements were made in $(1.5\text{–}2.0) \cdot 10^{-7}$ mm Hg vacuum and the vacuum was not allowed to deteriorate below 10^{-6} mm Hg during the heating. We found that the peak of s_0, observed after the heating at 700°K, disappeared completely and the value of s_0 decreased monotonically with increasing heating temperature (Fig. 30).

These results of the preliminary and qualitative measurements seemed to confirm that water was responsible for the surface recombination on the germanium. On the other hand, the possibility of altering s by a change in the atmosphere provided a convenient investigation method. Therefore, in spite of the technical difficulties encountered in the high-temperature heating, we carried out some measurements using the field effect.

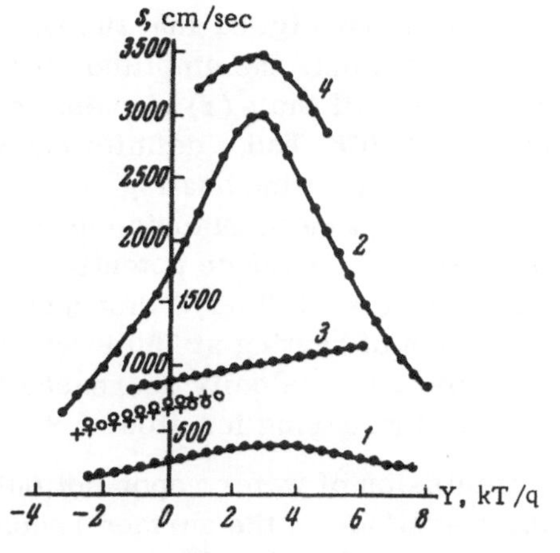

Fig. 31. Changes in the s (Y) curves after high-temperature heating in vacuum. Sample No. 103 (p-type). 1) Before heating; 2) after heating at 500°K; 3) after heating at 750°K for 5 min; 4) after admission of H_2O. Circles represent the results obtained after heating at 750°K for 1 h; crosses represent the results of a second heating at 750°K for 20 min after admission of water vapor.

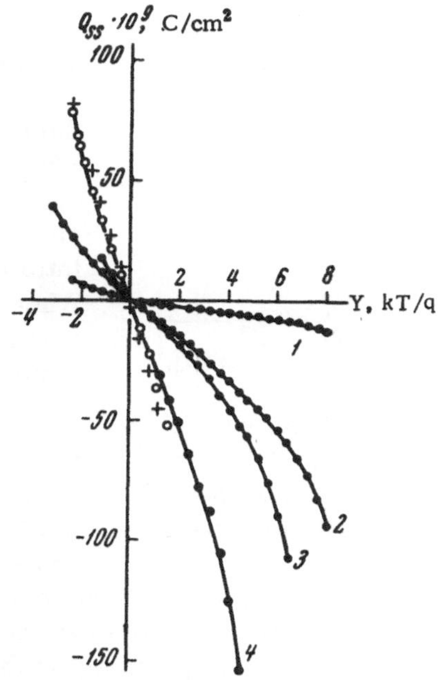

Fig. 32. Changes in the charge captured by the fast states caused by high-temperature heating in vacuum (p-type sample No. 103). Curves 1-4 have the same meaning as in Fig. 31.

All these experiments were carried out in high vacuum in order to avoid an uncontrolled influence of the water. (A sample was cooled to 300°K in 1-1.5 h and a small concentration of water vapor could have altered strongly the results.)

Figures 31 and 32 show the dependences s (Y) and Q_{ss} (Y), recorded for a p-type sample No. 103. It was very interesting to note the rapid rise of Q_{ss} (Y) after the heating in the 650-750°K range. Since all the Q_{ss} (Y) curves were similar, we could conclude that a change took place in the density of defects at the germanium–oxide interface and this conclusion was reached without the unconvincing resolution of the Q_{ss} (Y) dependences into "elementary" curves (the justification for this procedure will be considered in the next chapter).

The change in the surface recombination velocity followed almost exactly the change in the number of fast surface states after heating to 650°K [18]. Heating above 650°K produced different results: the density of the fast states increased monotonically and, after the heating at 750°K, this density became about 10 times higher than the density observed after the heating at 500°K; the surface recombination velocity decreased monotonically.

When a sample was exposed to water vapor (admitted through a capillary to produce a partial pressure of 10^{-2} mm Hg) after the sample had been heated at 750°K, the surface recombination velocity increased strongly but, within the limits of the experimental error, Q_{ss} (Y) remained unchanged, i.e., the number of defects at the germanium–oxide interface was not affected by the admission of water but some of these defects became recombination centers. The subsequent heating in vacuum destroyed completely these new recombination centers but the number of defects [Q_{ss} (Y)] again did not change.

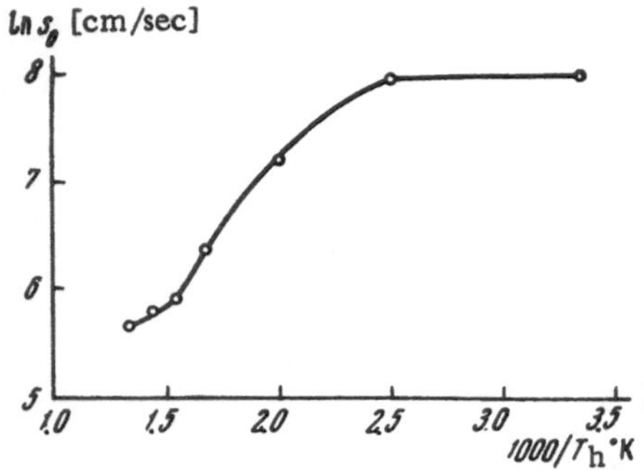

Fig. 33. Second heating cycle in vacuum after heating to 750°K in vacuum and admission of water vapor. Sample No. 108 (n-type).

It is evident from Fig. 31 that the heating of a sample at 750°K altered not only the amplitude but also the shape of the s (Y) curve. All the s (Y) curves, recorded after the heating up to 650°K, had a definite maximum at Y = +3.0 kT/q, whereas after the heating at 750°K the s (Y) dependence changed to a monotonic curve increasing slowly with increasing surface potential. The characteristic minimum at Y = + 3.0 kT/q disappeared completely. When the duration of heating at 750°K was increased, the surface recombination velocity decreased but s remained a monotonically increasing function of Y.

The admission of water vapor not only greatly increased the magnitude of the surface recombination velocity but also produced a characteristic maximum in the Y = + 3.0 kT/q region. Subsequent heating at 750°K destroyed completely this change. Thus, the admission of water vapor after a sample had been heated at 750°K in vacuum produced the same recombination centers which existed on the real surface of germanium after it had been etched and which became active after the pumping or the heating to temperatures up to 650°K.

We determined in detail the process of annihilation of the recombination centers introduced by the admission of water vapor: this was done by heating a sample at 750°K, admitting water vapor, and repeating the heating in vacuum. It is evident from Fig. 33 that the recombination centers, formed again as a result of the adsorption of water molecules, retained their concentration when the sample was heated to 400°K. In the 500-650°K range, these centers disappeared completely and further heating at 650-750°K had little influence on the surface recombination velocity. This observation provided a convincing proof that the reduction in the surface recombination velocity caused by heating in the temperature range 500-650°K was due to the desorption of chemisorbed water molecules from the surface of germanium.

It was interesting to consider the changes in the surface potential, which were observed in the absence of a transverse electric field (Y_0) after high-temperature heating in vacuum (Table 7). Beginning from 650°K, the surface potential shifted in the p-type direction. Exposure to water vapor converted the surface back to n-type conduction and subsequent heating at 750°K produced p-type conduction. When the duration of heating at 750°K was increased, the absolute value of the negative surface potential continued to increase. However, we were unable to determine this value since the field-effect curve shifted so much in the p-type direction that we could not reach the surface conductivity minimum.

TABLE 7. Changes in Y_0 Caused by Heating in Vacuum (n-type sample No. 113)

Treatment (T, °K)	Y_0	Treatment (T, °K)	Y_0
300	+1.2	700	+4.7
400	+1.2	720	+2.7
500	+1.7	750	−1.5
600	+4.4	H_2O ($p \cdot 10^{-3}$ mm Hg)	+0.9
650	+5.6	750 after H_2O	−2.7

7. Heating at High Temperatures in Oxygen and Ozone

The heating of germanium at high temperatures in high vacuum [111] shows that the principal contribution to the surface recombination velocity is made by defects at which water molecules are chemisorbed. Such heating can remove completely the water chemisorbed on the surface of germanium and reduce the surface recombination velocity to low values. However, such heating alters the structure of the oxide film on the germanium surface so that the density of the fast surface states increases considerably. Moreover, the oxide film has a fairly loose structure and the defects at the germanium–oxide interface are easily accessible to molecules from the atmosphere. Even a slight deterioration in the vacuum to 10^{-4}-10^{-3} mm Hg increases considerably the surface recombination velocity and the admission of air gives rise to surface recombination velocities of the order of 10^5 cm/sec.

On the other hand, experiments [111] indicate that ozone is capable of healing the defects at which the recombination centers are formed. It would seem that if high-temperature heating is carried out not in vacuum but in an atmosphere of oxygen or ozone, we should obtain low values of the surface recombination velocity which are not greatly affected by the ambient gaseous medium.

The results of heating up to 750-800°K in an atmosphere of dry oxygen (oxygen pressure 60-200 mm Hg) showed that the surface recombination velocity decreased monotonically with increasing temperature of heating, as in the case of heating in high vacuum. The most characteristic result was this: the subsequent admission of water vapor or atmospheric air and repeated cycles of heating in vacuum did not alter appreciably the surface recombination velocity of the samples heated at high temperatures in oxygen.

Heating in an ozone atmosphere was even more effective [113]. Figure 34 shows the results of experiments carried out on two p-type samples ($\rho \approx 30\ \Omega \cdot$cm). When a given temperature had been reached, the ozonizer was switched on throughout the heating period. An interesting point was that the surface recombination velocity, obtained after a sample had been heated at 750°K in an ozone atmosphere, was lower than the velocity on a freshly etched sample. After the heating in vacuum at 750°K, the velocity decreased on the average to 300 cm/sec, whereas heating in ozone at the same temperature gave the value s_0 = 60-70 cm/sec.

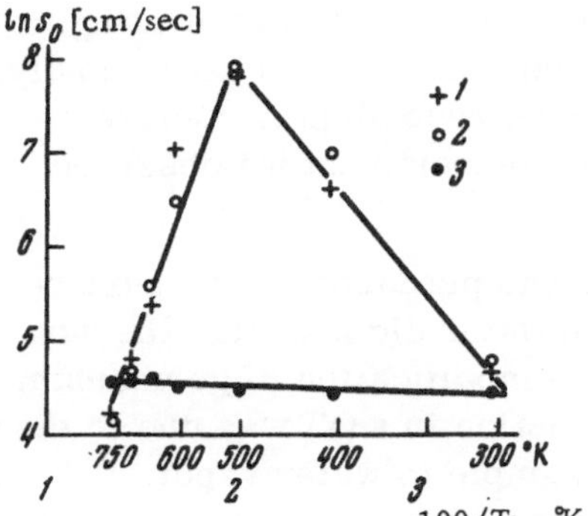

Fig. 34. Dependence of s_0 on the reciprocal of the temperture (T_h) of heating: 1) p-type sample No. 2 in ozone; 2) p-type sample No. 95 in ozone; 3) p-type sample No. 95 heated for the second time in vacuum after exposure to air.

After it had been heated in ozone, one of the samples was exposed to room atmosphere and then heated in vacuum (Fig. 34). The exposure to air increased the surface recombination velocity from 63 to 77 cm/sec. A subsequent heating cycle in vacuum had very little effect on the surface recombination velocity.

A second sample was heated in ozone and stored in air for several weeks. It was found that the surface recombination velocity did not change at all. Even when this sample was immersed in water, its effective lifetime remained in the 140-145 μsec range ($s_0 \approx$ 90 cm/sec).

The results of these experiments can be explained in the following different ways:

1) a considerably smaller number of defects at the germanium–oxide interface is produced during heating in oxygen or ozone than during heating in vacuum;

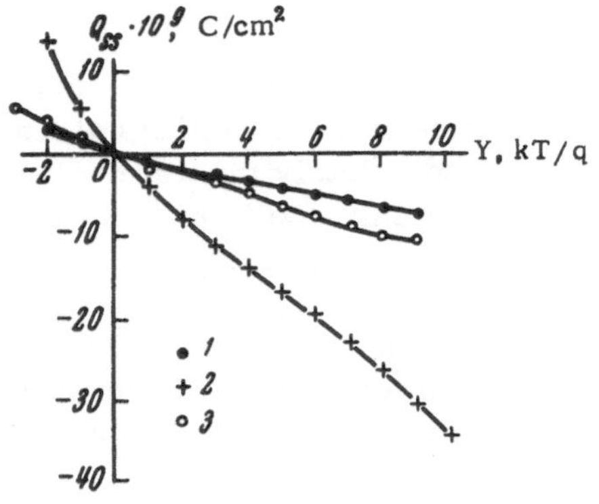

Fig. 35. Changes in the charge captured by the fast states caused by heating in oxygen. Sample No. 42 (p-type). 1) Before heating; 2), 3) after heatings 500 and 715°K, respectively.

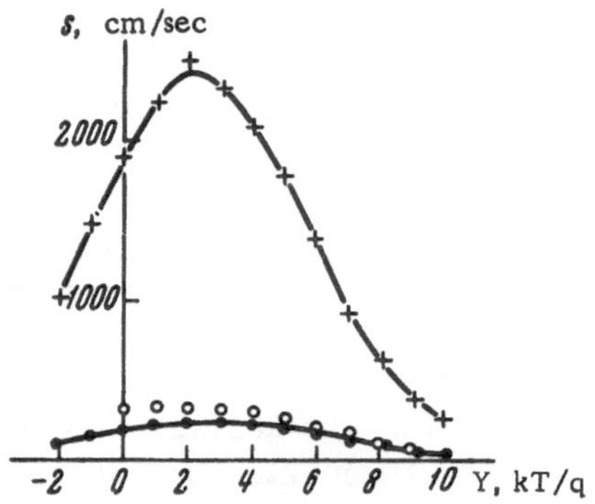

Fig. 36. Changes in s (Y) caused by heating in oxygen. Sample No. 42 (p-type). The symbols used for experimental points have the same meaning as in Fig. 35.

2) the heating in an oxygen atmosphere produces a stronger shift of the surface potential in the direction of high negative values than does the heating in vacuum;

3) oxygen hinders the adsorption of water molecules at defects.

The correct interpretation can be given only on the basis of field-effect experiments. In view of the high thermal conductivity of oxygen, these experiments are more difficult than in vacuum.

Figures 35 and 36 give the results obtained by heating a p-type sample to 715°K in an atmosphere of dry oxygen. The highest density of the fast states was observed after heating at 500°K. When the temperature was increased, the density of the fast states decreased rapidly, approaching the density on an unheated sample (Fig. 35). The surface recombination curve (Fig. 36) was also quite close to the original s (Y) curve, i.e., in this experiment we observed a qualitative correlation between the changes in Q_{ss} (Y) and s (Y).

The experimental results indicated that a perfect oxide film was produced on the surface of germanium which had been heated in ozone. Since immersion in water did not alter the surface characteristics, the oxide film was evidently a water-insoluble modification of germanium dioxide [114]. Thus, when a sample was heated in vacuum, the reduction in s (Y) was due to the removal of water from the germanium surface; the exposure of a sample to water vapor rapidly increased the surface recombination velocity to values of the order of 10^4 cm/sec. On the other hand, the heating of a sample in oxygen or ozone reduced the number of recombination centers and lowered very considerably the total number of fast states.

The high-temperature heating in ozone was, to some extent, similar to the chemical etching of a sample. Such a treatment reduced the number of defects (fast states) and the surface recombination velocity assumed a low value. However, there was an important difference between the ozone and etching treatments. The heating of a freshly etched sample in vacuum to 500°K increased the surface recombination velocity and the density of the fast states by about an order of magnitude, whereas the vacuum heating of a sample subjected to a high-temperature ozone treatment had practically no effect on the surface recombination velocity (Fig. 34).

CHAPTER IV

Discussion

1. Analysis of Experimental Data

The experimentally obtained dependences of the charge trapped by the fast states on the surface potential, $Q_{ss}(Y)$, can be described satisfactorily by various combinations of discrete levels as well as by continuous distributions of levels [68, 70, 115]. The field-effect data are clearly insufficient for a decision to be made in favor of one of the available models. The experimentally obtained curves have no characteristic points which could be used for level identification. The monotonic rise of the captured charge with increasing absolute value of the surface potential and the absence of inflection points in the experimental curves (such points could be used as "boundaries" between discrete levels) would seem to support the continuous distribution model. However, most of the investigators have interpreted their results using various combinations of discrete levels.

Attempts have been made to support the discrete model by interpreting all the experimental curves employing a single set of discrete levels [18]. Such an operation does not provide a convincing proof of the discrete nature of the levels. An experimental curve can be described by many sets of discrete levels. The number of such levels in a set increases as the range of the investigated surface potentials increases. It is usual to assume that levels are separated by 3-4 kT from each other. We can easily show that a set of such levels can be used to describe any experimental dependence $Q_{ss}(Y)$. Such an operation is analogous, to some extent, to the expansion of an arbitrary function as a series of Fermi functions. The most typical example of such an expansion as a Fermi series is given in [116]. The experimental curves $Q_{ss}(Y)$ obtained at 233.4°K (Fig. 37) has an extended linear region in the range 6-7 kT/q. This straight line can be described, to a high degree of accuracy, by a set of Fermi functions. The excellent agreement between the expansion and the actual curve is the best proof that the expansion of $Q_{ss}(Y)$ as a Fermi series is purely formal. Therefore, systems of discrete levels should be regarded only as convenient approximations. It is incorrect to attribute physical meaning to formally determined levels [3, 4, 7, 15, 19, 88].

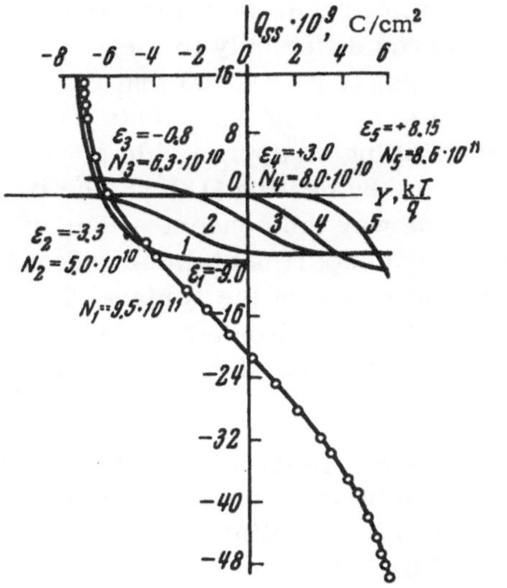

Fig. 37. Data taken from [116]. Circles represent the experimental points; continuous curves are theoretical.

It has been reported in [64, 65] that the surface recombination velocity and the capture by the fast states are altered in the same way by various surface treatments: a treatment which increases the captured charge also increases the surface recombination velocity; the converse is also true. On this basis, it has been concluded that the traps responsible for recombination are the same traps which are active in the field effect. Since this is true in many cases, the foregoing conclusion has been elevated to the rank of a postulate, which has been employed to show that the expansion of the $Q_{ss}(Y)$ curve as a Fermi series is unique. Using the data on the surface recombination velocity, one can then find a recombination level which can be used as the starting point in the selection of a system of discrete levels suitable for the description of the $Q_{ss}(Y)$ curve.

We shall not, at this stage, discuss the correctness of this type of analysis of the field effect but we shall point out that this approach can be used successfully only if there is only one discrete recombination level. Otherwise, the s (Y) curve can be described by an infinite number of variants and, therefore, it cannot be used to obtain a unique distribution of traps.

The experimental curves s (Y) cannot usually be approximated accurately by theoretical relationships. Frequently, particularly in the case of freshly etched surfaces, the s (Y) curves are not bell-shaped (Fig. 3) but the surface recombination velocity varies monotonically with the surface potential. However, a change in the surface treatment can often alter the s (Y) curve in such a way that the change can be attributed to a single discrete recombination level [4, 61, 62, 70].

On the other hand, in a great number of cases, the experimental values of the surface recombination velocity vary much more slowly than predicted by the theoretical formulas [18, 65, 68]. Moreover, the fact that an experimental curve can be approximated satisfactorily by a theoretical expression implying a single level cannot be regarded as a proof of the existence of just one discrete recombination level. The same dependence can also be obtained assuming the existence of several discrete levels and even a continuous distribution [68].

It is desired to separate the experimental curve s (Y) into a series of elementary curves. The shape of the experimental curve cannot by itself provide a unique set of elementary curves; this can be done only if additional experimental data are available [20].

Thus, the surface recombination velocity cannot be used as the starting point of a rigorous determination of the distribution of the fast states. Moreover, the assumption that the densities of the recombination and trapping levels vary in the same way, on which this method of analysis is based, is rather suspect. Heating in the 500-650°K range much reduces the surface recombination velocity, whereas the Q_{ss} (Y) curve remains practically unaffected. A similar result is obtained by vacuum treatment of a freshly etched sample: the surface recombination velocity increases severalfold but Q_{ss} (Y) remains unchanged [75].

Thus, neither the dependence Q_{ss}(Y) by itself nor the combined analysis of the Q_{ss}(Y) and s (Y) curves allows us to determine unambiguously the spectrum of the fast surface states (traps and recombination centers). Additional data are required. Such additional data are provided by an analysis of a family of s (Y) curves for a single sample, recorded after various heat treatments. As mentioned in the preceding chapter, these curves differ only in their amplitude. If all such curves are plotted on the reduced scale s (Y)/s_{max} it is found that the differences between them are within the limits of the experimental error (Fig. 25). This means that the concentrations of all the different recombination centers making an appreciable contribution to the recombination processes are altered in the same way by heating [18]. Since it is incorrect to assume that heating in vacuum increases in the same way the concentrations of surface centers of different origin, it follows that any deviation of the experimental curve from the theoretical dependence cannot be attributed to accidental contamination of the surface [77]. Retention of the same form by the s (Y) curve in spite of a change of its amplitude by a factor of 10-20 can be interpreted only in one way: heating alters the concentration of recombination centers and these centers are all of one type.

This experimental observation cannot be reconciled with the assumption that a range of discrete levels, or a continuous distribution of levels, makes an appreciable contribution to the s (Y) curve [117]. However, although the only possible explanation is that heating alters the concentration of recombination centers which are all of the same kind, the form of the s (Y) curve differs considerably from the theoretical dependence. Moreover, an analysis carried out by Rzhanov [67] shows that allowance for the finite carrier lifetime at the excited levels of a recombination center does not alter the theoretical dependence s (Y). It follows that we must find other causes for the deviation of the experimental curve from its theoretically predicted shape.

The formulas for the dependence of the surface recombination velocity on the surface potential are deduced making the following assumptions (which apply to the formula given in [58] as well as to the Rzhanov formula [67]):

a) the recombination centers present on the surface are all of the same type;

b) the surface is perfectly uniform;

c) the recombination centers are located on a geometrically smooth surface (no distribution with depth is permitted).

A violation of any one of these three conditions should result in the deviation of the experimental s (Y) curve from the theoretical dependence. However, as discussed in the preceding paragraphs, any such deviation is usually taken to indicate the violation of only the first assumption and the possibility that the second and third assumptions have been violated is ignored.

If the surface is not uniform, the experimentally determined dependence s (Y) is the result of the superposition of local curves representing regions of the same type which differ in the surface potential and in the amplitude. The nonuniformity of the surface may be due to various causes [118-121]. From our point of view, the most important is the emergence of different crystallographic faces on the surface [121]. In contrast to microcracks and other local defects, the presence of different crystallographic faces is responsible for the "patchy" structure [62, 65] which may cause broadening of the s (Y) curve. In this case, surface regions of comparable area can have different adsorption properties and regions of different orientation can differ in the specific charge and, consequently, in the surface potential.

This type of surface nonuniformity is manifested in the field-effect experiments, as indicated by the results of investigations of samples with large surfaces cut parallel to various crystallographic planes [76]. The results show that, whereas the changes in the concentration of the recombination centers caused by heating in vacuum are of the same kind for all the investigated orientations, the profiles of the s (Y) curves are different for samples with different crystallographic orientations.

We must recall that our experiments, like those reported in [76], were carried out mainly on samples etched in hydrogen peroxide mixed with an alkali. It is known that this etchant is selective and, therefore, etching of the (111) and (110) surfaces partly reveals other crystallographic faces. Only in the case when the large faces of a sample are parallel to the (100) plane does the etching not cause changes in the orientation of the surface so that it remains uniform in the crystallographic sense. It is found that in precisely this case the experimental s (Y) curves can frequently be approximated satisfactorily by the theoretical formula [20].

Apart from the surface uniformity caused by the action of a selective etchant [108], a uniformity may be induced in the measurements of the field effect. The capacitance of the system consisting of a sample and a field electrode is, on the average, half the value calculated on the assumption that the gap between the electrode and the sample is filled by mica. This shows that an air gap having an effective thickness of 1.5-2.0 μ exists in such a capacitor. Consequently, part of the surface is in contact with the mica and the rest is separated from the mica by the air gap. This gives rise to a nonuniformity in the specific capacitance and, consequently, of the charge induced by the application of a voltage. This inhomogeneity due to the nonuniform distribution of the transverse electric field is responsible for distortions of the field-effect curve in the region of a minimum when large static or pulsed fields are applied to the capacitor. This is observed as a change in the form of the $Q_{ss}(Y)$ curve and it has been interpreted as induced adsorption under the influence of the field [3]. However, the observation that the slopes of the branches of the field-effect curve remain unchanged and only the region near the conductivity minimum is broadened is evidence against the case for adsorption and any change in the effective distribution of the fast surface states. The observations can be interpreted in terms of the superposition of the field-effect curves of regions with different specific capacitance.

Similar broadening of the $Q_{ss}(Y)$ and $s(Y)$ curves is observed when recombination centers are distributed across the depth of the space-charge layer, which is a highly likely occurrence [62].

We can thus see that departures of the experimental curve $s(Y)$ from the theoretical form can be easily explained by the nonuniformity of the surface, and that the constancy of the shape of the experimental curve observed when the surface recombination velocity changes by a factor of 10-20 shows that there is a change in the concentration of surface recombination centers of one type only.

The lack of correlation between the changes in $Q_{ss}(Y)$ and $s(Y)$ is to be expected. It is unlikely that all the fast states are recombination centers. A recombination center is a very special case of a trap [122]. A center is active in recombination when the probability of trapping is similar for electrons and holes, i.e., when the ratio of the electron and hole capture cross sections does not differ greatly from unity. Naturally, the number of recombination centers is usually a small fraction of all the fast surface states.

We shall now show that the experimental results confirm these general conclusions. Thus, the change in the maximum surface recombination velocity by a factor of about 5, caused by heating at 500-650°K (Fig. 25), shows that the concentration of the recombination centers decreases by a factor of 5. Since such heating has little effect on $Q_{ss}(Y)$ [18], it follows that, in any energy interval, the concentration of the recombination centers is at least one order of magnitude lower than the concentration of the traps. A similar conclusion follows from the experiments in which samples are placed in an evacuated chamber after etching [75]. Finally, the most conclusive is the experiment in which a sample heated at 750°K is exposed to water vapor. We then observe the characteristic bell-shaped curve with the same parameters as the recombination curves obtained after the sample had been heated up to 650°K. The $Q_{ss}(Y)$ curve is not affected by such a treatment.

Thus, as assumed a priori and demonstrated experimentally, the $Q_{ss}(Y)$ curve cannot be used to determine the concentration of recombination centers. All that we can do is to estimate the upper limit of this concentration, bearing in mind that a change in the concentration of the recombination centers by a factor of 10 has little effect on the $Q_{ss}(Y)$ curve.

We can see that the method involving the combined use of the $Q_{ss}(Y)$ and $s(Y)$ curves does not provide the relevant information, and again we meet the problem of an unambiguous expansion of the $Q_{ss}(Y)$ curve as a Fermi series. Since, in practice, the $Q_{ss}(Y)$ curve represents a dependence not related to the recombination centers in view of their low concentration, there are no reasons for preferring any particular method for the expansion of this curve as a Fermi series. Therefore, in discussing changes in the $Q_{ss}(Y)$ curves due to surface treatments, we shall not assume either a set of discrete levels or a continuous distribution of traps. All that we say is that, from the physical point of view, it is more likely that the distribution of traps is continuous.

Fortunately, our experimental data on $Q_{ss}(Y)$ can be interpreted without the expansion of this curve into a series of elementary dependences. The $Q_{ss}(Y)$ curves are not affected by heating to 650°K. Figure 38 shows the curves obtained after various heat treatments: these curves are plotted on a relative scale with $Q_{ss}(Y)/Q_{ss}(+6)$ as the ordinate. The scatter is within the limits of the experimental error. This allows us to describe completely a change in the density of the fast surface levels, in the investigated range of the surface potential, by a change

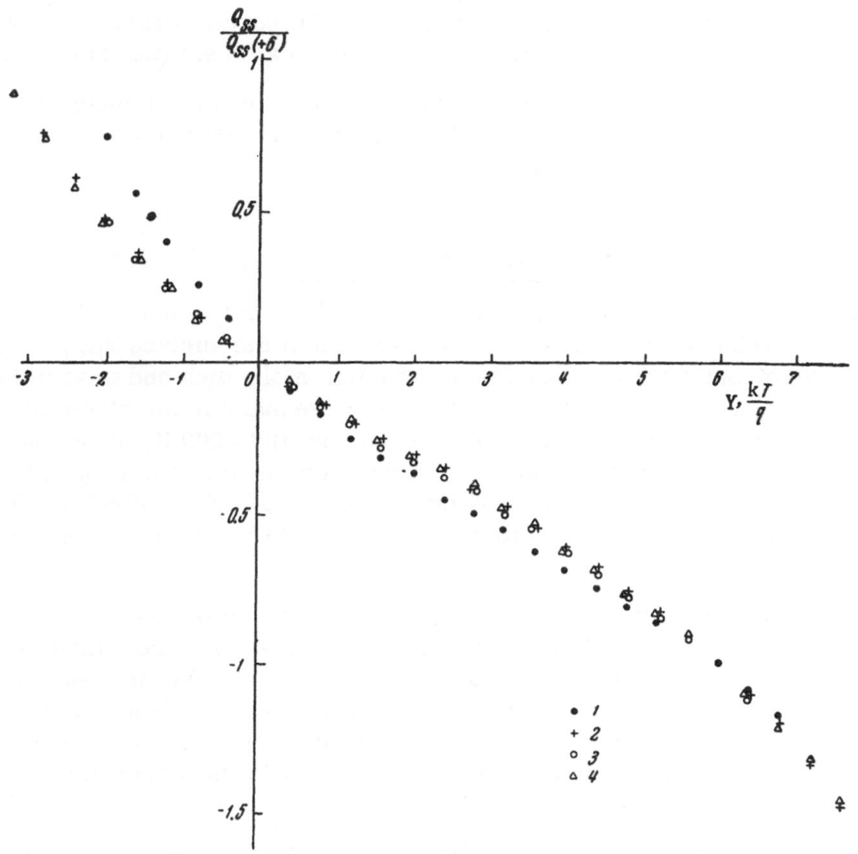

Fig. 38. Reproducibility of the $Q_{ss}(Y)$ curve during heating. Sample No. 3 (p-type): 1) before heating; 2), 3), 4) after heating at 400, 500, and 650°K, respectively.

in the value of Q_{ss} (+6) due to various surface treatments. Some deviations of the energy distribution of the fast surface states from the curve in Fig. 38 (these deviations have been observed in various experiments [20]) do not affect the results of the estimates of the change in the density of states based on the value of Q_{ss} at any fixed value of the surface potential.

In summarizing the discussion given in the present section, we can draw the following conclusions:

a) heating to 650°K produces recombination centers of one type, as indicated by the constancy of the profile of the s (Y) curve, which is not affected by such heating;

b) the deviation of the experimental dependence s (Y) from the theoretical formula should not be attributed to the presence of many recombination levels [18, 20, 65, 68, 75, 76, 117] but to the nonuniformity of the surface originating in the initial chemical treatment and induced in measurements of the field effect;

c) the concentration of the recombination centers is so low that changes in this concentration do not affect the $Q_{ss}(Y)$ curve, i.e., that opposite changes in Q_{ss} (Y) and s (Y) dependences caused by some treatments should not be regarded as surprising; on the other hand, we cannot say there is no correlation whatever between the s (Y) and $Q_{ss}(Y)$ curves (this point will be discussed later);

d) an interpretation of the $Q_{ss}(Y)$ curve by a set of discrete levels or a continuous distribution of levels is a purely formal device and has no physical basis (we shall not use this approach);

e) the constancy of the shape of the $Q_{ss}(Y)$ curve observed after many treatments suggests changes in the density of the fast surface states or, which is equivalent, in the concentration of defects at the germanium–oxide interface.

2. Nature of the Compensating Effect of Water

The results of experiments on the adsorption of the molecules of polar liquids on dehydrated germanium surfaces show that the neutralization of the surface recombination centers is related to the electrostatic interaction between polar molecules and recombination centers. No chemical changes in the oxide film or at the oxide–germanium interface take place during the pumping out in vacuum and the heating of germanium up to 500°K, or during the adsorption of water on the surface of the germanium after such treatments. The process of neutralization is associated with the establishment, by a polar molecule, of a local electric field near a surface recombination center. This field alters the energy of the center as well as its electron and hole capture cross sections.

In principle, we can explain the effect of polar molecules in two ways. First, they can give rise to new slow states which strongly alter the surface potential. Since we can assume that we are dealing with a nonuniform surface, it is likely that polar molecules are adsorbed on those regions of the surface where the recombination centers are located. Then, the considerable change in the surface potential in these regions would give rise to a strong shift of the energy level of the recombination centers with respect to the Fermi level, and this should markedly alter the surface recombination velocity.

Secondly, polar molecules may affect the fast states. However, to do this they must diffuse through the oxide film to the oxide–germanium interface.

The data on the kinetics of the neutralization of surface recombination centers by the adsorption of nitrobenzene point clearly to the second mechanism. After the exposure to nitrobenzene (or other) vapor, the equilibrium with the slow states is established practically instantaneously, as indicated by the rapid establishment of a new equilibrium value of the surface potential. The neutralization of the recombination centers is a much slower process than the establishment of a new equilibrium value of the surface potential.

The slow neutralization of the recombination centers by the nitrobenzene vapor indicates that the diffusion of nitrobenzene molecules across the oxide film is necessary in this process. However, the experimentally obtained power law for the time dependence of the maximum surface recombination velocity cannot be deduced on the basis of diffusion alone. This shows that the process of neutralization is limited not by diffusion but by the localization of dipoles near the recombination centers.

The retention of the original profile of the s (Y) curve and of the position of its maximum along the potential axis, as well as the reduction of its amplitude under the action of polar molecules, all show that these molecules simply reduce the concentration of the recombination centers. A different result is reported in [80, 81] for a sample of germanium heated in liquid ether. The shape of the s (Y) curve after such treatment differs markedly from the shape of the s (Y) curve for a sample that has been heated in vacuum. The s_{max} point in vacuum is observed at $Y = +3.0$ kT/q, whereas the position of s_{max} reported in [80, 81] is located at $Y = +6.0$ kT/q (Fig. 39). Since the position of s_{max} is governed by the ratio of the capture cross sections c_p/c_n, such a shift in s_{max} indicates an increase in c_p/c_n by a factor of 400. This was the interpretation given in [80, 81].

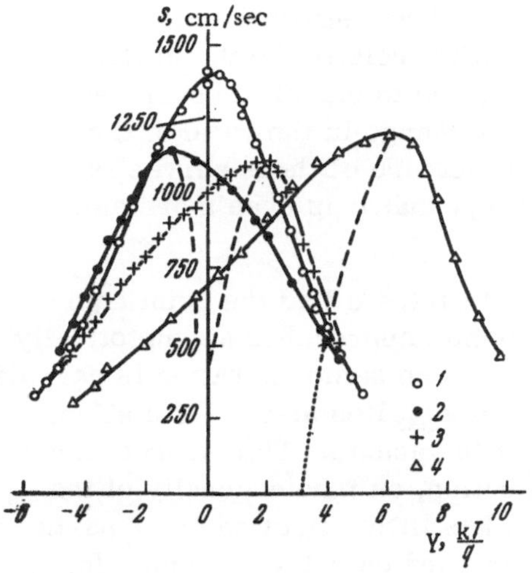

Fig. 39. Comparison of the results of a modeling experiment (curves 1-3) with the results reported in [80, 81] (curve 4): 1) in the absence of a static field; 2) in a negative field; 3) in a positive field. Sample No. 139 (n-type) after heating at 450°K. The "symmetry lines" are shown dashed.

However, according to Eq. (27), such a change in the cross sections should produce a simultaneous drop in the value of s_{max} by a factor of about 20. In fact, the experimentally observed amplitudes of s_{max} after the heating in liquid ether are equal, within the limits of experimental error, to the values of s_{max} observed for germanium samples heated in vacuum. This contradiction is apparently removed [20, 67] if we use Eq. (29). Since the numerator and the additional term in the denominator of Eq. (29) are proportional to $(c_p c_n)^{1/2}$, it follows that the amplitude of the surface recombination velocity should not change when the cross sections change provided the additional term in the denominator predominates over the first term.

We shall not analyze in detail Eq. (29) as applied to the case considered but we shall point out that, if we accept this interpretation, we should find that the s (Y) curve would shift along the potential axis without a change in shape. In fact, only the top of the curve shifts considerably (by 3 kT/q) so that the curve becomes markedly asymmetrical (Fig. 39).

The results reported in [80, 81] can be interpreted more simply: the shift of the s_{max} point along the potential axis is due to a strong nonuniformity of the surface after treatment in liquid ether. There are many reasons for this interpretation. First, the authors themselves report that in some experiments with ether a large shift along the potential axis is observed for the field-effect curves of the two large faces. In principle, both faces of the sample are subjected to identical conditions and this large difference between the potentials of the faces indicates the presence of very pronounced inhomogeneities of the potential on each face. Thus, even slight differences between the conditions on the faces give rise to differences in the average potential. Such macroscopic effects are not observed in other cases, because it is rare for two surfaces to differ very greatly in their homogeneity. Moreover, the authors themselves report that there are no differences between the shape of the s (Y) curves in such cases. This indicates that all these surfaces are strongly heterogeneous but that the number of regions of a given type on both faces is different in some cases but equal in others.

Secondly, the time dependences of the results require a different interpretation. It is evident from Fig. 22 that ether does not restore completely the initial properties of the surface even after tens of hours, whereas the reported measurements in liquid ether were carried out 15 min after the end of heating when only 30-40% of the recombination centers could have been neutralized.

The observation that $s_{max}(t)$ did not pass, as in the case of nitrobenzene, through the origin indicated that exposure to ether did not result simply in physical adsorption but gave rise to some chemical process. This is particularly true of treatment in ether at high temperatures.

These conclusions were checked by the following simple experiment, in which the influence of inhomogeneity on the s (Y) curve was simulated. A sample was heated in vacuum at 500°K. The same alternating field was applied to both surfaces of a sample and, in addition, a static positive or negative bias was applied to one of the surfaces. The results of this experiment are presented in Fig. 39. We can see that the point s_{max} shifts, relative to the initial position of s_{max}, to one side or the other when a static bias is applied to one of the surfaces. The traditional way to interpret such a shift is to attribute it to a change in the ratio of the capture cross sections by a factor of about 50. A characteristic feature of these curves is their asymmetry. The curve reported in [80, 81] is similar except that it is even more asymmetrical.

These curves have an interesting property. If, at each ordinate, we find the point equidistant from both branches of the curve, the line joining these points approaches asymptotically the true position of s_{max} when the curves shift to zero potential. If the same operation is carried out on the curve given in [80, 81], we find that the true position of s_{max} lies at $Y = +3.0\ kT/q$, which corresponds to the usual position of s_{max} after the heating in vacuum. This operation has an obvious physical meaning. As mentioned in the preceding section, an inhomogeneity of the surface strongly distorts the surface conductivity minimum but has little effect on other parts of the conductivity curve. The data reported in [80, 81] were obtained on p-type samples for which the position of s_{max} along the potential axis was almost identical with the position of the surface conductivity minimum. Therefore, an inhomogeneity of the surface produced a strong shift of the position of s_{max} and a weak shift of the rest of the s (Y) curve. The weaker effect in our experiment was due to the fact that s_{max} of the n-type samples was separated by a fairly large potential difference from the surface conductivity minimum.

The compensation of the recombination centers by water explains also the changes in the surface recombination velocity after various chemical treatments of the germanium samples. As mentioned earlier, a sample etched in hydrogen peroxide has a considerably lower surface recombination velocity than one etched in CP-4. Experiments involving heating in vacuum to 500°K show that this difference is not due to the different nature or concentration of the surface recombination centers. When physically adsorbed water is removed from the surface, the values of s_{max} are found to be practically the same for a sample etched in H_2O_2 and for one etched in CP-4.

Thus, the large differences between the initial surface recombination velocities observed after different chemical treatments are due to different amounts of water adsorbed on the surface of germanium. It is known [108] that a surface etched in H_2O_2 becomes hydrophilic and that one etched in CP-4 becomes hydrophobic so that much less water is adsorbed on the surface in the second case and, clearly, the amount adsorbed is insufficient for the complete compensation of the surface recombination centers.

3. Nature of Recombination Centers on Real Surface of Germanium

The results of the experiments in which we determined the effect of ozone on a dehydrated surface have shown clearly that ozone does not introduce additional recombination centers but destroys a considerable proportion of those already present.

Ozone is unstable and, after a time, decomposes into molecular and atomic oxygen. The atomic species reacts strongly with water [123]: we may therefore expect dehydration of the surface in the presence of a continuous supply of ozone molecules. If this process physically affects the adsorbed water, the surface recombination velocity increases; if a monolayer of chemisorbed water is affected, the recombination velocity decreases. When ozone interacts with a freshly etched sample whose surface carries a large amount of physically adsorbed water, we should observe both processes. Since ozone cannot reach some of the defects, it cannot remove the chemisorbed water completely and, consequently, the interaction of freshly etched and dehydrated surfaces with ozone produces saturation at a level close to $s_{max} = 800$ cm/sec.

The experiments involving repeated heating of sample No. 108 (Fig. 33) show that the concentration of the recombination centers is reduced appreciably after the sample had been heated only to 500°K. Therefore, the value s_{max} = 2000-2500 cm/sec, obtained after the heating at 500°K, does not correspond to the total concentration of the surface recombination centers but to a concentration which is approximately 40-50% lower. Thus, the reduction of the maximum surface recombination velocity to 700-800 cm/sec after prolonged exposure to ozone corresponds to the annihilation of about 80% of the recombination centers. It is quite reasonable to assume that about 20% of the defects at which surface recombination centers are formed are inaccessible to the ozone.

The effect of ozone is thus, in many respects, similar to the effect of heating. There are, of course, some differences. Multistage heating allows us to influence, in turn, the physically adsorbed water and the chemisorbed water monolayer. Ozone does not show this selectivity to the kind of adsorption and, therefore, the results of the interaction with ozone are not as easy to interpret. On the other hand, ozone interacts only with water. It is evident from Fig. 29 that the s (Y) curve fits very well the theoretical dependence (discrepancies at low values of s are due to the contribution of the lateral faces of a sample, which is ignored in the calculations; see Sec. 9 in Chap. II). Heating induces other processes which increase the constant component of the s (Y) curve (Fig. 28). Thus, the effects of ozone and of heating in vacuum are not completely identical but, in practice, they are the same in respect of the dominant recombination centers.

It is interesting to consider the relationship between the changes in s (Y) and Q_{ss} (Y) observed after the interaction of ozone with a sample dehydrated by heating at 500°K (Figs. 26 and 27). Ozone reduces s_{max} by a factor of about 2.5, i.e., it destroys two-thirds of the total number of recombination centers. The Q_{ss} (Y) curve is not affected, i.e., the distribution $N_t(E_t)$ of the fast states is retained within the limits of the experimental error. This can be interpreted in two ways:

1. When a water molecule is removed, a defect at the germanium–oxide interface loses its recombination properties but continues to act as a fast state (a hole trap) in the field effect. When water is removed from a defect, its energy level remains practically unaffected although the hole capture cross section decreases considerably. This assumption is reasonable because such modification of a defect does not alter its charged state (it remains an acceptor).

2. When a water molecule is removed, the values of c_p/c_n and E_t change greatly. The distribution of the fast states $N_t(E_t)$ is retained because the concentration of recombination centers is much lower than the total density of fast states in a given range of energies.

We can easily see that these two interpretations yield the same conclusion. The second interpretation means that the fast states with the same energy level E_t may be of the recombination or nonrecombination type, i.e., the transformation of a recombination center into a trap does not involve a change in the energy level. This confirms fully the conclusion, drawn in Sec. 1 of the present chapter, that the usual combined analysis of the Q_{ss} (Y) and s (Y) curves is incorrect.

The results reported in [75] can now be easily explained. Traps are converted into recombination centers when physically adsorbed water is removed from the surface of germanium by pumping out in vacuum. The surface recombination velocity increases severalfold and a maximum, characteristic of the dominant centers, appears in the s (Y) curve. The Q_{ss} (Y) curve is practically unaffected.

We may conclude that the lack of correlation between changes in the s (Y) and Q_{ss} (Y) curves is normal, whereas similarity of the changes in the density of fast states and the concentration of recombination centers, observed sometimes after the heating in vacuum to 500°K, is purely accidental.

This conclusion is too sweeping because there is a definite relationship between the density of fast states (or the number of defects at the germanium–oxide interface) and the concentration of recombination centers. The recombination centers are based on defects. Other conditions being equal, we find that the larger the number of defects at the germanium–oxide interface, the higher is the concentration of recombination centers at which the water molecules are localized. The conclusion that there is no relation between the changes in the concentration of recombination centers and those in the density of fast states does not apply to the hypothetical case of "other conditions being equal" but to actual experiments involving heating and interactions with ozone, in which the conditions are not equal. For example, the number of defects at the germanium–oxide interface remains constant when the samples are heated in the temperature range 500-650°K or when they interact with ozone. However, these defects become inaccessible to water molecules and, therefore, cannot become recombination centers. Hence, there is no correlation between changes in Q_{ss} (Y) and s (Y) in these experiments.

Heating in vacuum at 700-750°K, which appreciably alters the oxide film structure, is a case of "other conditions being equal." The number of defects increases strongly compared with the number after the heating at 500°K, and they become easily accessible to the water molecules. Therefore, we observe a fairly good correlation between the changes in the density of fast states and those in the concentration of recombination centers when we compare the results obtained after the heating at 500°K and after the interaction of water vapor with a sample heated to 750°K. In the last case, s_{max} reaches values of the order of 10^4 cm/sec (see Sec. 7 in Chap. III), i.e., values which are about 10 times higher than s_{max} after the heating at 500°K. The density of fast states also changes by a factor of 10.

This applies also to the heating in oxygen and ozone. A strong reduction in the concentration of recombination centers is correlated with a reduction in the total density of fast states.

When germanium samples are heated in vacuum at 750°K, the number of defects rises strongly when the heating period is increased (Fig. 32). This confirms that such heating alters the oxide film structure. The density of fast states changes only by 10-20% when the duration of heating at 400-600°K is increased from 5 min to 1 h. Obviously, heating at these relatively low temperatures does not affect the oxide film structure but simply influences the processes of adsorption and desorption (changes in the oxide film structure probably begin at 650°K; see Fig. 24).

The heating of germanium in high vacuum at 750°K reduces the surface recombination velocity in spite of a large increase in the density of fast states. Such heating destroys completely the recombination centers with $c_p/c_n = 10\text{-}30$, and this is manifested by the disappearance of the characteristic maximum in the s (Y) curve in the region of $Y = +3.0$ kT/q (Fig. 31). Since the subsequent interaction with water vapor restores these recombination centers, we may conclude that the defects generated by the heating in vacuum at 750°K include a large number of defects at which the recombination centers are formed. A monotonic rise of the s (Y) curve is observed up to $Y = +6.0$ kT/q. If we use the standard analysis of Eq. (27), we must attribute the result to the effect of the recombination centers with $c_p/c_n \gg 10^4$, i.e., to the effect of ordinary acceptors which do not have recombination properties. The appearance of a large number of acceptors after this treatment is also indicated by a strong shift of the surface potential in the negative direction (Table 7).

Thus, an acceptor defect at the germanium–oxide interface is the basis of a recombination center. When a water molecule (an electric dipole) is localized at such a defect, its electron capture cross section increases appreciably and the defect becomes an effective acceptor-like recombination center. This model of the surface recombination centers explains all the changes in the surface recombination velocity observed in experiments involving polar liquids, heating in vacuum, or in oxygen and ozone atmospheres, as well as in experiments involving interaction with ozone.

The etching of a sample leaves several layers of physically adsorbed water. The presence of this water gives rise to regions where double layers of water dipoles are formed at the germanium–oxide interface. This process destroys local fields of water dipoles near defects. Consequently, the surface recombination velocity falls to its minimum value. Different amounts of physically adsorbed water left after different chemical treatments are responsible for differences in the surface recombination velocity. Etching in CP-4, which produces a hydrophobic surface, results in a higher value of the surface recombination velocity than etching in hydrogen peroxide, which produces a hydrophilic surface.

The gradual removal of physically adsorbed water by simple pumping out in vacuum [75] or, more effectively, by pumping and heating, produces a maximum in the monotonic $s(\varphi_s)$ curve. This maximum occurs in the region of $q\varphi_s/kT = 5$ and its amplitude increases as the duration of pumping is extended, or the temperature is raised. This continues until 500°K is reached (at this temperature, all the physically adsorbed water is removed). The differences in the surface recombination velocity, observed after the samples have been etched in different solutions, are completely destroyed by such treatment.

Heating in the 500–650°K range removes the chemisorbed water layer, i.e., it removes those water molecules which are responsible for the surface recombination. Such heating reduces the surface recombination velocity. The reaction which accompanies the desorption of chemisorbed water [112] heals the defects on which the recombination centers are based. Such healing alters the characteristics irreversibly: the interaction with water vapor after the heating to 650°K has practically no effect on the surface recombination velocity. Some increase in the surface recombination velocity is observed after prolonged exposure to saturated water but this effect is weak (of the order of 30%). The increase may be due to the weakness of the freshly healed oxide, and the exposure to saturated water vapor may result in the formation of a liquid layer of water, which dissolves some of the freshly formed regions in the oxide film and thus increases somewhat the surface recombination velocity.

Heating above 650°K alters the film structure so that the number of defects at the germanium–oxide interface again increases strongly. These defects may attract water molecules and become recombination centers so that, depending on the concentration of water vapor in the ambient atmosphere, the surface recombination velocity may either decrease or increase.

It is interesting to note that the s(Y) curve obtained after the heating in vacuum to 750°K is similar to the s(Y) curve of the freshly etched samples. These two cases are identical from the point of view of the proposed model of recombination centers, and the similarity of these two curves is fully expected. In both cases, there is no local field of an electric dipole near a defect, which could convert a defect into an effective recombination center. In the case of the samples heated to 750°K, such electric dipoles are simply absent, whereas in the case of the etched samples, dipoles are present but located alongside other dipoles (molecules of physically adsorbed water or of other polar liquids) which screen the defect from the field of the electric dipole of a chemisorbed water molecule.

The fundamental problem in the theory of recombination is the explanation of the large carrier capture cross section of neutral centers, which is frequently comparable with the capture cross section of charged centers (for example, in our case, $c_p/c_n \approx 10$). The polarization potential suggested in [124] cannot explain this effect [125] because, in cases typical of deep traps in germanium and silicon, this potential cannot generate the number of excited levels of a center necessary for a cascade process. The suggestion that a permanent dipole moment must be located near a recombination center [126] marks a step forward to the solution of the problem. The calculations reported in [126, 127] have shown that the presence of a permanent dipole moment can, under certain conditions, give rise to a continuous spectrum of excited levels of a center. The results reported in the present paper may be regarded as a direct experimental proof of this suggestion.

Conclusions

The experiments reported in the present paper have enabled us to suggest a new model of the dominant surface recombination centers on germanium. The activation and deactivation of these centers are governed by the presence of an electric dipole near an acceptor-type defect. This dipole owes its origin to water molecules adsorbed on the surface of a sample, which diffuse through the oxide film. We shall not discuss the nature of the electric dipole, i.e., we shall leave open the question whether it is the electrostatic dipole of the water molecule or the quantum-mechanical dipole formed by the donor–acceptor binding of a water molecule to coordination-unsaturated germanium atoms [128]. We cannot pursue the question of the nature of the dipole further because, on the one hand, the estimates given in [127] suggest that the dipole moment necessary for a large capture cross section of a neutral center is not sufficiently large to be attributed to the usual electrostatic dipole of the water molecule; on the other hand, even the order of magnitude of the quantum-mechanical dipole has not yet been estimated. Information on this subject would most probably be given by investigations of the influence of the adsorption of molecules of various types (with different dipole moments) capable of the formation of donor–acceptor bonds but incapable of having an influence on the parameters of the dominant recombination centers).

The nature of primary defects, on which recombination centers are based, has not been considered in the present paper. All that we can deduce from a comparison of the results obtained by high-temperature heating in vacuum and oxygen is that these defects are associated with the oxide film. Therefore, investigations of the influence of the structure, composition, and thickness of an oxide film seem to be the most promising ways to determine the nature of these defects. The establishment of a relationship between the parameters of the surface states and the properties of the bulk of a crystal and its oxide film should help in the determination of the nature of the primary defects (nonadsorptive defects, whose modification in the processes of sorption, excitation, etc., alters the parameters of the surface states.

It is likely that success in investigations of the surface properties of semiconductors lies along the path of the establishment of relationships between the electrical parameters and adsorption processes. The main problem can be formulated as the determination of the "chemical" nature of the surface states, i.e, the determination of the origin of particular surface states, the establishment of a relationship between the parameters (and changes in the parameters) of the surface states, on the one hand, and the sorption processes, parameters of the oxide film and of the bulk of a semiconductor, on the other.

The ambient medium is the main instrument for the investigation of the surface states because changes in this medium can be used to alter the spectrum of the surface levels. Attention should be concentrated on the kinetics of the various processes. The rate of change of the spectrum of the surface states should be measured and the nature of the surface states (slow, fast, recombination) undergoing changes should be determined in order to obtain information on the location of the various surface states, on their interaction with external perturbations, and on the mechanisms of these interactions.

The steady-state methods described in the present paper are suitable for this purpose. It is paradoxical that the "kinetic" methods can be used to investigate only the steady-state parameters of the surface, whereas the steady-state methods can be used to study the kinetics of the processes occurring on the surface. On the other hand, the determination of plain numerical values of the parameters of the surface states does not help, in most cases, in finding the relationships between these parameters and the nature of defects on the surface of a semiconductor. It is more logical to investigate the kinetic parameters of the surface states only after the establishment of relationships between the nature of the surface states and the "chemistry" of the associated defects.

As discussed in detail in the preceding chapters, the field effect in gases and in vacuum can be used only as a semiquantitative method in comparisons of directions of changes in such quantities as Y_0, $Q_{ss}(Y)$, and s(Y) caused by various perturbations. Attempts to deduce accurate values of the parameters N_t, E_t, c_p, and c_n from measurements of the field effect meet with serious difficulties, associated with the inhomogeneity of the surface, which may be intrinsic or induced by the field effect. We are thus faced with two problems: 1) preparation of a uniform surface; 2) development of measurement methods which do not disturb the uniformity of the surface.

In the light of these problems, it seems imperative to study in greater detail the dependences of s(Y) and $Q_{ss}(Y)$ on the crystallographic orientation of the surface, as begun in [20], and to search for new measurement methods. In principle, the second task is already solved: one can use the field effect in MOS structures and in electrolyte-semiconductor systems (see, for example, [129]). In both cases, the inhomogeneity associated with the variable air gap in the usual field effect is eliminated. However, an MOS structure is not convenient for the solution of the problem outlined above since surface conditions can be changed only with great difficulty, if at all.

The electrolyte-semiconductor system is more promising in this respect. It is surprising to find that this system is not very popular compared with the usual gas-semiconductor system, although the use of an electrolyte has the following advantages: 1) a high degree of surface homogeneity or at least complete elimination of the inhomogeneity of the induced charge; 2) omnidirectional field effect; 3) the possibility of the experimental determination of $\mu_s(Y)$ by comparison of the C(Y) and $\delta G(Y)$ curves.

Moreover, the electrolyte-semiconductor system retains many advantages of the field effect in gases: it provides opportunities for changing the surface structure, oxidizing it, carrying out adsorption experiments, etc.

Magnetic fields, used in the same way as in the electric-field effect, may be found useful in such studies because they can be employed to alter the carrier density in the surface layer [130-132]. The advantage of magnetic fields would lie in measurements of the "magnetic field effect" on a free surface, which are very desirable in studies of adsorption and absolutely essential in investigations of clean surfaces.

Literature Cited

1. Tamm, I. E., Zh. Eksp. Teor. Fiz., 3:34 (1933).
2. Shockley, W., Phys. Rev., 56:317 (1939).
3. Litovchenko, V. G., and Lyashenko, V. I., Fiz. Tverd. Tela, 1:1609 (1959).
4. Primachenko, V. E., Litovchenko, V. G., Lyashenko, V. I., and Snitko, O. V., Ukr. Fiz. Zh., 5:345 (1960).
5. Schlier, R. E., and Farnsworth, H. E., J. Chem. Phys., 30:917 (1959).
6. Barnes, G. A., and Banbury, P. C., Proc. Phys. Soc. London, 71:1020 (1958).
7. Lyashenko, V. I., Chernaya, N. S., and Gerasimov, A. B., Fiz. Tverd. Tela, 2:2421 (1960).
8. Lax, M., Proc. Fifth Intern. Conf. on Physics of Semiconductors, Prague, 1960, publ. by Academic Press, New York (1961), p. 484.
9. Law, J. T., and Garrett, C. G. B., J. Appl. Phys., 27:656 (1956).
10. Pratt, Jr., G. W., Phys. Rev., 98:1543 (1955).
11. Kingston, R. H., and McWhorter, A. L., Phys. Rev., 103:534 (1956).
12. Montgomery, H. C., and Brown, W. L., Phys. Rev., 103:865 (1956).
13. Rzhanov, A. V., Pavlov, N. M., and Selezneva, M. A., Zh. Tekh. Fiz., 28:2645 (1958).

14. Rzhanov, A. V., Novototskii-Vlasov, Yu. F., and Neizvestnyi, I. G., Fiz. Tverd. Tela, 1:1471 (1959).
15. Lyashenko, V. I., and Chernaya, N. S., Fiz. Tverd. Tela, Sbornik, 1:110 (1959).
16. Lyashenko, V. I., Snitko, O. V., and Sytenko, T. N., Fiz. Tverd. Tela, Sbornik, 1:3 (1959).
17. Margoninski, Y., J. Chem. Phys., 32:1791 (1960).
18. Rzhanov, A. V., Novototskii-Vlasov, Yu. F., Neizvestnyi, I. G., Pokrovskaya, S. V., and Galkina, T. I., Fiz. Tverd. Tela, 3:822 (1961).
19. Balzarotti, A., Chiarotti, G., and Frova, A., Nuovo Cimento, 26:1205 (1962).
20. Rzhanov, A. V., Tr. Fiz. Inst. Akad. Nauk SSSR, 20:3 (1963).
21. Kingston, R. H., J. Appl. Phys., 27:101 (1956).
22. Statz, H., de Mars, G. A., Davis, Jr., L., and Adams, Jr., A., Phys. Rev., 101:1272 (1956).
23. Grimley, T. B., J. Phys. Chem. Solids, 14:227 (1960).
24. Morrison, S. R., in: Semiconductor Surface Physics (Proc. Conf. Philadelphia, 1956), ed. by R. H. Kingston et al., Univ. of Pennsylvania Press, Philadelphia (1957), p. 169.
25. Morrison, S. R., Advan. Catal. Relat. Subj., 7:259 (1955).
26. Vol'kenshtein, F. F., Zh. Fiz. Khim., 32:2383 (1958).
27. Vol'kenshtein, F. F., and Kogan, Sh. M., Zh. Fiz. Khim., 34:1996 (1960).
28. Brattain, W. H., and Bardeen, J., Bell System Tech. J., 32:1 (1953).
29. Pratt, Jr., G. W., and Kolm, H. H., in: Semiconductor Surface Physics (Proc. Conf. Philadelphia, 1956), ed. by R. H. Kingston et al., Univ. of Pennsylvania Press, Philadelphia (1957), p. 297.
30. Low, G. G. E., Proc. Phys. Soc. London, 68B:10 (1955).
31. Bardeen, J., Phys. Rev., 71:717 (1947).
32. Meyerhof, W. E., Phys. Rev., 71:727 (1947).
33. Benzer, S., Phys. Rev., 71:141 (1947).
34. Kingston, R. H., and Neustadter, S. F., J. Appl. Phys., 26:718 (1955).
35. Garrett, C. G. B., and Brattain, W. H., Phys. Rev., 99:376 (1955).
36. Kurskii, Yu. A., Fiz. Tverd. Tela, 3:212 (1961).
37. Schrieffer, J. R., Phys. Rev., 97:641 (1955).
38. Litovchenko, V. G., Lyashenko, V. I., and Frolov, O. S., in: Surface Properties of Semiconductors (ed. by A. N. Frumkin), Consultants Bureau, New York (1964), p. 103.
39. Millea, M. F., and Hall, T. C., Phys. Rev. Lett., 1:276 (1958).
40. Many, A., Grover, N. B., Goldstein, Y., and Harnik, E., J. Phys. Chem. Solids, 14:186 (1960).
41. Greene, R. F., Frankl, D. R., and Zemel, J., Phys. Rev., 118:967 (1960).
42. Grover, N. B., Goldstein, Y., and Many, A., J. Appl. Phys., 32:2538 (1961).
43. Goldstein, Y., Grover, N. B., Many, A., and Greene, R. F., J. Appl. Phys., 32:2540 (1961).
44. Zemel, J. N., and Petriz, R. L., J. Phys. Chem. Solids, 8:102 (1959).
45. Morrison, S. R., Phys. Rev., 102:1297 (1956).
46. McWhorter, A. L., Phys. Rev., 98:1191 (1955).
47. Koc, S., Status Solidi, 2:1304 (1962).
48. Lasser, M., Wysocki, C., and Bernstein, B., Phys. Rev., 105:491 (1957).
49. Lasser, M., Wysocki, C., and Bernstein, B., in: Semiconductor Surface Physics (Proc. Conf. Philadelphia, 1956), ed. by R. H. Kingston et al., Univ. of Pennsylvania Press, Philadelphia (1957), p. 197.
50. Rzhanov, A. V., Fiz. Tverd. Tela, 3:1718 (1961).
51. Novototskii-Vlasov, Yu. F., and Sinyukov, M. P., in: Surface Properties of Semiconductors (ed. by A. N. Frumkin), Consultants Bureau, New York (1964), p. 45.
52. Brown, W. L., Phys. Rev., 100:590 (1955).
53. Bardeen, J., Coovert, R. E., Morrison, S. R., Schrieffer, J. R., and Sun, R., Phys. Rev., 104:47 (1956).

54. Shockley, W., Electrons and Holes in Semiconductors, Van Nostrand, New York (1950).
55. Bir, G. L., Fiz. Tverd. Tela, 1:67 (1959).
56. Petrusevich, V. A., Sorokin, O. V., and Kruglov, V. I., Fiz. Tverd. Tela, 3:2023 (1961).
57. Shockley, W., and Read, Jr., W. T., Phys. Rev., 87:835 (1952).
58. Stevenson, D. T., and Keyes, R. J., Physica, 20:1041 (1954).
59. Harnik, E., Many, A., Margoninski, Y., and Alexander, E., Phys. Rev., 101:1434 (1956).
60. Rzhanov, A. V., Novototskii-Vlasov, Yu. F., and Neizvestnyi, I. G., Zh. Tekh. Fiz., 27:2440 (1957).
61. Many, A., Harnik, E., and Margoninski, Y., Semiconductor Surface Physics (Proc. Conf. Philadelphia, 1956), ed. by R. H. Kingston et al., Univ. of Pennsylvania Press, Philadelphia (1957), p. 85.
62. Dousmanis, G. C., Phys. Rev., 112:369 (1958).
63. Wang, S., and Wallis, G., Phys. Rev., 107:947 (1957).
64. Brattain, W. H., and Garrett, C. G. B., Bell System Tech. J., 35:1019 (1956).
65. Garrett, C. G. B., and Brattain, W. H., Bell System Tech. J., 35:1041 (1956).
66. Rzhanov, A. V., Fiz. Tverd. Tela, 3:3691 (1961).
67. Rzhanov, A. V., Fiz. Tverd. Tela, 3:3698 (1961).
68. Rzhanov, A. V., Pavlov, N. M., and Selezneva, M. A., Fiz. Tverd. Tela, 3:832 (1961).
69. Henisch, H. K., Reynolds, W. N., and Tipple, P. M., Physica, 20:1033 (1954).
70. Many, A., and Gerlich, D., Phys. Rev., 107:404 (1957).
71. Rzhanov, A. V., Radiotekh. Elektron., 1:1086 (1956).
72. Wang, S., and Wallis, G., Phys. Rev., 105:1459 (1957).
73. Lyashenko, V. I., Snitko, O. V., and Litovchenko, V. G., Proc. Fifth Intern. Conf. on Physics of Semiconductors, Prague, 1960, publ. by Academic Press, New York (1961), p. 515.
74. Albers, Jr., W. A., and Rickel, A. M., J. Electrochem. Soc., 109:582 (1962).
75. Rzhanov, A. V., and Arkhipova, I. A., Fiz. Tverd. Tela, 4:1274 (1962).
76. Rzhanov, A. V., in: Surface Properties of Semiconductors (ed. by A. N. Frumkin), Consultants Bureau, New York (1964), p. 70.
77. Morrison, S. R., J. Phys. Chem. Solids, 14:214 (1960).
78. Keyes, R. J., and Maple, T. G., Phys. Rev., 94:1416 (1954).
79. Smirnov, G. V., Polukarov, Yu. M., and Arslambekov, V. A., in: Surface Properties of Semiconductors (ed. by A. N. Frumkin), Consultants Bureau, New York (1964), p. 62.
80. Neizvestnyi, I. G., in: Surface Properties of Semiconductors (ed. by A. N. Frumkin), Consultants Bureau, New York (1964), p. 51.
81. Rzhanov, A. V., and Neizvestnyi, I. G., Fiz. Tverd. Tela, 3:3317 (1961).
82. Rzhanov, A. V., Novototskii-Vlasov, Yu. F., Neizvestnyi, I. G., Galkina, T. I., and Pokrovskaya, S. V., Fifth Intern. Conf. on Physics of Semiconductors, Prague, 1960, publ. by Academic Press, New York (1961), p. 506.
83. Burshtein, R. Kh., and Sergeev, S. I., Dokl. Akad. Nauk SSSR, 139:134 (1961).
84. Boddy, P. J., and Brattain, W. H., J. Electrochem. Soc., 109:812 (1962).
85. Low, G. G. E., Proc. Phys. Soc. London, 68B:1154 (1955).
86. Litovchenko, V. G., Fiz. Tverd. Tela, Sbornik, 2:83 (1959).
87. Litovchenko, V. G., and Lyashenko, V. I., Fiz. Tverd. Tela, 2:1592 (1960).
88. Litovchenko, V. G., and Snitko, O. V., Fiz. Tverd. Tela, 2:591 (1960).
89. Harnik, E., Goldstein, Y., Grover, N. B., and Many, A., J. Phys. Chem. Solids, 14:193 (1960).
90. Navon, D., Bray, R., and Fan, H. Y., Proc. IRE, 40:1342 (1952).
91. Many, A., Proc. Phys. Soc. London, 67B:9 (1954).
92. Bogdanov, S. V., and Kopylovskii, B. D., Prib. Tekh. Eksp., No. 1, p. 66 (1956).
93. Many, A., Goldstein, Y., Grover, N. B., and Harnik, E., Proc. Fifth Intern. Conf. on Physics of Semiconductors, Prague, 1960, publ. by Academic Press, New York (1961), p. 498.

94. Galkina, T. I., Fiz. Tverd. Tela, 1:216 (1959).
95. Zhuze, V. P., Pikus, G. E., and Sorokin, O. V., Fiz. Tverd. Tela, 1:1420 (1959).
96. Zhuze, V. P., Pikus, G. E., and Sorokin, O. V., Zh. Tekh. Fiz., 27:1167 (1957).
97. Rzhanov, A. V., Arkhipova, I. A., and Bidulya, V. N., Zh. Tekh. Fiz., 28:1051 (1958).
98. Many, A., Margoninski, Y., Harnik, E., and Alexander, E., Phys. Rev., 101:1433 (1956).
99. Lyashenko, V. I., Snitko, O. V., and Sytenko, T. N., Fiz. Tverd. Tela, Sbornik, 1:11 (1959).
100. van Roosbroeck, W., Phys. Rev., 101:1713 (1956).
101. Rzhanov, A. V., Fiz. Tverd. Tela, 2:2431 (1960).
102. Sim, A. C., J. Electron. Control, 10:117 (1961).
103. Curcio, J. A., and Petty, C. C., J. Opt. Soc. Amer., 41:302 (1951).
104. Brooks, H., Advan. Electron. Electron Phys., 7:87 (1955).
105. Ryvkin, S. M., Zh. Tekh. Fiz., 24:2136 (1954).
106. Morin, F. J., and Maita, J. P., Phys. Rev., 94:1525 (1954).
107. Rzhanov, A. V., and Plotnikov, A. F., Fiz. Tverd. Tela, 3:1557 (1961).
108. Ellis, S. G., J. Appl. Phys., 28:1262 (1957).
109. Novototskii-Vlasov, Yu. F., and Neizvestnyi, I. G., Prib. Tekh. Eksp., No. 4, p. 127 (1961).
110. Hagstrum, H. D., J. Appl. Phys., 32:1020 (1961).
111. Novototskii-Vlasov, Yu. F., Fiz. Tverd. Tela, 7:1086 (1965).
112. Law, J. T., J. Phys. Chem., 59:67 (1955).
113. Novototskii-Vlasov, Yu. F., Fiz. Tverd. Tela, 6:3500 (1964).
114. Clarke, E. N., Phys. Rev., 98:1178 (1955).
115. Yunovich, A. É., and Anokhin, B. G., Dokl. Vyssh. Shkoly, Ser. Fiz.-Mat., 5:176 (1958).
116. Flietner, H., and Hundt, G., Phys. Status Solidi, 3:108 (1963).
117. Rzhanov, A. V., Fiz. Tverd. Tela, 4:1279 (1962).
118. Roginskii, S. Z., in: Surface Properties of Semiconductors (ed. by A. N. Frumkin), Consultants Bureau, New York (1964), p. 2.
119. Law, J. T., Phys. Chem., 59:543 (1955).
120. Rideal, E. K., Introduction to Surface Chemistry, 2nd ed., Cambridge University Press (1930).
121. Adam, N. K., The Physics and Chemistry of Surfaces, 3rd ed., Oxford University Press (1942).
122. Rose, A., Phys. Rev., 97:322 (1955).
123. Waters, W. A., Chemistry of Free Radicals, 2nd ed., Oxford University Press (1948).
124. Lax, M., Phys. Rev., 119:1502 (1960).
125. Bonch-Bruevich, V. L., and Glasko, V. B., Fiz. Tverd. Tela, 4:510 (1962).
126. Kurskii, Yu. A., Fiz. Tverd. Tela, 4:2620 (1962).
127. Kurskii, Yu. A., Fiz. Tverd. Tela, 6:1485 (1964).
128. Prudnikov, R. V., Novototskii-Vlasov, Yu. F., and Kiselev, V. F., Fiz. Tverd. Tela, 8:2458 (1966).
129. Myamlin, V. A., and Pleskov, Yu. V., Electrochemistry of Semiconductors, Plenum Press, New York (1967).
130. Aerts, E., Amelinckx, S., and Vennik, J., J. Electron. Control, 7:497 (1959).
131. Aerts, E., J. Electron. Control, 9:217 (1960).
132. Einspruch, N. G., J. Phys. Chem. Solids, 23:1743 (1962).

THEORY OF ELECTROMAGNETIC WAVE ABSORPTION IN IDEAL AND NONIDEAL IONIC CRYSTAL LATTICES*

V. S. Vinogradov

A theoretical analysis is given of the influence of the interaction of phonons with charged impurities and with other phonons on the absorption of infrared and microwave electromagnetic waves in ionic crystals. The calculation is made by the double-time retarded Green's function method. The theoretical expressions obtained are compared with experimental results.

Introduction

This paper contains a theoretical treatment of two physical phenomena observed in ionic crystals, namely the absorption and dispersion of electromagnetic waves in the infrared region of the spectrum, and absorption at microwave frequencies, which depends on the defect content of the crystal.

The infrared absorption and dispersion in ionic crystals have been studied in many experiments with various crystals, semiconductors, and dielectrics, because of the great interest of these phenomena for both pure and applied science. The classical theory has long existed, but the quantum theory is not yet complete.

The object of the present work is to eliminate the deficiencies in previous discussions on the quantum theory of infrared absorption and dispersion in ionic crystals, and to arrive at a more satisfactory theory.

The phenomenon of dielectric losses at microwave frequencies has also been much studied experimentally, but has been the subject of hardly any theoretical work.

In this paper, an attempt is made to give a theoretical elucidation of certain aspects of this phenomenon.

The introduction is followed by four chapters, conclusions, and Appendices A-C.

The appendices contain the lengthy subsidiary calculations. Expressions taken from the appendices are indicated in the body of the paper by the letter A, B, or C.

*Thesis for the degree of Candidate of Physico-Mathematical Sciences, defended at the Lebedev Institute on January 11, 1965. Supervisor, V. A. Chuenkov.

CHAPTER I

Review of Work on the Theory of Electromagnetic Wave Absorption by Ideal and Nonideal Crystal Lattices

1. Classical and Quantum Theories of Infrared Absorption and Dispersion in an Ideal Ionic Crystal Lattice

Elementary theory gives as the complex permittivity of an ionic crystal

$$\varepsilon(\omega) = \varepsilon_\infty + (\varepsilon_0 - \varepsilon_\infty)/\left[1 - \left(\frac{\omega}{\omega_0}\right)^2 - i\left(\frac{\omega}{\omega_0}\right)\frac{\gamma}{\omega_0}\right], \tag{I.1}$$

where ε_0 and ε_∞ are the static and high-frequency permittivities, ω_0 the frequency of transverse long-wave optical vibrations of the ions (the dispersion frequency), and γ a constant describing the damping of the dispersion vibration, whose value in this theory is obtained from experiment.

If the reflectivity $R(\omega)$ is calculated from Eq. (I.1), the result shows that it has a single maximum at a frequency close to ω_0.

In 1930-1932, Czerny and his co-workers [1-3] measured the reflection and transmission coefficients of alkali-halide crystals, and found additional maxima of the reflectivity, which are not given by Eq. (I.1). This showed that the elementary theory, which describes the damping by a frequency-independent quantity γ, is not satisfactory.

In 1933, Born and Blackman used classical lattice dynamics to construct a microscopic theory of infrared absorption [4, 5], which led to a function $\gamma(\omega)$. They took as basis the work of Pauli [6] on the same subject, published as early as 1925. Pauli considered a linear sequence of ions having alternating charges $+e$ and $-e$ and equal masses.

In the potential energy of this configuration, Pauli took account of cubic as well as quadratic terms in the displacements of the ions. This made it possible to set up a coupling between the dispersion vibrations built up by the electric field and the remaining vibrations in a state close to thermodynamic equilibrium, which led to damping of the dispersion vibrations.

For $\varepsilon(\omega)$, Pauli obtained an expression similar to Eq. (I.1), with γ proportional to the absolute temperature T and independent of the frequency. (He overlooked that this γ must vanish at frequencies $\omega > \sqrt{2}\omega_0$.) In Pauli's treatment, the frequency-independence of γ in the range $0 < \omega < \sqrt{2}\omega_0$ is a consequence of the fact that the positive and negative ions have equal masses, as assumed in his crystal model.

In [4, 5], Born and Blackman dispensed with this limitation, and considered more general models: a linear sequence of ions with different masses, and a three-dimensional crystal lattice of the NaCl type. They obtained a function $\gamma(\omega)$ similar to that which is observed experimentally.

Their calculations show that the damping of dispersion vibrations built up by an electric field having frequency ω is due to energy transfer to two other vibrations related to the field frequency ω by

$$\begin{aligned} \omega &= \omega_j(\mathbf{y}) + \omega_{j'}(\mathbf{y}), \\ \omega &= \omega_j(\mathbf{y}) - \omega_{j'}(\mathbf{y}), \end{aligned} \tag{I.2}$$

where j and j' are branches of the vibrational spectrum of the crystal, and $\mathbf{y}$ the wave vectors of the vibrations.

The expressions (I.2) become clearer if regarded from the viewpoint of quantum mechanics. The first equation then signifies that a photon with energy $\hbar\omega$ creates two phonons with energies $\hbar\omega_j(\mathbf{y})$ and $\hbar\omega_{j'}(\mathbf{y})$. The second equation describes a process in which a photon is absorbed to create one phonon and absorb another. The equality of the arguments in the functions $\omega_j(\mathbf{y})$ and $\omega_{j'}(\mathbf{y})$ follows from the conservation of quasimomentum and the relations $\omega_j(-\mathbf{y}) = \omega_j(\mathbf{y})$, $\nu \ll \mathbf{y}$, where ν is the wave vector of the radiation.

The damping constant $\gamma(\omega)$ relating to these processes is determined by the number of waves whose frequencies and wave vectors satisfy Eqs. (I.2) for a given frequency ω, i.e., by the density of states which corresponds to the functions $\omega_j(\mathbf{y}) \pm \omega_{j'}(\mathbf{y})$.

The question of the density of states in the spectrum was analyzed by Blackman in 1936 [7], as regards its effect on $\gamma(\omega)$, in greater detail than in [4, 5]. Blackman's paper also corrected certain errors in the expressions for $\varepsilon(\omega)$ given in [4, 5].

Thus, in [4-7], the classical theory of infrared absorption and dispersion in ionic crystals was basically completed.

Among later publications, mention should be made of the work of Neuberger and Hatcher [8] in 1961 and of Jepsen and Wallis [9] in 1962. The former authors repeated the calculations given in [5, 7], and found agreement with the results of [7]. They also made a numerical evaluation of $\gamma(\omega)$ and $\varepsilon(\omega)$ for NaCl. A comparison of the calculated optical constants with the experimental values showed satisfactory agreement.

In [9], Born and Blackman's methods were used to calculate $\varepsilon(\omega)$ with allowance for fourth-order anharmonic terms in the potential energy. It was found that these terms, though less important than the cubic terms, are not negligible, especially in accounting for the temperature dependence of the absorption at a frequency $\omega \approx \omega_0$.

Let us next consider the use of quantum-mechanical methods to calculate the infrared absorption. The first such work was that of Barnes, Brattain, and Seitz [10] in 1935, where in fact the transition probabilities between states of an anharmonic lattice in an alternating electric field were calculated. From the condition for these probabilities to be nonzero, relations were obtained between the field frequency and the frequencies of the phonons concerned in the absorption process. As well as the Eqs. (I.2), other relations were obtained, including

$$\hbar\,[\pm\omega_1(\mathbf{y}_1)\pm\omega_2(\mathbf{y}_2)\pm\omega_3(\mathbf{y}_3)\pm\omega_0(\mathbf{K})]=\hbar\omega \qquad \text{(I.3)}$$

for example. Here, $\mathbf{y}_1 \pm \mathbf{y}_2 \pm \mathbf{y}_3 = \mathbf{K}'$; $\mathbf{K}$, $\mathbf{K}'$ are reciprocal-lattice vectors.

For this reason, the authors [10] supposed that the frequency dependence of the absorption must be more complicated than is indicated in [4, 5, 7].

The appearance of the relations (I.3) in [10] is due to the fact that the authors calculated only transition probabilities and not the more complicated quantities $\varepsilon(\omega)$ and $\gamma(\omega)$, which are combinations of these probabilities. In the calculation of $\varepsilon(\omega)$ and $\gamma(\omega)$, it is found that the terms which contain a contribution from the processes (I.3) vanish, leaving only those which contain a contribution from the processes (I.2).

Other deficiencies of [10] include the use of perturbation theory in order to take account of anharmonicity; this does not allow a consideration of absorption at frequencies close to ω_0. No account, moreover, was taken of the temperature dependence of the absorption.

In 1954, Born and Huang [11] made an attempt to overcome these deficiencies and create a more satisfactory theory. They used the Weisskopf–Wigner method to calculate $\varepsilon(\omega)$. Their conclusion was that $\gamma(\omega) \propto T^3$ at high temperatures, in contradiction with the results of the classical theory [4, 5, 7], which give $\gamma(\omega) \propto T$.

Born and Huang's calculations were analyzed by Mitskevich [12], Maradudin and Wallis [13], and Vinogradov [14] in 1960 and 1961, and a number of inaccuracies were revealed.

One was the incorrect formulation of the initial conditions in solving the Schrödinger equation. The initial conditions in [11] were taken to be that, at $t = t_0$, $a_0 = 1$, and $a_s = 0$ for $s \neq 0$, where a_0 and a_s are the coefficients in the expansion of the complete wave function in eigenfunctions of the Hamiltonian in the harmonic approximation; a_0 corresponds to the state in which the numbers of photons are equal to their values in thermodynamic equilibrium. These initial conditions are in contradiction with the statistical result that, at the initial instant, we must have $|a_s|^2 \propto \exp(-E_s/kT) \neq 0$. In short, Born and Huang did not correctly allow for the distribution of the system between states.

Another inaccuracy is the neglect of certain intermediate states in calculating the transition probabilities.

In [12-14] and in the work of Kleinman [15], Szigeti [16], and Vinogradov [17], a more satisfactory theory of the effect was given; perturbation theory was used in [13-16], the density-matrix method in [12], and the retarded double-time Green's function method in [17]. The calculations in [12] and [17], unlike the others, are valid when the radiation frequency is the same as the dispersion frequency. The expressions for $\varepsilon(\omega)$ derived in these two papers become the classical expressions [7, 8] when $\hbar$ tends to zero.

Maradudin and Wallis [18] in 1962 calculated $\varepsilon(\omega)$ by summation of diagrams, but the results were less satisfactory than in [12, 17]. In calculating $\gamma(\omega)$, they neglected processes in which one phonon is absorbed and another created, and their frequency dependence $\varepsilon(\omega)$ does not tend to the classical form when $\hbar \to 0$. In his last paper on this subject [19], in 1963, Maradudin obtained an expression for $\varepsilon(\omega)$ identical with those given in [12, 17]. A similar expression was derived by Lax [20] in 1963.

Mitskevich [21, 22] in 1961 and 1962 made detailed calculations of the functions $\gamma(\omega)$ and $\varepsilon(\omega)$ in the classical limit and of the optical constants k and n for NaCl, LiF, and MgO, taking into account the contribution to $\varepsilon(\omega)$ from the deformation of the electron shells. Satisfactory agreement was obtained between the calculated and experimental values of k and n.

The papers by Gurevich and Ipatova [23] and Kashcheev [24, 25] appeared in 1963. The former authors avoided lengthy numerical calculations and estimated the relative role of the third- and fourth-order anharmonic terms in the potential energy, as regards absorption. They concluded that these terms give comparable contributions to $\gamma(\omega)$ at $\omega = \omega_0$.

Kashcheev [24] used the method of double-time temperature-dependent Green's functions, calculating $\varepsilon(\omega)$ for a crystal having several optically active branches and including the effect of their interaction. His calculation involves a number of errors, however, which will be discussed in detail in chapter III, section 3.

In [25], Kashcheev included the effect of fourth-order anharmonic terms in the potential energy on $\gamma(\omega)$. This paper has the same shortcomings as [24].

In describing below the quantum theory of infrared absorption, we shall use principally the author's paper [17]. New features as compared with [17] are the allowance for fourth-order anharmonic terms in the potential energy and also the treatment of a crystal having several optically active branches.

2. Theories Which Take Account of the Influence of Charged Defects on Electromagnetic Wave Absorption by a Crystal Lattice

There has been much theoretical work on the problem of how the phonon spectrum of a crystal is influenced by defects of various kinds, and how these changes affect various physical phenomena.

First of all, Lifshits [26, 27] (the latter paper giving a long list of this author's publications), Montroll, Potts, Maradudin, and co-workers [28-30], and the Japanese workers Takeno, Hori, and others deserve mention.

The influence of defects on electromagnetic wave absorption by a crystal lattice was studied by Lifshits [26], Lax and Burstein [32], Wallis and Maradudin [33], and Dawber and Elliott [34]. These papers were concerned mainly with isotopic defects, i.e., those due to a change of mass of the particles forming the lattice.

There has been very little discussion of charged defects. One paper is that by Lysenko [35] in 1938 on work done at the suggestion of Frenkel'.

Lysenko considered the absorption of electromagnetic radiation with frequency $\omega \ll \omega_{max}$ (where ω_{max} is the maximum frequency in the lattice spectrum) by a cubic lattice consisting of a atoms with equal mass, one atom having an electric charge. Absorption occurs in this system because the charged atom acquires energy from the electric field and emits acoustic waves.

A similar problem for a linear chain was solved by Lifshits in 1942 [26].

The problem of microwave absorption in an ionic lattice of the NaCl type containing charged defects was solved by Vinogradov in 1960-1963 [36-38]. The essentially new contribution was the allowance for the Coulomb interaction between the charged defect and the lattice ions.

The calculations [37, 38] were done in order to account for the frequency and temperature dependence of dielectric microwave losses in $SrTiO_3$ and other crystals.

CHAPTER II

The Hamiltonian and the Equations of Motion. Expression of the Complex Polarizability in Terms of Retarded Green's Functions

1. The Hamiltonian for a Lattice Containing Defects. Properties of Its Coefficients

In this paper, we shall consider only those infrared dispersion and absorption effects which are not due to deformation of the electron shells of the ions during their vibrations. Such effects have been quite fully discussed in [11, 16, 32], and we shall therefore regard the ions as rigid.

Let us therefore consider an infinite lattice consisting of rigid point ions. We take a volume of $N\,(= L^3)$ cells and suppose it to be periodically repeated. Periodicity conditions are to be imposed on the displacement of the ions.

Let $U = \Phi + \delta\Phi$ be the potential energy of the crystal lattice, $\delta\Phi$ being due to the presence of defects. In an ideal lattice, the ions are customarily labeled by indices $\binom{l}{k}$, where l is the set of three integers which give the position of the cell, and k the number of the ion in the cell.

We shall label the ions in a defective lattice in the same way, assuming that, when $\delta\Phi = 0$, a given ion is at the position of the ion $\binom{l}{k}$ in the periodic lattice.

Let $u_\alpha\binom{l}{k}$ denote the (time-dependent) displacement of ion $\binom{l}{k}$ from its equilibrium position in the defective lattice. We assume the $u_\alpha\binom{l}{k}$ to be small and expand U in series:

$$U = U(0) + \sum_{lk\alpha} U_\alpha\binom{l}{k} u_\alpha\binom{l}{k} + {}^1/_2 \sum_{lk\alpha}\sum_{l'k'\beta} U_{\alpha\beta}\binom{ll'}{kk'} u_\alpha\binom{l}{k} u_\beta\binom{l'}{k'} +$$
$$+ \frac{1}{3!}\sum_{lk\alpha}\sum_{l'k'\beta}\sum_{l''k''\gamma} U_{\alpha\beta\gamma}\binom{ll'l''}{kk'k''} u_\alpha\binom{l}{k} u_\beta\binom{l'}{k'} u_\gamma\binom{l''}{k''} +$$
$$+ \frac{1}{4!}\sum_{lk\alpha}\sum_{l'k'\beta}\sum_{l''k''\gamma}\sum_{l'''k'''\delta} U_{\alpha\beta\gamma\delta}\binom{ll'l''l'''}{kk'k''k'''} u_\alpha\binom{l}{k} u_\beta\binom{l'}{k'} u_\gamma\binom{l''}{k''} u_\delta\binom{l'''}{k'''} + \dots . \quad \text{(II.1)}$$

The coefficients $U_{\alpha\dots}\binom{l\,\dots}{k\,\dots}$ which are the derivatives of U with respect to the $u_\alpha\binom{l}{k}$, change when the entire lattice is displaced through a lattice vector, since they are taken at the equilibrum positions of the ions of a nonperiodic (defective) lattice. These coefficients can be, in part, expressed in terms of translation-invariant quantities if they are expanded in terms of quantities $v_\alpha\binom{l}{k}$, which describe the equilibrium positions of the ions in the nonperiodic lattice relative to the equilibrium positions in the periodic lattice (i.e., the displacements of the ion equilibrium positions in the lattice when the potential $\delta\Phi$ is added). The subscript zero will denote the values of coefficients at the ion equilibrium positions in the ideal lattice. If the $v_\alpha\binom{l}{k}$ are small, we have

$$U_\alpha\binom{l}{k} = U_\alpha\binom{l}{k}_0 + \sum_{l'k'\beta} U_{\alpha\beta}\binom{ll'}{kk'}_0 v_\beta\binom{l'}{k'} + \dots , \quad \text{(II.2)}$$

$$U_{\alpha\beta}\binom{ll'}{kk'} = U_{\alpha\beta}\binom{ll'}{kk'}_0 + \sum_{l''k''\gamma} U_{\alpha\beta\gamma}\binom{ll'l''}{kk'k''}_0 v_\gamma\binom{l''}{k''} + \dots , \quad \text{(II.3)}$$

and so on. Since, in equilibrium, $U_\alpha\binom{l}{k} = 0$ and $\Phi_\alpha\binom{l}{k}_0 = 0$, we obtain from Eq. (II.2) equations which give the displacements $v_\alpha\binom{l}{k}$ of the equilibrium positions:

$$0 = \delta\Phi_\alpha\binom{l}{k}_0 + \sum_{l'k'\beta}\left(\Phi_{\alpha\beta}\binom{ll'}{kk'}_0 + \delta\Phi_{\alpha\beta}\binom{ll'}{kk'}_0\right) v_\beta\binom{l'}{k'} + \dots . \quad \text{(II.4)}$$

Substituting Eqs. (II.2), (II.3), etc., in Eq. (II.1) and adding the kinetic energy, we get as the lattice Hamiltonian (ignoring defects due to a change of mass of the ions)

$$H = {}^1/_2 \sum_{lk\alpha} m_k \dot{u}_\alpha^2\binom{l}{k} + {}^1/_2 \sum_{lk\alpha}\sum_{l'k'\beta} \Phi_{\alpha\beta}\binom{ll'}{kk'}_0 u_\alpha\binom{l}{k} u_\beta\binom{l'}{k'} +$$
$$+ {}^1/_2 \sum_{lk\alpha}\sum_{l'k'\beta}\left\{\delta\Phi_{\alpha\beta}\binom{ll'}{kk'}_0 + \sum_{l''k''\gamma}\left(\Phi_{\alpha\beta\gamma}\binom{ll'l''}{kk'k''}_0 + \delta\Phi_{\alpha\beta\gamma}\binom{ll'l''}{kk'k''}_0\right) v_\gamma\binom{l''}{k''} + \dots\right\} \times$$
$$\times u_\alpha\binom{l}{k} u_\beta\binom{l'}{k'} + \frac{1}{3!}\sum_{lk\alpha}\sum_{l'k'\beta}\sum_{l''k''\gamma}\left\{\Phi_{\alpha\beta\gamma}\binom{ll'l''}{kk'k''}_0 + \delta\Phi_{\alpha\beta\gamma}\binom{ll'l''}{kk'k''}_0 +\right.$$
$$\left. + \sum_{l'''k'''\delta}\left(\Phi_{\alpha\beta\gamma\delta}\binom{ll'l''l'''}{kk'k''k'''}_0 + \delta\Phi_{\alpha\beta\gamma\delta}\binom{ll'l''l'''}{kk'k''k'''}_0\right) v_\delta\binom{l'''}{k'''} + \dots\right\} u_\alpha\binom{l}{k} u_\beta\binom{l'}{k'} \times$$

$$\times u_\gamma\begin{pmatrix}l''\\k''\end{pmatrix}+\frac{1}{4!}\sum_{lk\alpha}\sum_{l'k'\beta}\sum_{l''k''\gamma}\sum_{l'''k'''\delta}\left\{\Phi_{\alpha\beta\gamma\delta}\begin{pmatrix}ll'l''l'''\\kk'k''k'''\end{pmatrix}_0+\delta\Phi_{\alpha\beta\gamma\delta}\begin{pmatrix}ll'l''k'''\\kk'k''k'''\end{pmatrix}_0\right\}u_\alpha\begin{pmatrix}l\\k\end{pmatrix}u_\beta\begin{pmatrix}l'\\k'\end{pmatrix}u_\gamma\begin{pmatrix}l''\\k''\end{pmatrix}u_\delta\begin{pmatrix}l'''\\k'''\end{pmatrix}. \tag{II.5}$$

The first two terms form the harmonic part of the Hamiltonian, which can be represented as the sum of the energies of the oscillators by using the Fourier transformation. We shall regard this as the unperturbed part of the Hamiltonian, and the remaining terms at the perturbation. In the expression for the Hamiltonian, we shall retain only terms containing derivatives of $\Phi + \delta\Phi$ of not higher than the fourth order, neglecting all other terms in the expansion.

Since $u_\alpha\begin{pmatrix}l\\k\end{pmatrix}$ and $v_\alpha\begin{pmatrix}l\\k\end{pmatrix}$ are functions of $\mathbf{x}\begin{pmatrix}l\\k\end{pmatrix}$, the positions of the ions in the periodic lattice, they can be written as expansions in plane waves:

$$u_\alpha\begin{pmatrix}l\\k\end{pmatrix}=\frac{1}{\sqrt{Nm_k}}\sum_{\mathbf{y}j}Q\begin{pmatrix}\mathbf{y}\\j\end{pmatrix}w_\alpha\left(k\middle|\begin{matrix}\mathbf{y}\\j\end{matrix}\right)\exp\left[2\pi i\mathbf{y}\mathbf{x}\begin{pmatrix}l\\k\end{pmatrix}\right], \tag{II.6}$$

$$v_\alpha\begin{pmatrix}l\\k\end{pmatrix}=\frac{1}{\sqrt{Nm_k}}\sum_{\mathbf{y}j}q\begin{pmatrix}\mathbf{y}\\j\end{pmatrix}w_\alpha\left(k\middle|\begin{matrix}\mathbf{y}\\j\end{matrix}\right)\exp\left[2\pi i\mathbf{y}\mathbf{x}\begin{pmatrix}l\\k\end{pmatrix}\right], \tag{II.7}$$

where

$$\mathbf{y}=\mathbf{y}\left(\frac{h}{L}\right)=\frac{h_1}{L}\mathbf{b}_1+\frac{h_2}{L}\mathbf{b}_2+\frac{h_3}{L}\mathbf{b}_3,$$

$\mathbf{b}_1$, $\mathbf{b}_2$, $\mathbf{b}_3$ are reciprocal-lattice vectors, and h_1, h_2, h_3 take integral values in the range $-L/2 \le h_1, h_2, h_3 < L/2$ (for further details, see [11]).

Since the displacements are real, we have

$$Q\begin{pmatrix}\mathbf{y}\\j\end{pmatrix}^*w_\alpha\left(k\middle|\begin{matrix}\mathbf{y}\\j\end{matrix}\right)^*=Q\begin{pmatrix}-\mathbf{y}\\j\end{pmatrix}w_\alpha\left(k\middle|\begin{matrix}-\mathbf{y}\\j\end{matrix}\right). \tag{II.8}$$

The quantities $w_\alpha\left(k\middle|\begin{matrix}\mathbf{y}\\j\end{matrix}\right)$ are the solutions of the oscillation equations with the harmonic Hamiltonian. These equations are (for further details, see [11])

$$\omega^2\begin{pmatrix}\mathbf{y}\\j\end{pmatrix}w_\alpha\left(k\middle|\begin{matrix}\mathbf{y}\\j\end{matrix}\right)=\sum_{k'\beta}C_{\alpha\beta}\begin{pmatrix}\mathbf{y}\\kk'\end{pmatrix}w_\beta\left(k'\middle|\begin{matrix}\mathbf{y}\\j\end{matrix}\right), \tag{II.9}$$

where

$$C_{\alpha\beta}\begin{pmatrix}\mathbf{y}\\kk'\end{pmatrix}=\sum_{l-l'}\frac{1}{\sqrt{m_km_{k'}}}\Phi_{\alpha\beta}\begin{pmatrix}ll'\\kk'\end{pmatrix}_0\exp\left[-2\pi i\mathbf{y}\mathbf{x}\begin{pmatrix}ll'\\kk'\end{pmatrix}\right]. \tag{II.10}$$

The quantities $C_{\alpha\beta}\begin{pmatrix}\mathbf{y}\\kk'\end{pmatrix}$ have the following properties:

$$\begin{aligned}C_{\alpha\beta}\begin{pmatrix}\mathbf{y}\\kk'\end{pmatrix}^*&=C_{\beta\alpha}\begin{pmatrix}\mathbf{y}\\k'k\end{pmatrix},\\C_{\alpha\beta}\begin{pmatrix}\mathbf{y}\\kk'\end{pmatrix}^*&=C_{\alpha\beta}\begin{pmatrix}-\mathbf{y}\\kk'\end{pmatrix}.\end{aligned} \tag{II.11}$$

From Eqs. (II.9)-(II.11), it is seen that

$$\omega\binom{\mathbf{y}}{j} = \omega\binom{-\mathbf{y}}{j},\; w_\alpha^*\left(k\middle|\begin{matrix}\mathbf{y}\\ j\end{matrix}\right) = w_\alpha\left(k\middle|\begin{matrix}-\mathbf{y}\\ j\end{matrix}\right).$$

Then, from Eq. (II.8), we find

$$Q\binom{\mathbf{y}}{j}^* = Q\binom{-\mathbf{y}}{j} \tag{II.12}$$

A similar equation is valid for $q\binom{\mathbf{y}}{j}$. The quantities $w_\alpha\left(k\middle|\begin{matrix}\mathbf{y}\\ j\end{matrix}\right)$ satisfy the normalization conditions

$$\sum_{k\alpha} w_\alpha^*\left(k\middle|\begin{matrix}\mathbf{y}\\ j\end{matrix}\right) w_\alpha\left(k\middle|\begin{matrix}\mathbf{y}\\ j'\end{matrix}\right) = \delta_{jj'}. \tag{II.13}$$

Substituting Eqs. (II.6) and (II.7) in Eq. (II.5) and using Eqs. (II.9), (II.12), and (II.13), we can bring the Hamiltonian H to the form

$$H = \frac{1}{2}\sum_{\mathbf{y}j}\left\{\dot{Q}\binom{\mathbf{y}}{j}\dot{Q}\binom{-\mathbf{y}}{j} + \omega^2\binom{\mathbf{y}}{j} Q\binom{\mathbf{y}}{j} Q\binom{-\mathbf{y}}{j}\right\} + \frac{1}{2}\sum_{\substack{\mathbf{y}\mathbf{y}'\\ jj'}} A\binom{\mathbf{y}\mathbf{y}'}{jj'} Q\binom{\mathbf{y}}{j} Q\binom{\mathbf{y}'}{j'} +$$

$$+ \frac{1}{3!}\sum_{\substack{\mathbf{y}\mathbf{y}'\mathbf{y}''\\ jj'j''}} A\binom{\mathbf{y}\mathbf{y}'\mathbf{y}''}{jj'j''} Q\binom{\mathbf{y}}{j} Q\binom{\mathbf{y}'}{j'} Q\binom{\mathbf{y}''}{j''} + \frac{1}{4!}\sum_{\substack{\mathbf{y}\mathbf{y}'\mathbf{y}''\mathbf{y}'''\\ jj'j''j'''}} A\binom{\mathbf{y}\mathbf{y}'\mathbf{y}''\mathbf{y}'''}{jj'j''j'''} Q\binom{\mathbf{y}}{j} Q\binom{\mathbf{y}'}{j'} Q\binom{\mathbf{y}''}{j''} Q\binom{\mathbf{y}'''}{j'''}. \tag{II.14}$$

The coefficients $A\binom{\mathbf{y}\ldots}{j\ldots}$ are

$$A\binom{\mathbf{y}\mathbf{y}'}{jj'} = \delta\Phi\binom{\mathbf{y}\mathbf{y}'}{jj'}_0 + \sum_{\mathbf{y}''j''}\left(\Phi\binom{\mathbf{y}\mathbf{y}'\mathbf{y}''}{jj'j''}_0 + \delta\Phi\binom{\mathbf{y}\mathbf{y}'\mathbf{y}''}{jj'j''}_0\right) q\binom{\mathbf{y}''}{j''} + \frac{1}{2}\sum_{\substack{\mathbf{y}''\mathbf{y}'''\\ j''j'''}}\left(\Phi\binom{\mathbf{y}\mathbf{y}'\mathbf{y}''\mathbf{y}'''}{jj'j''j'''}_0 + \delta\Phi\binom{\mathbf{y}\mathbf{y}'\mathbf{y}''\mathbf{y}'''}{jj'j''j'''}_0\right) q\binom{\mathbf{y}''}{j''} q\binom{\mathbf{y}'''}{j'''}, \tag{II.15}$$

$$A\binom{\mathbf{y}\mathbf{y}'\mathbf{y}''}{jj'j''} = \Phi\binom{\mathbf{y}\mathbf{y}'\mathbf{y}''}{jj'j''}_0 + \delta\Phi\binom{\mathbf{y}\mathbf{y}'\mathbf{y}''}{jj'j''}_0 + \sum_{\mathbf{y}'''j'''}\left(\Phi\binom{\mathbf{y}\mathbf{y}'\mathbf{y}''\mathbf{y}'''}{jj'j''j'''}_0 + \delta\Phi\binom{\mathbf{y}\mathbf{y}'\mathbf{y}''\mathbf{y}'''}{jj'j''j'''}_0\right) q\binom{\mathbf{y}'''}{j'''},$$

$$A\binom{\mathbf{y}\mathbf{y}'\mathbf{y}''\mathbf{y}'''}{jj'j''j'''} = \Phi\binom{\mathbf{y}\mathbf{y}'\mathbf{y}''\mathbf{y}'''}{jj'j''j'''}_0 + \delta\Phi\binom{\mathbf{y}\mathbf{y}'\mathbf{y}''\mathbf{y}'''}{jj'j''j'''}_0,$$

where

$$\delta\Phi\binom{\mathbf{y}\mathbf{y}'\ldots\mathbf{y}^{(n)}}{jj'\ldots j^{(n)}} = \frac{1}{N^{n/2}}\sum_{\substack{lk\alpha\\ l'k'\beta\\ \ldots\\ l^{(n)}k^{(n)}\alpha^{(n)}}} \frac{\delta\Phi_{\alpha\beta\ldots\alpha^{(n)}}\binom{ll'\ldots l^{(n)}}{kk'\ldots k^{(n)}}_0}{\sqrt{m_k m_{k'}\ldots m_{k^{(n)}}}} \times$$

$$\times w_\alpha\left(k\middle|\begin{matrix}\mathbf{y}\\ j\end{matrix}\right) w_\beta\left(k'\middle|\begin{matrix}\mathbf{y}'\\ j'\end{matrix}\right)\ldots w_{\alpha^{(n)}}\left(k^{(n)}\middle|\begin{matrix}\mathbf{y}^{(n)}\\ j^{(n)}\end{matrix}\right)\exp\left\{2\pi i\left[\mathbf{y}\mathbf{x}\binom{l}{k} + \mathbf{y}'\mathbf{x}\binom{l'}{k'} + \ldots + \mathbf{y}^{(n)}\mathbf{x}\binom{l^{(n)}}{k^{(n)}}\right]\right\}. \tag{II.16}$$

The coefficients $\Phi\binom{\mathbf{y}\mathbf{y}'\ldots\mathbf{y}^{(n)}}{jj'\ldots j^{(n)}}$ differ from the $\delta\Phi\binom{\mathbf{y}\mathbf{y}'\ldots\mathbf{y}^{(n)}}{jj'\ldots j^{(n)}}$ in that they contain the function $\Phi_{\alpha\beta\ldots\alpha^{(n)}}\binom{ll'\ldots l^{(n)}}{kk'\ldots k^{(n)}}$, which is unchanged by translation through a lattice vector, i.e.,

$$\Phi_{\alpha\beta\gamma\ldots\alpha^{(n)}}\binom{ll'\ldots l^{(n)}}{kk'\ldots k^{(n)}}_0 = \Phi_{\alpha\beta\gamma\ldots\alpha^{(n)}}\binom{0\;l'-l\ldots l^{(n)}-l}{k\;\;k'\ldots k^{(n)}}_0.$$

Hence, the coefficients $\Phi_{\alpha\beta\gamma\ldots\alpha^{(n)}}\begin{pmatrix}\mathbf{y}\mathbf{y}' & \cdots & \mathbf{y}^{(n)} \\ jj' & \cdots & j^{(n)}\end{pmatrix}_0$ may be written

$$\Phi\begin{pmatrix}\mathbf{y}\mathbf{y}' & \cdots & \mathbf{y}^{(n)} \\ jj' & \cdots & j^{(n)}\end{pmatrix}_0 = \frac{1}{N^{\frac{n}{2}-1}}\Delta(\mathbf{y}+\mathbf{y}'+\ldots+\mathbf{y}^{(n)}) \sum_{\substack{k\alpha \\ l'k'\beta \\ \cdots\cdots \\ l^{(n)}k^{(n)}\alpha^{(n)}}} \frac{\Phi_{\alpha\beta\ldots\alpha^{(n)}}\begin{pmatrix}0\; l'-l & l^{(n)}-l \\ k\; k' & k^{(n)}\end{pmatrix}_0}{\sqrt{m_k m_{k'} \ldots m_{k^{(n)}}}} \times$$

$$\times w_\alpha\left(k\middle|\begin{matrix}\mathbf{y}\\ j\end{matrix}\right) w_\beta\left(k'\middle|\begin{matrix}\mathbf{y}'\\ j'\end{matrix}\right)\ldots w_{\alpha^{(n)}}\left(k^{(n)}\middle|\begin{matrix}\mathbf{y}^{(n)}\\ j^{(n)}\end{matrix}\right)\exp\left\{2\pi i\left[\mathbf{y}\mathbf{x}\begin{pmatrix}0\\ k\end{pmatrix}+\mathbf{y}'\mathbf{x}\begin{pmatrix}l'-l\\ k'\end{pmatrix}+\ldots+\mathbf{y}^{(n)}\mathbf{x}\begin{pmatrix}l^{(n)}-l\\ k^{(n)}\end{pmatrix}\right]\right\}, \quad \text{(II.17)}$$

where the function $\Delta(\mathbf{y}) = \frac{1}{N}\sum_l e^{2\pi i\mathbf{y}\mathbf{x}(l)}$ is nonzero and equal to unity if the wave vector $\mathbf{y}$ is a reciprocal-lattice vector, including zero.

It is seen immediately from Eq. (II.17) that $\Phi\begin{pmatrix}\mathbf{y}\mathbf{y}' & \cdots & \mathbf{y}^{(n)} \\ jj' & \cdots & j^{(n)}\end{pmatrix}_0$ has the properties

$$\text{a)}\quad \Phi\begin{pmatrix}\mathbf{y}\mathbf{y}' & \cdots & \mathbf{y}^{(n)} \\ jj' & \cdots & j^{(n)}\end{pmatrix}_0^* = \Phi\begin{pmatrix}-\mathbf{y} & -\mathbf{y}' & \cdots & -\mathbf{y}^{(n)} \\ j & j' & \cdots & j^{(n)}\end{pmatrix}_0,$$

$$\text{b)}\quad \Phi\begin{pmatrix}\mathbf{y}\mathbf{y}' & \cdots & \mathbf{y}^{(n)} \\ jj' & \cdots & j^{(n)}\end{pmatrix}_0 = \Phi\begin{pmatrix}\mathbf{y}'\mathbf{y} & \cdots & \mathbf{y}^{(n)} \\ j'j & \cdots & j^{(n)}\end{pmatrix}_0 = \Phi\begin{pmatrix}\mathbf{y}^{(n)}\mathbf{y}' & \cdots & \mathbf{y} \\ j^{(n)}j' & \cdots & j\end{pmatrix}_0, \ldots, \quad \text{(II.18)}$$

i.e., the indices $\begin{pmatrix}\mathbf{y}\\ j\end{pmatrix}, \begin{pmatrix}\mathbf{y}'\\ j'\end{pmatrix}, \ldots, \begin{pmatrix}\mathbf{y}^{(n)}\\ j^{(n)}\end{pmatrix}$ can be interchanged. The same properties are valid for the coefficients $\delta\Phi\begin{pmatrix}\mathbf{y}\mathbf{y}' & \cdots & \mathbf{y}^{(n)} \\ jj' & \cdots & j^{(n)}\end{pmatrix}_0$ and $A\begin{pmatrix}\mathbf{y}\mathbf{y}' & \cdots & \mathbf{y}^{(n)} \\ jj' & \cdots & j^{(n)}\end{pmatrix}$, as is seen from the definitions (II.16) and (II.15).

Expressions for coefficients of the form (II.16) and (II.17), for the case where the forces acting between the ions are between pairs of ions, are given in Appendix A.

2. Expression for the Complex Polarizability of the Lattice in Terms of Retarded Green's Functions

In this paper, we shall use the method of double-time retarded Green's functions (see [39, 40]). We therefore need an expression for the complex polarizability of the lattice in terms of the Green's functions. To derive this, we shall first find an expression for the mean dipole moment per unit volume of the crystal.

The ion dipole moment is

$$p_\alpha\begin{pmatrix}l\\ k\end{pmatrix} = e_k^l u_\alpha\begin{pmatrix}l\\ k\end{pmatrix} = (e_k + \delta e_k^l)\, u_\alpha\begin{pmatrix}l\\ k\end{pmatrix}, \quad \text{(II.19)}$$

where δe_k^l is the change of charge at the point $\begin{pmatrix}l\\ k\end{pmatrix}$ caused by the presence of the defect, and e_k is the charge of the kth ion in a defect-free lattice. We expand the dipole moment as a Fourier series in $\mathbf{y}$ and integral in ω:

$$p_\alpha\begin{pmatrix}l\\ k\end{pmatrix} = \sum_{\mathbf{y}}\int p_\alpha\left(\begin{matrix}\mathbf{y}\\ k\end{matrix}, \omega\right)\exp\left[2\pi i\mathbf{y}\mathbf{x}\begin{pmatrix}l\\ k\end{pmatrix} - i\omega t\right]d\omega, \quad \text{(II.20)}$$

where

$$p_\alpha\left(\begin{matrix}\nu\\k\end{matrix},\Omega\right)=\frac{1}{2\pi N}\sum_l\int p_\alpha\left(\begin{matrix}l\\k\end{matrix}\right)\exp\left[-2\pi i\nu\mathbf{x}\left(\begin{matrix}l\\k\end{matrix}\right)+i\Omega t\right]dt. \tag{II.21}$$

The property of realness shows that

$$p_\alpha\left(\begin{matrix}\mathbf{y}\\k\end{matrix},\omega\right)=p_\alpha\left(\begin{matrix}-\mathbf{y}\\k\end{matrix},-\omega\right)^*. \tag{II.22}$$

The quantity that is of interest to us is the mean dipole moment per unit volume,

$$\langle p_\alpha(\nu,\Omega)\rangle=\frac{1}{2\pi v_a N}\sum_{lk}\int\left\langle p_\alpha\left(\begin{matrix}l\\k\end{matrix}\right)\exp\left[-2\pi i\nu\mathbf{x}\left(\begin{matrix}l\\k\end{matrix}\right)+i\Omega t\right]\right\rangle dt, \tag{II.23}$$

where the brackets $\langle\,\rangle$ denote averaging with the density matrix ρ and over the distributions of defects, and v_a is the volume of the unit cell.

In order to obtain an expression for the mean value, we solve the equation for the density matrix (see also [39]):

$$i\hbar\frac{d\rho(t)}{dt}=[H+U_E,\rho(t)], \tag{II.24}$$

where H is the Hamiltonian (II.14), and U_E takes account of the interaction between the lattice and the macroscopic electric field.

We shall solve this equation by perturbation theory, assuming that U_E is a perturbation and writing $\rho(t)=\rho+\Delta\rho(t)$. As the initial condition, we take $\rho(t)|_{t\to\infty}=\rho=Q^{-1}e^{-\frac{H}{kT}}$ (assuming that the macroscopic field is applied adiabatically).

The expression for the mean value of a quantity A is

$$\langle A\rangle=Sp\{\rho(t)A(t)\}. \tag{II.25}$$

For $\Delta\rho(t)$, we have from Eq. (II.24), in the first-order approximation of perturbation theory,

$$\Delta\rho(t)=\frac{1}{i\hbar}\int_{-\infty}^{+\infty}\exp\left[\frac{iH(\tau-t)}{\hbar}\right]\left[U_E,\rho\right]\exp\left[-\frac{iH(\tau-t)}{\hbar}\right]d\tau. \tag{II.26}$$

Using this, we get for the time-dependent part of the mean value

$$\langle A\rangle=\frac{1}{i\hbar}\int_{-\infty}^{+\infty}\theta(t-\tau)\langle[A^h(t),U_E^h(\tau)]\rangle d\tau. \tag{II.27}$$

Here, the operators $A^h(t)$ and $U_E^h(\tau)$ in the commutator are taken to be Heisenberg operators, i.e.,

$$A^h(t)=e^{\frac{iHt}{\hbar}}Ae^{-\frac{iHt}{\hbar}}, \tag{II.28}$$

and the averaging is effected by means of the equilibrium density matrix $\rho=Q^{-1}e^{-\frac{H}{kT}}$.

The function $\theta(t)$ is such that

$$\theta(t)=\begin{cases}1 & \text{for } t>0\\ 0 & \text{for } t<0\end{cases}\qquad \frac{d\theta(t)}{dt}=\delta(t). \tag{II.29}$$

In order to find the required quantity $\langle p_\alpha(\nu,\Omega)\rangle$, we must substitute for A in Eq. (II.27) the expression $p_\alpha\binom{l}{k}\exp\left[-2\pi i\nu\mathbf{x}\binom{l}{k}+i\Omega t\right]$, and then use the result in Eq. (II.23).

The energy of interaction with the macroscopic electric field may be written

$$U_E=-\sum_{l'k'}p_\alpha\binom{l'}{k'}E_\alpha\binom{l'}{k'}. \tag{II.30}$$

The macroscopic electric field is the sum of the field due to the external sources and the part of the lattice field which varies slowly in space. In order to avoid including the macroscopic field due to the lattice twice, it must be subtracted from the Hamiltonian (II.14); for further details, see [11, 41].

The macroscopic field may be written as an expansion

$$E_\alpha\binom{l'}{k'}=\sum_{\mathbf{y}}\int E_\alpha(\mathbf{y},\omega)\exp\left[2\pi i\mathbf{y}\mathbf{x}\binom{l'}{k'}-i\omega t\right]d\omega, \tag{II.31}$$

where

$$E_\alpha(\mathbf{y},\omega)^*=E_\alpha(-\mathbf{y},-\omega). \tag{II.32}$$

Using Eqs. (II.23), (II.27), (II.30), and (II.31), we find

$$\langle p_\alpha(\nu,\Omega)\rangle=-\frac{1}{2\pi v_a N}\sum_{\substack{\mathbf{y}ll'\\kk'}}\iint\frac{1}{i\hbar}\theta(t-\tau)\left\langle\left[p_\alpha\binom{l}{k},p'_\beta\binom{l'}{k'}\right]E_\beta(\mathbf{y},\omega)\times\right.$$
$$\left.\times\exp\left[2\pi i\mathbf{y}\mathbf{x}\binom{l'}{k'}-2\pi i\nu\mathbf{x}\binom{l}{k}\right]\right\rangle\exp(i\Omega t-i\omega\tau)\,d\Omega\,dt\,d\tau. \tag{II.33}$$

Here and henceforth, we shall omit the superscript h to the Heisenberg-representation operators. The prime to the operator $p_\beta\binom{l'}{k'}$ signifies that it depends on the time τ. The unprimed operator $p_\alpha\binom{l}{k}$ depends on t.

Since the averaging in Eq. (II.33) is taken over the Gibbs distribution and the spatial distribution of defects, which we assume homogeneous, the expression in the angular brackets $\langle\ \rangle$ must depend on the differences $l-l'$, $t-\tau$. Hence, we find

$$\langle p_\alpha(\nu,\omega)\rangle=-\frac{1}{v_a\hbar}\sum_{\substack{l-l'\\kk'}}\int\frac{1}{i}\theta(t-\tau)\left\langle\left[p_\alpha\binom{l}{k},p'_\beta\binom{l'}{k'}\right]E_\beta(\nu,\Omega)\times\right.$$
$$\left.\times\exp\left[-2\pi i\nu\mathbf{x}\binom{ll'}{kk'}\right]\right\rangle\exp[i\Omega(t-\tau)]\,d(t-\tau). \tag{II.34}$$

Using the definition of the Fourier transform

$$G(\Omega)=\frac{1}{2\pi}\int G(t)\,e^{i\Omega t}dt$$

and taking the macroscopic field amplitude outside the averaging, we get the following expression for the complex polarizability of the lattice:

$$\chi_{\alpha\beta}(\mathbf{v}, \Omega) = -\frac{2\pi}{Nv_a\hbar} \sum_{ll'kk'} \left\{ \frac{\theta(t-\tau)}{i} \left\langle \left[p_\alpha \begin{pmatrix} l \\ k \end{pmatrix}, p'_\alpha \begin{pmatrix} l' \\ k' \end{pmatrix} \right] \exp\left[-2\pi i \mathbf{v} \mathbf{x} \begin{pmatrix} ll' \\ kk' \end{pmatrix} \right] \right\rangle \right\}_\Omega . \tag{II.35}$$

The subscript Ω to the braces signifies that the Fourier component of the expression within them is taken.

Substitution of Eqs. (II.6) and (II.19) in Eq. (II.35) gives

$$\chi_{\alpha\beta}(\mathbf{v}, \Omega) = \chi^{(1)}_{\alpha\beta}(\mathbf{v}, \Omega) + \chi^{(2)}_{\alpha\beta}(\mathbf{v}, \Omega) + \chi^{(3)}_{\alpha\beta}(\mathbf{v}, \Omega) + \chi^{(4)}_{\alpha\beta}(\mathbf{v}, \Omega),$$

where

$$\chi^{(1)}_{\alpha\beta}(\mathbf{v}, \Omega) = -\frac{2\pi}{v_a\hbar} \sum_{kk'jj'} \frac{e_k e_{k'} w_\alpha\left(k \middle| \begin{matrix} \mathbf{v} \\ j \end{matrix}\right) w_\beta\left(k' \middle| \begin{matrix} -\mathbf{v} \\ j' \end{matrix}\right)}{\sqrt{m_k m_{k'}}} \left\{ \frac{\theta(t-\tau)}{i} \left\langle \left[Q\begin{pmatrix} \mathbf{v} \\ j \end{pmatrix}, Q'\begin{pmatrix} -\mathbf{v} \\ j' \end{pmatrix} \right] \right\rangle \right\}_\Omega ,$$

$$\chi^{(2)}_{\alpha\beta}(\mathbf{v}, \Omega) = -\frac{2\pi}{Nv_a\hbar} \sum_{\substack{l-l',\, kk' \\ \mathbf{y}jj'}} \frac{w_\alpha\left(k \middle| \begin{matrix} \mathbf{y} \\ j \end{matrix}\right) w_\beta\left(k' \middle| \begin{matrix} \mathbf{y} \\ j' \end{matrix}\right)}{\sqrt{m_k m_{k'}}} \left\{ \frac{\theta(t-\tau)}{i} \left\langle \left[Q\begin{pmatrix} \mathbf{y} \\ j \end{pmatrix}, Q'\begin{pmatrix} -\mathbf{y} \\ j' \end{pmatrix} \right] \delta e^l_k \delta e^{l'}_{k'} \exp\left[2\pi i (\mathbf{y}-\mathbf{v}) \mathbf{x}\begin{pmatrix} ll' \\ kk' \end{pmatrix} \right] \right\rangle \right\}_\Omega ,$$

$$\chi^{(3)}_{\alpha\beta}(\mathbf{v}, \Omega) = -\frac{2\pi}{Nv_a\hbar} \sum_{\substack{lkk' \\ \mathbf{y}jj'}} \frac{e_k w_\alpha\left(k \middle| \begin{matrix} \mathbf{v} \\ j \end{matrix}\right) w_\beta\left(k' \middle| \begin{matrix} \mathbf{y}' \\ j' \end{matrix}\right)}{\sqrt{m_k m_{k'}}} \left\{ \frac{\theta(t-\tau)}{i} \left\langle \left[Q\begin{pmatrix} \mathbf{v} \\ j \end{pmatrix}, Q'\begin{pmatrix} \mathbf{y}' \\ j' \end{pmatrix} \right] \delta e^{l'}_{k'} \exp\left[2\pi i (\mathbf{y}'+\mathbf{v}) \mathbf{x}\begin{pmatrix} l' \\ k' \end{pmatrix} \right] \right\rangle \right\}_\Omega , \tag{II.36}$$

$$\chi^{(4)}_{\alpha\beta}(\mathbf{v}, \Omega) = -\frac{2\pi}{Nv_a\hbar} \sum_{\substack{lkk' \\ \mathbf{y}jj'}} \frac{e_{k'} w_\alpha\left(k \middle| \begin{matrix} \mathbf{y} \\ j \end{matrix}\right) w_\beta\left(k' \middle| \begin{matrix} -\mathbf{v} \\ j' \end{matrix}\right)}{\sqrt{m_k m_{k'}}} \left\{ \frac{\theta(t-\tau)}{i} \times \right.$$

$$\left. \times \left\langle \left[Q\begin{pmatrix} \mathbf{y} \\ j \end{pmatrix}, Q'\begin{pmatrix} -\mathbf{v} \\ j' \end{pmatrix} \right] \delta e^l_k \exp\left[2\pi i (\mathbf{y}-\mathbf{v}) \mathbf{x}\begin{pmatrix} l \\ k \end{pmatrix} \right] \right\rangle \right\}_\Omega .$$

Thus, we obtain equations which give the complex polarizability in terms of the retarded double-time Green's functions (the expressions in the braces).

3. The Phonon Creation and Annihilation Operators. The Hamiltonian and the Polarizability in Terms of These Operators. Equations of Motion for Operators

We can define the phonon creation and annihilation operators b_+ and b_- by

$$Q\begin{pmatrix} \mathbf{y} \\ j \end{pmatrix} = \sqrt{\frac{\hbar}{2\omega\begin{pmatrix} \mathbf{y} \\ j \end{pmatrix}}} \left(b_+\begin{pmatrix} -\mathbf{y} \\ j \end{pmatrix} + b_-\begin{pmatrix} \mathbf{y} \\ j \end{pmatrix} \right). \tag{II.37}$$

These operators have the properties

$$\begin{aligned} b^+_-\begin{pmatrix} \mathbf{y} \\ j \end{pmatrix} &= b_+\begin{pmatrix} \mathbf{y} \\ j \end{pmatrix} \\ b^+_+\begin{pmatrix} \mathbf{y} \\ j \end{pmatrix} &= b_-\begin{pmatrix} \mathbf{y} \\ j \end{pmatrix} \end{aligned} \tag{II.38}$$

(the superscript plus denoting the conjugate operation), and satisfy the commutation relations

$$b_-\binom{\mathbf{y}}{j}b_+\binom{\mathbf{y}'}{j'} - b_+\binom{\mathbf{y}'}{j'}b_-\binom{\mathbf{y}}{j} = \delta_{\mathbf{y}\mathbf{y}'}\,\delta_{jj'}. \tag{II.39}$$

Substitution of Eq. (II.37) in Eq. (II. 14) gives

$$\begin{aligned} H = {} & \sum_{\mathbf{y}j} \hbar\omega\binom{\mathbf{y}}{j} b_+\binom{\mathbf{y}}{j} b_-\binom{\mathbf{y}}{j} + \frac{1}{2}\sum_{\substack{\mathbf{y}\mathbf{y}'\\ jj'}} B\binom{\mathbf{y}\mathbf{y}'}{jj'}\left(b_+\binom{-\mathbf{y}}{j} + b_-\binom{\mathbf{y}}{j}\right)\left(b_+\binom{-\mathbf{y}'}{j'} + \right. \\ & \left. + b_-\binom{\mathbf{y}'}{j'}\right) + \frac{1}{3}\sum_{\substack{\mathbf{y}\mathbf{y}'\mathbf{y}''\\ jj'j''}} B\binom{\mathbf{y}\mathbf{y}'\mathbf{y}''}{jj'j''}\left(b_+\binom{-\mathbf{y}}{j} + b_-\binom{\mathbf{y}}{j}\right)\left(b_+\binom{-\mathbf{y}'}{j'} + \right. \\ & \left. + b_-\binom{\mathbf{y}'}{j'}\right)\left(b_+\binom{-\mathbf{y}''}{j''} + b_-\binom{\mathbf{y}''}{j''}\right) + \frac{1}{4}\sum_{\substack{\mathbf{y}\mathbf{y}'\mathbf{y}''\mathbf{y}'''\\ jj'j''j'''}} B\binom{\mathbf{y}\mathbf{y}'\mathbf{y}''\mathbf{y}'''}{jj'j''j'''}\left(b_+\binom{-\mathbf{y}}{j} + \right. \\ & \left. + b_-\binom{\mathbf{y}}{j}\right)\left(b_+\binom{-\mathbf{y}'}{j'} + b_-\binom{\mathbf{y}'}{j'}\right)\left(b_+\binom{-\mathbf{y}''}{j''} + b_-\binom{\mathbf{y}''}{j''}\right)\left(b_+\binom{-\mathbf{y}'''}{j'''} + b_-\binom{\mathbf{y}'''}{j'''}\right), \end{aligned} \tag{II.40}$$

where

$$\begin{aligned} B\binom{\mathbf{y}\mathbf{y}'}{jj'} &= A\binom{\mathbf{y}\mathbf{y}'}{jj'}\frac{\hbar}{\left(4\omega\binom{\mathbf{y}}{j}\omega\binom{\mathbf{y}'}{j'}\right)^{1/2}}, \\ B\binom{\mathbf{y}\mathbf{y}'\mathbf{y}''}{jj'j''} &= \frac{1}{2}A\binom{\mathbf{y}\mathbf{y}'\mathbf{y}''}{jj'j''}\frac{\hbar^{3/2}}{\left(8\omega\binom{\mathbf{y}}{j}\omega\binom{\mathbf{y}'}{j'}\omega\binom{\mathbf{y}''}{j''}\right)^{1/2}}, \\ B\binom{\mathbf{y}\mathbf{y}'\mathbf{y}''\mathbf{y}'''}{jj'j''j'''} &= \frac{1}{6}A\binom{\mathbf{y}\mathbf{y}'\mathbf{y}''\mathbf{y}'''}{jj'j''j'''}\frac{\hbar^2}{\left(16\omega\binom{\mathbf{y}}{j}\omega\binom{\mathbf{y}'}{j'}\omega\binom{\mathbf{y}''}{j''}\omega\binom{\mathbf{y}'''}{j'''}\right)^{1/2}}. \end{aligned} \tag{II.41}$$

These coefficients have the properties (II.18).

In deriving Eq. (II.40), we have used the relation

$$i\hbar\frac{d}{dt}\left(b_+\binom{-\mathbf{y}}{j} + b_-\binom{\mathbf{y}}{j}\right) = -\hbar\omega\binom{\mathbf{y}}{j}\left(b_+\binom{-\mathbf{y}}{j} + b_-\binom{\mathbf{y}}{j}\right), \tag{II.42}$$

which is independent of the interaction energy.

From Eq. (II.28) we have

$$i\hbar\frac{dA^{\mathrm{h}}}{dt} = [A^{\mathrm{h}}, H]. \tag{II.43}$$

Using Eqs. (II.43), (II.40), and (II.39), we get the equations of motion for the operators:

$$\begin{aligned} i\hbar\frac{db_+\binom{\mathbf{y}^0}{j^0}}{dt} = {} & -\hbar\omega\binom{\mathbf{y}^0}{j^0}b_+\binom{\mathbf{y}^0}{j^0} - \sum_{\mathbf{y}j} B\binom{\mathbf{y}^0\mathbf{y}}{j^0 j}\left(b_+\binom{-\mathbf{y}}{j} + b_-\binom{\mathbf{y}}{j}\right) - \\ & - \sum_{\substack{\mathbf{y}\mathbf{y}'\\ jj'}} B\binom{\mathbf{y}^0\mathbf{y}\mathbf{y}'}{j^0 jj'}\left(b_+\binom{-\mathbf{y}}{j} + b_-\binom{\mathbf{y}}{j}\right)\left(b_+\binom{-\mathbf{y}'}{j'} + b_-\binom{\mathbf{y}'}{j'}\right) - \\ & - \sum_{\substack{\mathbf{y}\mathbf{y}'\mathbf{y}''\\ jj'j''}} B\binom{\mathbf{y}^0\mathbf{y}\mathbf{y}'\mathbf{y}''}{j^0 jj'j''}\left(b_+\binom{-\mathbf{y}}{j} + b_-\binom{\mathbf{y}}{j}\right)\left(b_+\binom{-\mathbf{y}'}{j'} + b_-\binom{\mathbf{y}'}{j'}\right)\left(b_+\binom{-\mathbf{y}''}{j''} + b_-\binom{\mathbf{y}''}{j''}\right), \end{aligned} \tag{II.44}$$

$$i\hbar\frac{db_-\binom{\mathbf{y}^0}{j^0}}{dt}=\hbar\omega\binom{\mathbf{y}^0}{j^0}b_-\binom{\mathbf{y}^0}{j^0}+\sum_{\mathbf{y}j}B\left(-\frac{\mathbf{y}^0\mathbf{y}}{j^0 j}\right)\left(b_+\binom{-\mathbf{y}}{j}+b_-\binom{\mathbf{y}}{j}\right)+$$

$$+\sum_{\substack{\mathbf{y}\mathbf{y}'\\jj'}}B\left(-\frac{\mathbf{y}^0\mathbf{y}\mathbf{y}'}{j^0 j j'}\right)\left(b_+\binom{-\mathbf{y}}{j}+b_-\binom{\mathbf{y}}{j}\right)\left(b_+\binom{-\mathbf{y}'}{j'}+b_-\binom{\mathbf{y}'}{j'}\right)+$$

$$+\sum_{\substack{\mathbf{y}\mathbf{y}'\mathbf{y}''\\jj'j''}}B\left(-\frac{\mathbf{y}^0\mathbf{y}\mathbf{y}'\mathbf{y}''}{j^0 j j' j''}\right)\left(b_+\binom{-\mathbf{y}}{j}+b_-\binom{\mathbf{y}}{j}\right)\left(b_+\binom{-\mathbf{y}'}{j'}+b_-\binom{\mathbf{y}'}{j'}\right)\left(b_+\binom{-\mathbf{y}''}{j''}+b_-\binom{\mathbf{y}''}{j''}\right). \quad \text{(II.45)}$$

These show that Eq. (II.42) is satisfied.

If we change to the operators b_+ and b_- in Eqs. (II.36) (taking $\overline{\delta e_k^l \delta e_{k'}^{l'}}$ outside the averaging in $\chi^{(2)}_{\alpha\beta}(\nu, \Omega)$, since the latter will be calculated only with accuracy $\overline{\delta e_k^l \delta e_{k'}^{l'}}$ and higher-order correlations will be neglected), the result is

$$\chi^{(1)}_{\alpha\beta}(\nu,\Omega)=-\frac{\pi}{v_a}\sum_{\substack{kk'\\jj'}}e_k e_{k'}p_\alpha\left(k\,\Big|\,\begin{matrix}\nu\\ j\end{matrix}\right)p_\beta\left(k'\,\Big|\,\begin{matrix}-\nu\\ j'\end{matrix}\right)\mathcal{G}\begin{pmatrix}\nu & -\nu\\ j & j'\end{pmatrix}_\Omega,$$

$$\chi^{(2)}_{\alpha\beta}(\nu,\Omega)=-\frac{\pi}{v_a}\sum_{\substack{l-l',\mathbf{y}\\kk'jj'}}p_\alpha\left(k\,\Big|\,\begin{matrix}\mathbf{y}\\ j\end{matrix}\right)p_\beta\left(k'\,\Big|\,\begin{matrix}-\mathbf{y}\\ j'\end{matrix}\right)\overline{\delta e_k^l\,\delta e_{k'}^{l'}}\exp\left[2\pi i(\mathbf{y}-\nu)\mathbf{x}\binom{ll'}{kk'}\right]\mathcal{G}\begin{pmatrix}\mathbf{y} & -\mathbf{y}\\ j & j'\end{pmatrix}_\Omega,$$

$$\chi^{(3)}_{\alpha\beta}(\nu,\Omega)=-\frac{\pi}{v_aN}\sum_{\substack{l'\ \mathbf{y}'\\kk'jj'}}e_k p_\alpha\left(k\,\Big|\,\begin{matrix}\nu\\ j\end{matrix}\right)p_\beta\left(k'\,\Big|\,\begin{matrix}\mathbf{y}'\\ j'\end{matrix}\right)\mathcal{G}_{\substack{l'\\k'}}\begin{pmatrix}\nu & \mathbf{y}'\\ j & j'\end{pmatrix}_\Omega, \quad \text{(II.46)}$$

$$\chi^{(4)}_{\alpha\beta}(\nu,\Omega)=-\frac{\pi}{v_aN}\sum_{\substack{l\ \mathbf{y}\\kk'jj'}}e_k p_\alpha\left(k\,\Big|\,\begin{matrix}\mathbf{y}\\ j\end{matrix}\right)p_\beta\left(k'\,\Big|\,\begin{matrix}-\nu\\ j'\end{matrix}\right)\mathcal{G}_{\substack{l\\k}}\begin{pmatrix}\mathbf{y} & -\nu\\ j & j'\end{pmatrix}_\Omega,$$

where

$$p_\alpha\left(k\,\Big|\,\begin{matrix}\mathbf{y}\\ j\end{matrix}\right)=\frac{w_\alpha\left(k\,\Big|\,\begin{matrix}\mathbf{y}\\ j\end{matrix}\right)}{\sqrt{m_k\omega\binom{\mathbf{y}}{j}}},$$

$$\mathcal{G}_{\substack{l\\k}}\begin{pmatrix}\mathbf{y}\mathbf{y}'\\ jj'\end{pmatrix}=G_{\substack{l\\k}}\begin{pmatrix}\mathbf{y}\mathbf{y}'\\ jj'\end{pmatrix}+\Gamma_{\substack{l\\k}}\begin{pmatrix}\mathbf{y}\mathbf{y}'\\ jj'\end{pmatrix}+G^+_{\substack{l\\k}}\begin{pmatrix}-\mathbf{y} & -\mathbf{y}'\\ j & j'\end{pmatrix}+\Gamma^+_{\substack{l\\k}}\begin{pmatrix}-\mathbf{y} & -\mathbf{y}'\\ j & j'\end{pmatrix}, \quad \text{(II.47)}$$

$$G_{\substack{l\\k}}\begin{pmatrix}\mathbf{y}\mathbf{y}'\\ jj'\end{pmatrix}=\frac{\theta(t-\tau)}{i}\left\langle\delta e_k^l\exp\left[2\pi i(\mathbf{y}+\mathbf{y}')\mathbf{x}\binom{l}{k}\right]\left[b_+\binom{-\mathbf{y}}{j},\ b'_-\binom{\mathbf{y}'}{j'}\right]\right\rangle, \quad \text{(II.48)}$$

$$\Gamma_{\substack{l\\k}}\begin{pmatrix}\mathbf{y}\mathbf{y}'\\ jj'\end{pmatrix}=\frac{\theta(t-\tau)}{i}\left\langle\delta e_k^l\exp\left[2\pi i(\mathbf{y}+\mathbf{y}')\mathbf{x}\binom{l}{k}\right]\left[b_-\binom{\mathbf{y}}{j},\ b'_-\binom{\mathbf{y}'}{j'}\right]\right\rangle. \quad \text{(II.49)}$$

The Green's functions $\mathcal{G}$, G, and Γ without the index $\binom{l}{k}$ differ from the same functions with the index by not containing the factor $\delta e_k^l \exp\left[2\pi i(\mathbf{y}+\mathbf{y}')\mathbf{x}\binom{l}{k}\right]$.

Thus, we have expressed the complex polarizability $\chi_{\alpha\beta}(\nu, \Omega)$ in terms of the retarded Green's functions of the operators b_+ and b_-.

CHAPTER III

The Quantum Theory of Infrared Absorption and Dispersion in an Ideal Ionic Lattice

In this chapter, we shall calculate the complex polarizability of an ideal ionic lattice, including anharmonic terms of the third and foruth order in its Hamiltonian. The calculation will be made by the method used in the work of Bogolyubov, Bonch-Bruevich, Zubarev, and Tyablikov, constructing and approximately solving a series of linked equations for the retarded Green's functions (see [39, 40]).

1. Equations for the Retarded Green's Functions

In the absence of charged defects, the complex polarizability of the lattice is given by [see Eqs. (II.46)]

$$\chi_{\alpha\beta}(\mathbf{v},\Omega) = -\frac{\pi}{v_a}\sum_{\substack{kk'\\jj'}} e_k e_{k'} p_\alpha\left(k\,\middle|\,\begin{matrix}\mathbf{v}\\ j\end{matrix}\right) p_\beta\left(k'\,\middle|\,\begin{matrix}-\mathbf{v}\\ j'\end{matrix}\right) \mathcal{G}\begin{pmatrix}\mathbf{v} & -\mathbf{v}\\ j & j'\end{pmatrix}_\Omega, \tag{III.1}$$

where

$$\mathcal{G}\begin{pmatrix}\mathbf{v} & -\mathbf{v}\\ j & j'\end{pmatrix} = G\begin{pmatrix}\mathbf{v} & -\mathbf{v}\\ j & j'\end{pmatrix} + \Gamma\begin{pmatrix}\mathbf{v} & -\mathbf{v}\\ j & j'\end{pmatrix} + G^+\begin{pmatrix}-\mathbf{v} & \mathbf{v}\\ j & j'\end{pmatrix} + \Gamma^+\begin{pmatrix}-\mathbf{v} & \mathbf{v}\\ j & j'\end{pmatrix}, \tag{III.2}$$

$$G\begin{pmatrix}\mathbf{v} & -\mathbf{v}\\ j & j'\end{pmatrix} = \frac{\theta(t-\tau)}{i}\left\langle\left[b_+\begin{pmatrix}-\mathbf{v}\\ j\end{pmatrix},\; b'_-\begin{pmatrix}-\mathbf{v}\\ j'\end{pmatrix}\right]\right\rangle, \tag{III.3}$$

$$\Gamma\begin{pmatrix}\mathbf{v} & -\mathbf{v}\\ j & j'\end{pmatrix} = \frac{\theta(t-\tau)}{i}\left\langle\left[b_-\begin{pmatrix}\mathbf{v}\\ j\end{pmatrix},\; b'_-\begin{pmatrix}-\mathbf{v}\\ j'\end{pmatrix}\right]\right\rangle. \tag{III.4}$$

The operators which appear in the Green's functions (III.3) and (III.4) satisfy the equations of motion (II.44) and (II.45).

For an ideal lattice, the coefficients $B\begin{pmatrix}\mathbf{y}^0\mathbf{y}\\ j^0 j\end{pmatrix}$ are zero, and the coefficients $B\begin{pmatrix}\mathbf{y}^0\mathbf{y}\mathbf{y}'\\ j^0 j j'\end{pmatrix}$ and $B\begin{pmatrix}\mathbf{y}^0\mathbf{y}\mathbf{y}'\mathbf{y}''\\ j^0 j j' j''\end{pmatrix}$ contain only the functions $\Phi\begin{pmatrix}\mathbf{y}^0\mathbf{y}\mathbf{y}'\\ j^0 j j'\end{pmatrix}_0$ and $\Phi\begin{pmatrix}\mathbf{y}^0\mathbf{y}\mathbf{y}'\mathbf{y}''\\ j^0 j j' j''\end{pmatrix}_0$ [see Eqs. (II.41)].

Using the commutation relations (II.39) and the properties of the coefficients $B\begin{pmatrix}\mathbf{y}^0\mathbf{y}\mathbf{y}'\\ j^0 j j'\end{pmatrix}$ and $B\begin{pmatrix}\mathbf{y}^0\mathbf{y}\mathbf{y}'\mathbf{y}''\\ j^0 j j' j''\end{pmatrix}$ [see Eqs. (II.18) and (II.41)], we can bring the equations of motion (II.44) and (II.45) into the more convenient form

$$\begin{aligned}
-i\hbar\frac{db_+\begin{pmatrix}\mathbf{y}^0\\ j^0\end{pmatrix}}{dt} &= \hbar\omega\begin{pmatrix}\mathbf{y}^0\\ j^0\end{pmatrix} b_+\begin{pmatrix}\mathbf{y}^0\\ j^0\end{pmatrix} + 3\sum_{\substack{\mathbf{y}\mathbf{y}'\\ jj'}} B\begin{pmatrix}\mathbf{y}^0\mathbf{y}' & -\mathbf{y}'\mathbf{y}\\ j^0 j' & j' j\end{pmatrix}\left(b_+\begin{pmatrix}-\mathbf{y}\\ j\end{pmatrix} + b_-\begin{pmatrix}\mathbf{y}\\ j\end{pmatrix}\right) + \\
&+ \sum_{\substack{\mathbf{y}\mathbf{y}'\\ jj'}} B\begin{pmatrix}\mathbf{y}^0\mathbf{y}\mathbf{y}'\\ j^0 j j'\end{pmatrix}\left\{b_+\begin{pmatrix}-\mathbf{y}\\ j\end{pmatrix} b_+\begin{pmatrix}-\mathbf{y}'\\ j'\end{pmatrix} + 2b_+\begin{pmatrix}-\mathbf{y}\\ j\end{pmatrix} b_-\begin{pmatrix}\mathbf{y}'\\ j'\end{pmatrix} + \right.\\
&\left. + b_-\begin{pmatrix}\mathbf{y}\\ j\end{pmatrix} b_-\begin{pmatrix}\mathbf{y}'\\ j'\end{pmatrix} + \delta_{-\mathbf{y}',\mathbf{y}}\delta_{j'j}\right\} + \sum_{\substack{\mathbf{y}\mathbf{y}'\mathbf{y}''\\ jj'j''}} B\begin{pmatrix}\mathbf{y}^0\mathbf{y}\mathbf{y}'\mathbf{y}''\\ j^0 j j' j''\end{pmatrix}\left\{b_+\begin{pmatrix}-\mathbf{y}\\ j\end{pmatrix} b_+\begin{pmatrix}-\mathbf{y}'\\ j'\end{pmatrix} b_+\begin{pmatrix}-\mathbf{y}''\\ j''\end{pmatrix} + \right.\\
&\left. + b_-\begin{pmatrix}\mathbf{y}\\ j\end{pmatrix} b_-\begin{pmatrix}\mathbf{y}'\\ j'\end{pmatrix} b_-\begin{pmatrix}\mathbf{y}''\\ j''\end{pmatrix} + 3b_+\begin{pmatrix}-\mathbf{y}\\ j\end{pmatrix} b_+\begin{pmatrix}-\mathbf{y}'\\ j'\end{pmatrix} b_-\begin{pmatrix}\mathbf{y}''\\ j''\end{pmatrix} + 3b_+\begin{pmatrix}-\mathbf{y}\\ j\end{pmatrix} b_-\begin{pmatrix}\mathbf{y}'\\ j'\end{pmatrix} b_-\begin{pmatrix}\mathbf{y}''\\ j''\end{pmatrix}\right\},
\end{aligned} \tag{III.5}$$

$$i\hbar\,\frac{db_-\binom{y^0}{j^0}}{dt}=\hbar\omega\binom{y^0}{j^0}b_-\binom{y^0}{j^0}+3\sum_{\substack{yy'\\jj'}}B\begin{pmatrix}-y^0y' & -y'y\\ j^0j' & j'j\end{pmatrix}\left(b_+\binom{-y}{j}+b_-\binom{y}{j}\right)+$$

$$+\sum_{\substack{yy'\\jj'}}B\begin{pmatrix}-y^0yy'\\ j^0jj'\end{pmatrix}\left\{b_+\binom{-y}{j}b_+\binom{-y'}{j'}+2b_+\binom{-y}{j}b_-\binom{y'}{j'}+\right.$$

$$\left.+b_-\binom{y}{j}b_-\binom{y'}{j'}+\delta_{-y',y}\delta_{j',j}\right\}+\sum_{\substack{yy'y''\\jj'j''}}B\begin{pmatrix}-y^0yy'y''\\ j^0jj'j''\end{pmatrix}\left\{b_+\binom{-y}{j}b_+\binom{-y'}{j'}\times\right.$$

$$\left.\times b_+\binom{-y''}{j''}+b_-\binom{y}{j}b_-\binom{y'}{j'}b_-\binom{y''}{j''}+3b_+\binom{-y}{j}b_+\binom{-y'}{j'}b_-\binom{y''}{j''}+3b_+\binom{-y}{j}b_-\binom{y'}{j'}b_-\binom{y''}{j''}\right\}. \quad \text{(III.6)}$$

Using the equations of motion (III.5) and (III.6), we can now derive equations for the Green's functions $G\begin{pmatrix}\nu & -\nu\\ j & j'\end{pmatrix}$, $\Gamma\begin{pmatrix}\nu & -\nu\\ j & j'\end{pmatrix}$. For $G\begin{pmatrix}\nu & -\nu\\ j & j'\end{pmatrix}$, we get, on differentiating with respect to t,

$$-i\hbar\,\frac{dG\begin{pmatrix}\nu & -\nu\\ j & j'\end{pmatrix}}{dt}=-\hbar\delta(t-\tau)\left\langle\left[b_+\binom{-\nu}{j},\ b_-\binom{-\nu}{j'}\right]\right\rangle+\frac{\theta(t-\tau)}{i}\left\langle\left[-i\hbar\,\frac{db_+\binom{\nu}{j}}{dt},\ b'_-\binom{-\nu}{j}\right]\right\rangle. \quad \text{(III.7)}$$

In the first commutator, we use the commutation relations (II.39), since the times in the operators concerned are the same. In the second commutator, we substitute Eq. (III.5), and also make use of the following fact. When ν (the wave number of the radiation) is small, the coefficients $B\begin{pmatrix}-\nu y' & -y'y\\ j^0j' & j'j\end{pmatrix}$ reduce to $B\begin{pmatrix}-\nu\, y' & -y'\,\nu\\ j^0j' & j'\,j\end{pmatrix}$ since the sum of the wave vectors in the coefficients for an ideal lattice must reduce to a reciprocal-lattice vector, and the wave vectors themselves lie in the first Brillouin zone [see Eqs. (II.17) and (II.7)]. The result is

$$-i\hbar\,\frac{dG\begin{pmatrix}\nu & -\nu\\ j & j'\end{pmatrix}}{dt}=\hbar\delta(t-\tau)\delta_{jj'}+\hbar\omega\binom{\nu}{j}G\begin{pmatrix}\nu & -\nu\\ j & j'\end{pmatrix}+3\sum_{y'\tilde{j}\tilde{\tilde{j}}}B\begin{pmatrix}-\nu\, y' & -y'\,\nu\\ j\,\tilde{j} & \tilde{j}\,\tilde{\tilde{j}}\end{pmatrix}\left\{G\begin{pmatrix}\nu & -\nu\\ \tilde{\tilde{j}} & j'\end{pmatrix}+\Gamma\begin{pmatrix}\nu & -\nu\\ \tilde{\tilde{j}} & j'\end{pmatrix}\right\}+$$

$$+\sum_{\substack{yy'\\jj'}}B\begin{pmatrix}-\nu\, y\, y'\\ j\,\tilde{j}\,\tilde{\tilde{j}}\end{pmatrix}\left\{\frac{\theta(t-\tau)}{i}\left\langle\left[b_+\binom{-y}{\tilde{j}}b_+\binom{-y'}{\tilde{\tilde{j}}},\ b'_-\binom{-\nu}{j'}\right]\right\rangle+2\,\frac{\theta(t-\tau)}{i}\left\langle\left[b_+\binom{-y}{\tilde{j}}b_-\binom{y'}{\tilde{\tilde{j}}},\ b'_-\binom{-\nu}{j'}\right]\right\rangle+\right.$$

$$\left.+\frac{\theta(t-\tau)}{i}\left\langle\left[b_-\binom{y}{\tilde{j}}b_-\binom{y'}{\tilde{\tilde{j}}},\ b'_-\binom{-\nu}{j'}\right]\right\rangle\right\}+\sum_{\substack{yy'y''\\ \tilde{j}\tilde{\tilde{j}}\tilde{\tilde{\tilde{j}}}}}B\begin{pmatrix}-\nu\, y\, y'\, y''\\ j\,\tilde{j}\,\tilde{\tilde{j}}\,\tilde{\tilde{\tilde{j}}}\end{pmatrix}\left\{\frac{\theta(t-\tau)}{i}\left\langle\left[b_+\binom{-y}{\tilde{j}}b_+\binom{-y'}{\tilde{\tilde{j}}}b_+\binom{-y''}{\tilde{\tilde{\tilde{j}}}}\right.\right.\right.$$

$$\left.\left.b'_-\binom{-\nu}{j'}\right]\right\rangle+\frac{\theta(t-\tau)}{i}\left\langle\left[b_-\binom{y}{\tilde{j}}b_-\binom{y'}{\tilde{\tilde{j}}}b_-\binom{y''}{\tilde{\tilde{\tilde{j}}}},\ b'_-\binom{-\nu}{j'}\right]\right\rangle+$$

$$\left.+3\,\frac{\theta(t-\tau)}{i}\left\langle\left[b_+\binom{-y}{\tilde{j}}b_+\binom{-y'}{\tilde{\tilde{j}}}b_-\binom{y''}{\tilde{\tilde{\tilde{j}}}},\ b'_-\binom{-\nu}{j'}\right]\right\rangle+3\,\frac{\theta(t-\tau)}{i}\left\langle\left[b_+\binom{-y}{\tilde{j}}b_-\binom{y'}{\tilde{\tilde{j}}}b_-\binom{y''}{\tilde{\tilde{\tilde{j}}}},\ b'_-\binom{-\nu}{j'}\right]\right\rangle\right\}. \quad \text{(III.8)}$$

From the sum

$$3\sum_{\substack{yy'y''\\ \tilde{j}\tilde{\tilde{j}}\tilde{\tilde{\tilde{j}}}}}B\begin{pmatrix}-\nu\, y\, y'\, y''\\ j\,\tilde{j}\,\tilde{\tilde{j}}\,\tilde{\tilde{\tilde{j}}}\end{pmatrix}\left\{\frac{\theta(t-\tau)}{i}\left\langle\left[b_+\binom{-y}{\tilde{j}}b_+\binom{-y'}{\tilde{\tilde{j}}}b_-\binom{y''}{\tilde{\tilde{\tilde{j}}}},\ b'_-\binom{-\nu}{j'}\right]\right\rangle+\right.$$

$$\left.+\frac{\theta(t-\tau)}{i}\left\langle\left[b_+\binom{y}{\tilde{j}}b_-\binom{y'}{\tilde{\tilde{j}}}b_-\binom{y''}{\tilde{\tilde{\tilde{j}}}},\ b'_-\binom{-\nu}{j'}\right]\right\rangle\right\}$$

we can approximately separate expressions proportional to the Green's functions $G\begin{pmatrix}\nu & -\nu\\ \tilde{\tilde{j}} & j'\end{pmatrix}$ and $\Gamma\begin{pmatrix}\nu & -\nu\\ \tilde{j} & j'\end{pmatrix}$. For example, the first term in the sum includes terms in which $-\mathbf{y}'=\mathbf{y}''$, $\tilde{\tilde{j}}=\tilde{\tilde{\tilde{j}}}$. Then, replacing $b_+\begin{pmatrix}-\mathbf{y}'\\ \tilde{\tilde{j}}\end{pmatrix}b_-\begin{pmatrix}-\mathbf{y}'\\ \tilde{\tilde{j}}\end{pmatrix}$ by $n\begin{pmatrix}\mathbf{y}'\\ \tilde{\tilde{j}}\end{pmatrix}=\left(\exp\left[\frac{\hbar\omega\begin{pmatrix}\mathbf{y}'\\ \tilde{\tilde{j}}\end{pmatrix}}{kT}\right]-1\right)^{-1}$, we can express these in terms of the Green's function $G\begin{pmatrix}\mathbf{y} & -\nu\\ \tilde{j} & j'\end{pmatrix}$. Proceeding in this way, we find

$$
\begin{aligned}
&3\sum_{\substack{\mathbf{y}\mathbf{y}'\mathbf{y}''\\ \tilde{j}\tilde{\tilde{j}}\tilde{\tilde{\tilde{j}}}}} B\begin{pmatrix}-\nu & \mathbf{y} & \mathbf{y}' & \mathbf{y}''\\ j & \tilde{j} & \tilde{\tilde{j}} & \tilde{\tilde{\tilde{j}}}\end{pmatrix}\Big\{\frac{\theta(t-\tau)}{i}\Big\langle\Big[b_+\begin{pmatrix}-\mathbf{y}\\ \tilde{j}\end{pmatrix}b_+\begin{pmatrix}-\mathbf{y}'\\ \tilde{\tilde{j}}\end{pmatrix}b_-\begin{pmatrix}\mathbf{y}''\\ \tilde{\tilde{\tilde{j}}}\end{pmatrix},\ b'_-\begin{pmatrix}-\nu\\ j'\end{pmatrix}\Big]\Big\rangle+\\
&\quad+\frac{\theta(t-\tau)}{i}\Big\langle\Big[b_+\begin{pmatrix}-\mathbf{y}\\ \tilde{j}\end{pmatrix}b_-\begin{pmatrix}\mathbf{y}'\\ \tilde{\tilde{j}}\end{pmatrix}b_-\begin{pmatrix}\mathbf{y}''\\ \tilde{\tilde{\tilde{j}}}\end{pmatrix},\ b'_-\begin{pmatrix}-\nu\\ j'\end{pmatrix}\Big]\Big\rangle\Big\}=\\
&=6\sum_{\mathbf{y}'\tilde{j}\tilde{\tilde{j}}} B\begin{pmatrix}-\nu & \mathbf{y}' & -\mathbf{y}' & \nu\\ j & \tilde{j} & \tilde{j} & \tilde{\tilde{j}}\end{pmatrix} n\begin{pmatrix}\mathbf{y}'\\ \tilde{j}\end{pmatrix}\Big\{G\begin{pmatrix}\nu & -\nu\\ \tilde{\tilde{j}} & j'\end{pmatrix}+\Gamma\begin{pmatrix}\nu & -\nu\\ \tilde{\tilde{j}} & j'\end{pmatrix}\Big\}+\\
&+3\sum_{\substack{\mathbf{y}\mathbf{y}'\mathbf{y}''\\ \tilde{j}\tilde{\tilde{j}}\tilde{\tilde{\tilde{j}}}}}{}' B\begin{pmatrix}-\nu & \mathbf{y} & \mathbf{y}' & \mathbf{y}''\\ j & \tilde{j} & \tilde{\tilde{j}} & \tilde{\tilde{\tilde{j}}}\end{pmatrix}\Big\{\frac{\theta(t-\tau)}{i}\Big\langle\Big[b_+\begin{pmatrix}-\mathbf{y}\\ \tilde{j}\end{pmatrix}b_+\begin{pmatrix}-\mathbf{y}'\\ \tilde{\tilde{j}}\end{pmatrix}b_-\begin{pmatrix}\mathbf{y}''\\ \tilde{\tilde{\tilde{j}}}\end{pmatrix},\ b'_-\begin{pmatrix}-\nu\\ j'\end{pmatrix}\Big]\Big\rangle+\\
&\quad+\frac{\theta(t-\tau)}{i}\Big\langle\Big[b_+\begin{pmatrix}-\mathbf{y}\\ \tilde{j}\end{pmatrix}b_-\begin{pmatrix}\mathbf{y}'\\ \tilde{\tilde{j}}\end{pmatrix}b_-\begin{pmatrix}\mathbf{y}''\\ \tilde{\tilde{\tilde{j}}}\end{pmatrix},\ b'_-\begin{pmatrix}-\nu\\ j'\end{pmatrix}\Big]\Big\rangle\Big\}.
\end{aligned}
\tag{III.9}
$$

The prime to the last sum means that the operators b_+ and b_- do not have the same indices.

The following notation will be used for the Green's functions which contain more than two operators:

$$
\begin{aligned}
G^a&=\frac{\theta(t-\tau)}{i}\Big\langle\Big[b_+\begin{pmatrix}-\mathbf{y}\\ \tilde{j}\end{pmatrix}b_+\begin{pmatrix}-\mathbf{y}'\\ \tilde{\tilde{j}}\end{pmatrix},\ b'_-\begin{pmatrix}-\nu\\ j'\end{pmatrix}\Big]\Big\rangle,\\
G^b&=\frac{\theta(t-\tau)}{i}\Big\langle\Big[b_+\begin{pmatrix}-\mathbf{y}\\ \tilde{j}\end{pmatrix}b_-\begin{pmatrix}\mathbf{y}'\\ \tilde{\tilde{j}}\end{pmatrix},\ b'_-\begin{pmatrix}-\nu\\ j'\end{pmatrix}\Big]\Big\rangle,\\
G^c&=\frac{\theta(t-\tau)}{i}\Big\langle\Big[b_-\begin{pmatrix}\mathbf{y}\\ \tilde{j}\end{pmatrix}b_-\begin{pmatrix}\mathbf{y}'\\ \tilde{\tilde{j}}\end{pmatrix},\ b'_-\begin{pmatrix}-\nu\\ j'\end{pmatrix}\Big]\Big\rangle,\\
G^1&=\frac{\theta(t-\tau)}{i}\Big\langle\Big[b_+\begin{pmatrix}-\mathbf{y}\\ \tilde{j}\end{pmatrix}b_+\begin{pmatrix}-\mathbf{y}'\\ \tilde{\tilde{j}}\end{pmatrix}b_+\begin{pmatrix}-\mathbf{y}''\\ \tilde{\tilde{\tilde{j}}}\end{pmatrix},\ b'_-\begin{pmatrix}-\nu\\ j'\end{pmatrix}\Big]\Big\rangle,\\
G^2&=\frac{\theta(t-\tau)}{i}\Big\langle\Big[b_-\begin{pmatrix}\mathbf{y}\\ \tilde{j}\end{pmatrix}b_-\begin{pmatrix}\mathbf{y}'\\ \tilde{\tilde{j}}\end{pmatrix}b_-\begin{pmatrix}\mathbf{y}''\\ \tilde{\tilde{\tilde{j}}}\end{pmatrix},\ b'_-\begin{pmatrix}-\nu\\ j'\end{pmatrix}\Big]\Big\rangle,\\
G^3&=\frac{\theta(t-\tau)}{i}\Big\langle\Big[b_+\begin{pmatrix}-\mathbf{y}\\ \tilde{j}\end{pmatrix}b_+\begin{pmatrix}-\mathbf{y}'\\ \tilde{\tilde{j}}\end{pmatrix}b_-\begin{pmatrix}\mathbf{y}''\\ \tilde{\tilde{\tilde{j}}}\end{pmatrix},\ b'_-\begin{pmatrix}-\nu\\ j'\end{pmatrix}\Big]\Big\rangle,\\
G^4&=\frac{\theta(t-\tau)}{i}\Big\langle\Big[b_+\begin{pmatrix}-\mathbf{y}\\ \tilde{j}\end{pmatrix}b_-\begin{pmatrix}\mathbf{y}'\\ \tilde{\tilde{j}}\end{pmatrix}b_-\begin{pmatrix}\mathbf{y}''\\ \tilde{\tilde{\tilde{j}}}\end{pmatrix},\ b'_-\begin{pmatrix}-\nu\\ j'\end{pmatrix}\Big]\Big\rangle.
\end{aligned}
\tag{III.10}
$$

With Eqs. (III.9) and (III.10), Eq. (III.8) becomes

$$
\begin{aligned}
&-i\hbar\frac{dG}{dt}\begin{pmatrix}\nu & -\nu\\ j & j'\end{pmatrix}=\hbar\delta(t-\tau)\,\delta_{jj'}+\hbar\omega\begin{pmatrix}\nu\\ j\end{pmatrix}G\begin{pmatrix}\nu & -\nu\\ j & j'\end{pmatrix}+\\
&+\sum_{\tilde{\tilde{j}}}C\begin{pmatrix}-\nu & \nu\\ j & \tilde{\tilde{j}}\end{pmatrix}\Big\{G\begin{pmatrix}\nu & -\nu\\ \tilde{\tilde{j}} & j'\end{pmatrix}+\Gamma\begin{pmatrix}\nu & -\nu\\ \tilde{\tilde{j}} & j'\end{pmatrix}\Big\}+\sum_{\substack{\mathbf{y}\mathbf{y}'\\ \tilde{j}\tilde{\tilde{j}}}}B\begin{pmatrix}-\nu & \mathbf{y} & \mathbf{y}'\\ j & \tilde{j} & \tilde{\tilde{j}}\end{pmatrix}\{G^a+2G^b+G^c\}+\\
&+\sum_{\substack{\mathbf{y}\mathbf{y}'\mathbf{y}''\\ \tilde{j}\tilde{\tilde{j}}\tilde{\tilde{\tilde{j}}}}}{}' B\begin{pmatrix}-\nu & \mathbf{y} & \mathbf{y}' & \mathbf{y}''\\ j & \tilde{j} & \tilde{\tilde{j}} & \tilde{\tilde{\tilde{j}}}\end{pmatrix}\{G^1+G^2+3G^3+3G^4\},
\end{aligned}
\tag{III.11}
$$

where

$$C\begin{pmatrix}-\nu & \nu \\ j & \tilde{\tilde{j}}\end{pmatrix} = 3\sum_{\tilde{j}y'} B\begin{pmatrix}-\nu & \nu & y' & -y' \\ j & \tilde{\tilde{j}} & \tilde{j} & \tilde{j}\end{pmatrix}\left(2n\begin{pmatrix}y' \\ \tilde{j}\end{pmatrix}+1\right). \quad \text{(III.12)}$$

The derivation of the equation for $\Gamma\begin{pmatrix}\nu & -\nu \\ j & j'\end{pmatrix}$ is similar to the above. This equation is

$$i\hbar\frac{d\Gamma\begin{pmatrix}\nu & -\nu \\ j & j'\end{pmatrix}}{dt} = \hbar\omega\begin{pmatrix}\nu \\ j\end{pmatrix}\Gamma\begin{pmatrix}\nu & -\nu \\ j & j'\end{pmatrix} + \sum_{\tilde{\tilde{j}}} C\begin{pmatrix}-\nu & \nu \\ j & \tilde{\tilde{j}}\end{pmatrix}\left\{G\begin{pmatrix}\nu & -\nu \\ \tilde{\tilde{j}} & j'\end{pmatrix} + \Gamma\begin{pmatrix}\nu & -\nu \\ \tilde{\tilde{j}} & j'\end{pmatrix}\right\} +$$

$$+\sum_{\substack{yy' \\ \tilde{j}\tilde{\tilde{j}}}} B\begin{pmatrix}-\nu & y & y' \\ j & \tilde{j} & \tilde{\tilde{j}}\end{pmatrix}\{G^a + 2G^b + G^c\} + \sum_{\substack{yy'y'' \\ \tilde{j}\tilde{\tilde{j}}\tilde{\tilde{\tilde{j}}}}}' B\begin{pmatrix}-\nu & y & y' & y'' \\ j & \tilde{j} & \tilde{\tilde{j}} & \tilde{\tilde{\tilde{j}}}\end{pmatrix}\{G^1 + G^2 + 3G^3 + 3G^4\}. \quad \text{(III.13)}$$

In order to obtain a closed set of equations for the Green's functions $G\begin{pmatrix}\nu & -\nu \\ j & j'\end{pmatrix}$, $\Gamma\begin{pmatrix}\nu & -\nu \\ j & j'\end{pmatrix}$, and (III.10), we must also derive equations for the functions (III.10). These are found in the same way as Eqs. (III.11) and (III.13).

For this purpose, we differentiate Eqs. (III.10) with respect to t. The resulting higher-order Green's functions (involving a larger number of operators) can be expressed approximately in terms of the known functions $G\begin{pmatrix}\nu & -\nu \\ j & j'\end{pmatrix}$, $\Gamma\begin{pmatrix}\nu & -\nu \\ j & j'\end{pmatrix}$, and (III.10). The Green's functions which cannot be so expressed are omitted.

These equations will not be derived in detail, since the calculations are simple though laborious. We shall give only an example of how the higher-order Green's functions can be expressed in terms of the known lower-order ones. In obtaining the equation for G^a, the following expression occurs:

$$2\sum_{\substack{y''y''' \\ j''j'''}} B\begin{pmatrix}y & y'' & y''' \\ \tilde{j} & j'' & j'''\end{pmatrix}\frac{\theta(t-\tau)}{i}\left\langle\left[b_+\begin{pmatrix}-y'' \\ j''\end{pmatrix}b_-\begin{pmatrix}y''' \\ j'''\end{pmatrix}b_+\begin{pmatrix}-y' \\ \tilde{\tilde{j}}\end{pmatrix}, b_-'\begin{pmatrix}-\nu \\ j'\end{pmatrix}\right]\right\rangle.$$

Replacing the operator product $b_-\begin{pmatrix}y''' \\ j'''\end{pmatrix}b_+\begin{pmatrix}-y' \\ \tilde{\tilde{j}}\end{pmatrix}$ by its mean value $\delta_{j'''\tilde{\tilde{j}}}\delta_{-y'''y'}\left(n\begin{pmatrix}y' \\ \tilde{\tilde{j}}\end{pmatrix}+1\right)$, we find

$$2\sum_{y''j''} B\begin{pmatrix}-y & y'' & y' \\ \tilde{j} & j'' & \tilde{\tilde{j}}\end{pmatrix}\left(n\begin{pmatrix}y' \\ \tilde{\tilde{j}}\end{pmatrix}+1\right)\frac{\theta(t-\tau)}{i}\left\langle\left[b_+\begin{pmatrix}-y'' \\ j''\end{pmatrix}, b_-'\begin{pmatrix}-\nu \\ j'\end{pmatrix}\right]\right\rangle.$$

Retaining in this sum only the term with $y'' = \nu$, we arrive at the expression

$$2\sum_{j''} B\begin{pmatrix}\nu & -y & -y' \\ j'' & \tilde{j} & \tilde{\tilde{j}}\end{pmatrix}\left(n\begin{pmatrix}y' \\ \tilde{\tilde{j}}\end{pmatrix}+1\right)G\begin{pmatrix}\nu & -\nu \\ j'' & j'\end{pmatrix}.$$

Transformations similar to the above-mentioned lead to the equations for G^a, G^b, and G^c

$$i\hbar\frac{dG^a}{dt} = -\hbar\left(\Omega\begin{pmatrix}y \\ \tilde{j}\end{pmatrix} + \Omega\begin{pmatrix}y' \\ \tilde{\tilde{j}}\end{pmatrix}\right)G^a - C\begin{pmatrix}-y' & y' \\ \tilde{\tilde{j}} & \tilde{\tilde{j}}\end{pmatrix}G^b -$$

$$-\sum_{j''} 2B\begin{pmatrix}\nu & -y & -y' \\ j'' & \tilde{j} & \tilde{\tilde{j}}\end{pmatrix}\left(n\begin{pmatrix}y \\ \tilde{j}\end{pmatrix} + n\begin{pmatrix}y' \\ \tilde{\tilde{j}}\end{pmatrix} + 1\right)\left\{G\begin{pmatrix}\nu & -\nu \\ j'' & j'\end{pmatrix} + \Gamma\begin{pmatrix}\nu & -\nu \\ j'' & j'\end{pmatrix}\right\},$$

$$
\begin{aligned}
i\hbar \frac{dG^b}{dt} &= \hbar\left(\Omega\binom{\mathbf{y}'}{\tilde{\tilde j}} - \Omega\binom{\mathbf{y}}{\tilde j}\right)G^b + C\begin{pmatrix} -\mathbf{y}'\mathbf{y}' \\ \tilde{\tilde j}\,\tilde{\tilde j}\end{pmatrix}G^a - C\begin{pmatrix} -\mathbf{y}\,\mathbf{y} \\ \tilde j\,\tilde j\end{pmatrix}G^c + \\
&+ \sum_{j''} 2B\begin{pmatrix} \boldsymbol{\nu} - \mathbf{y} - \mathbf{y}' \\ j''\,\tilde j\,\tilde{\tilde j}\end{pmatrix}\left(n\binom{\mathbf{y}}{\tilde j} - n\binom{\mathbf{y}'}{\tilde{\tilde j}}\right)\left\{G\begin{pmatrix} \boldsymbol{\nu} - \boldsymbol{\nu} \\ j''\,j'\end{pmatrix} + \Gamma\begin{pmatrix} \boldsymbol{\nu} - \boldsymbol{\nu} \\ j''\,j'\end{pmatrix}\right\}, \\
i\hbar \frac{dG^c}{dt} &= \hbar\left(\Omega\binom{\mathbf{y}}{\tilde j} + \Omega\binom{\mathbf{y}'}{\tilde{\tilde j}}\right)G^c + C\begin{pmatrix} -\mathbf{y}\mathbf{y} \\ \tilde j\,\tilde j\end{pmatrix}G^b + \\
&+ \sum_{j''} 2B\begin{pmatrix} \boldsymbol{\nu} - \mathbf{y} - \mathbf{y}' \\ j''\,\tilde j\,\tilde{\tilde j}\end{pmatrix}\left(n\binom{\mathbf{y}}{\tilde j} + n\binom{\mathbf{y}'}{\tilde{\tilde j}} + 1\right)\left\{G\begin{pmatrix} \boldsymbol{\nu} - \boldsymbol{\nu} \\ j''\,j'\end{pmatrix} + \Gamma\begin{pmatrix} \boldsymbol{\nu} - \boldsymbol{\nu} \\ j''\,j'\end{pmatrix}\right\},
\end{aligned}
\tag{III.14}
$$

where

$$
\Omega\binom{\mathbf{y}}{j} = \omega\binom{\mathbf{y}}{j} + \frac{C\begin{pmatrix} -\mathbf{y}\mathbf{y} \\ j\,j\end{pmatrix}}{\hbar}. \tag{III.15}
$$

Equations for G^1, G^2, G^3, and G^4 are found similarly. They are

$$
\begin{aligned}
i\hbar \frac{dG^1}{dt} &= -\hbar\left[\Omega\binom{\mathbf{y}}{\tilde j} + \Omega\binom{\mathbf{y}'}{\tilde{\tilde j}} + \Omega\binom{\mathbf{y}''}{\tilde{\tilde{\tilde j}}}\right]G^1 - C\begin{pmatrix} -\mathbf{y}''\mathbf{y}'' \\ \tilde{\tilde{\tilde j}}\,\tilde{\tilde{\tilde j}}\end{pmatrix}G^3 - \\
&- \sum_{j''} 6B\begin{pmatrix} \boldsymbol{\nu} - \mathbf{y} - \mathbf{y}' - \mathbf{y}'' \\ j''\,\tilde j\,\tilde{\tilde j}\,\tilde{\tilde{\tilde j}}\end{pmatrix}N_1\left\{G\begin{pmatrix} \boldsymbol{\nu} - \boldsymbol{\nu} \\ j''\,j'\end{pmatrix} + \Gamma\begin{pmatrix} \boldsymbol{\nu} - \boldsymbol{\nu} \\ j''\,j'\end{pmatrix}\right\}, \\
i\hbar \frac{dG^2}{dt} &= \hbar\left[\Omega\binom{\mathbf{y}}{\tilde j} + \Omega\binom{\mathbf{y}'}{\tilde{\tilde j}} + \Omega\binom{\mathbf{y}''}{\tilde{\tilde{\tilde j}}}\right]G^2 + C\begin{pmatrix} -\mathbf{y}\mathbf{y} \\ \tilde j\,\tilde j\end{pmatrix}G^4 + \\
&+ \sum_{j''} 6B\begin{pmatrix} \boldsymbol{\nu} - \mathbf{y} - \mathbf{y}' - \mathbf{y}'' \\ j''\,\tilde j\,\tilde{\tilde j}\,\tilde{\tilde{\tilde j}}\end{pmatrix}N_1\left\{G\begin{pmatrix} \boldsymbol{\nu} - \boldsymbol{\nu} \\ j''\,j'\end{pmatrix} + \Gamma\begin{pmatrix} \boldsymbol{\nu} - \boldsymbol{\nu} \\ j''\,j'\end{pmatrix}\right\}, \\
i\hbar \frac{dG^3}{dt} &= -\hbar\left[\Omega\binom{\mathbf{y}}{\tilde j} + \Omega\binom{\mathbf{y}'}{\tilde{\tilde j}} - \Omega\binom{\mathbf{y}''}{\tilde{\tilde{\tilde j}}}\right]G^3 - C\begin{pmatrix} -\mathbf{y}'\mathbf{y}' \\ \tilde{\tilde j}\,\tilde{\tilde j}\end{pmatrix}G^4 + C\begin{pmatrix} -\mathbf{y}''\mathbf{y}'' \\ \tilde{\tilde{\tilde j}}\,\tilde{\tilde{\tilde j}}\end{pmatrix}G^1 - \\
&- \sum_{j''} 6B\begin{pmatrix} \boldsymbol{\nu} - \mathbf{y} - \mathbf{y}' - \mathbf{y}'' \\ j''\,\tilde j\,\tilde{\tilde j}\,\tilde{\tilde{\tilde j}}\end{pmatrix}N_2\left\{G\begin{pmatrix} \boldsymbol{\nu} - \boldsymbol{\nu} \\ j''\,j'\end{pmatrix} + \Gamma\begin{pmatrix} \boldsymbol{\nu} - \boldsymbol{\nu} \\ j''\,j'\end{pmatrix}\right\}, \\
i\hbar \frac{dG^4}{dt} &= -\hbar\left[\Omega\binom{\mathbf{y}}{\tilde j} - \Omega\binom{\mathbf{y}'}{\tilde{\tilde j}} - \Omega\binom{\mathbf{y}''}{\tilde{\tilde{\tilde j}}}\right]G^4 - C\begin{pmatrix} -\mathbf{y}\mathbf{y} \\ \tilde j\,\tilde j\end{pmatrix}G^2 + C\begin{pmatrix} -\mathbf{y}'\mathbf{y}' \\ \tilde{\tilde j}\,\tilde{\tilde j}\end{pmatrix}G^3 - \\
&- \sum_{j''} 6B\begin{pmatrix} \boldsymbol{\nu} - \mathbf{y} - \mathbf{y}' - \mathbf{y}'' \\ j''\,\tilde j\,\tilde{\tilde j}\,\tilde{\tilde{\tilde j}}\end{pmatrix}N_3\left\{G\begin{pmatrix} \boldsymbol{\nu} - \boldsymbol{\nu} \\ j''\,j'\end{pmatrix} + \Gamma\begin{pmatrix} \boldsymbol{\nu} - \boldsymbol{\nu} \\ j''\,j'\end{pmatrix}\right\},
\end{aligned}
\tag{III.16}
$$

where

$$
\begin{aligned}
N_1 &= \left(n\binom{\mathbf{y}}{\tilde j} + 1\right)\left(n\binom{\mathbf{y}'}{\tilde{\tilde j}} + 1\right)\left(n\binom{\mathbf{y}''}{\tilde{\tilde{\tilde j}}} + 1\right) - n\binom{\mathbf{y}}{\tilde j}n\binom{\mathbf{y}'}{\tilde{\tilde j}}n\binom{\mathbf{y}''}{\tilde{\tilde{\tilde j}}}, \\
N_2 &= \left(n\binom{\mathbf{y}}{\tilde j} + 1\right)\left(n\binom{\mathbf{y}'}{\tilde{\tilde j}} + 1\right)n\binom{\mathbf{y}''}{\tilde{\tilde{\tilde j}}} - n\binom{\mathbf{y}}{\tilde j}n\binom{\mathbf{y}'}{\tilde{\tilde j}}\left(n\binom{\mathbf{y}''}{\tilde{\tilde{\tilde j}}} + 1\right), \\
N_3 &= n\binom{\mathbf{y}}{\tilde j}\left(n\binom{\mathbf{y}'}{\tilde{\tilde j}} + 1\right)\left(n\binom{\mathbf{y}''}{\tilde{\tilde{\tilde j}}} + 1\right) - \left(n\binom{\mathbf{y}}{\tilde j} + 1\right)n\binom{\mathbf{y}'}{\tilde{\tilde j}}n\binom{\mathbf{y}''}{\tilde{\tilde{\tilde j}}}.
\end{aligned}
\tag{III.17}
$$

2. Solution of the Equations for the Green's Functions

In order to solve Eqs. (III.11), (III.13), (III.14), and (III.16), we take the Fourier transform by means of the relations

$$
G(t) = \int G(\omega)\, e^{-i\omega t} d\omega, \tag{III.18}
$$

$$\delta(t-\tau)=\frac{1}{2\pi}\int e^{-i\omega(t-\tau)}d\omega. \tag{III.19}$$

Let $G^{+}(t)$ be the function conjugate to $G(t)$, and $G^{+}(\omega)$ the Fourier component of $G^{+}(t)$, while $G(\omega)^{+}$ denotes the function conjugate to $G(\omega)$. Then, using Eq. (III.18), we find

$$G^{+}(\omega)=G(-\omega)^{+}. \tag{III.20}$$

Thus, in order to obtain the Fourier component of the conjugate function, we must take the Fourier component of the original function with argument $-\omega$ and then take the complex conjugate. It is clear that, if the Fourier component is real, then

$$G(-\omega)^{+}=G(-\omega).$$

After substituting Eqs. (III.18) and (III.19) in Eqs. (III.11), (III.13), (III.14), and (III.16), we can express the Fourier components of the functions G^a, G^b, G^c, and G^1 through G^4 in terms of those of the functions $G\binom{\nu\ -\nu}{j''\ j'}$ by means of Eqs. (III.14) and (III.16), neglecting quantities of the type $C\binom{-\mathbf{y}\ \mathbf{y}}{\tilde{j}\ \tilde{j}}$. The expressions thus obtained for the Fourier components of G^a, G^b, G^c, and G^1 through G^4 are substituted in Eqs. (III.11) and (III.13) (or rather in the Fourier transforms of these equations).

This leads to the equations

$$-\left[\omega+\omega\binom{\nu}{j}+\Gamma_{i}(\nu,\omega)\right]G\binom{\nu\ -\nu}{j\ j'}_{\omega}-\Gamma_{j}(\nu,\omega)\,\Gamma\binom{\nu\ -\nu}{j\ j'}_{\omega}=\frac{1}{2\pi}\delta_{jj'}+\sum_{j''\neq j}\Gamma_{jj''}(\nu,\omega)\left\{G\binom{\nu\ -\nu}{j''\ j'}_{\omega}+\Gamma\binom{\nu\ -\nu}{j''\ j'}_{\omega}\right\}, \tag{III.21}$$

$$\left[\omega-\omega\binom{\nu}{j}-\Gamma_{j}(\nu,\omega)\right]\Gamma\binom{\nu\ -\nu}{j\ j'}_{\omega}-\Gamma_{j}(\nu,\omega)\,G\binom{\nu\ -\nu}{j\ j'}_{\omega}=\sum_{j''\neq j}\Gamma_{jj''}(\nu,\omega)\left\{G\binom{\nu\ -\nu}{j''\ j'}_{\omega}+\Gamma\binom{\nu\ -\nu}{j''\ j'}_{\omega}\right\}, \tag{III.22}$$

where

$$\begin{aligned}\Gamma_{jj''}(\nu,\omega)=\;&\frac{2}{\hbar^{2}}\sum_{\substack{\mathbf{y}\mathbf{y}'\\ \tilde{j}\tilde{\tilde{j}}}}B^{*}\begin{pmatrix}\nu & -\mathbf{y} & -\mathbf{y}'\\ j & \tilde{j} & \tilde{\tilde{j}}\end{pmatrix}B\begin{pmatrix}\nu & -\mathbf{y} & -\mathbf{y}'\\ j'' & \tilde{j} & \tilde{\tilde{j}}\end{pmatrix}\left\{\left(n\binom{\mathbf{y}}{\tilde{j}}+n\binom{\mathbf{y}'}{\tilde{\tilde{j}}}+1\right)\times\right.\\ &\times\left(\frac{1}{\omega-\Omega\binom{\mathbf{y}}{\tilde{j}}-\Omega\binom{\mathbf{y}'}{\tilde{\tilde{j}}}}-\frac{1}{\omega+\Omega\binom{\mathbf{y}}{\tilde{j}}+\Omega\binom{\mathbf{y}'}{\tilde{\tilde{j}}}}\right)+\left.\frac{2\left(n\binom{\mathbf{y}}{\tilde{j}}-n\binom{\mathbf{y}'}{\tilde{\tilde{j}}}\right)}{\omega+\Omega\binom{\mathbf{y}}{\tilde{j}}-\Omega\binom{\mathbf{y}'}{\tilde{\tilde{j}}}}\right\}+\\ &+\frac{3}{\hbar}\sum_{\tilde{j}\mathbf{y}'}B\begin{pmatrix}-\nu & \nu & \mathbf{y}' & -\mathbf{y}'\\ j & j'' & \tilde{j} & \tilde{j}\end{pmatrix}\left(2n\binom{\mathbf{y}'}{\tilde{j}}+1\right)+\frac{6}{\hbar^{2}}\sum_{\substack{\mathbf{y}\mathbf{y}'\mathbf{y}''\\ \tilde{j}\tilde{\tilde{j}}\tilde{\tilde{\tilde{j}}}}}{}'\,B^{*}\begin{pmatrix}\nu & -\mathbf{y} & -\mathbf{y}' & -\mathbf{y}''\\ j & \tilde{j} & \tilde{\tilde{j}} & \tilde{\tilde{\tilde{j}}}\end{pmatrix}B\begin{pmatrix}\nu & -\mathbf{y} & -\mathbf{y}' & -\mathbf{y}''\\ j'' & \tilde{j} & \tilde{\tilde{j}} & \tilde{\tilde{\tilde{j}}}\end{pmatrix}\times\\ &\times\left\{N_{1}\left(\frac{1}{\omega-\Omega\binom{\mathbf{y}}{\tilde{j}}-\Omega\binom{\mathbf{y}'}{\tilde{\tilde{j}}}-\Omega\binom{\mathbf{y}''}{\tilde{\tilde{\tilde{j}}}}}-\frac{1}{\omega+\Omega\binom{\mathbf{y}}{\tilde{j}}+\Omega\binom{\mathbf{y}'}{\tilde{\tilde{j}}}+\Omega\binom{\mathbf{y}''}{\tilde{\tilde{\tilde{j}}}}}\right)+\right.\\ &\left.+3N_{3}\left(\frac{1}{\omega+\Omega\binom{\mathbf{y}}{\tilde{j}}-\Omega\binom{\mathbf{y}'}{\tilde{\tilde{j}}}-\Omega\binom{\mathbf{y}''}{\tilde{\tilde{\tilde{j}}}}}-\frac{1}{\omega-\Omega\binom{\mathbf{y}}{\tilde{j}}+\Omega\binom{\mathbf{y}'}{\tilde{\tilde{j}}}+\Omega\binom{\mathbf{y}''}{\tilde{\tilde{\tilde{j}}}}}\right)\right\},\end{aligned}$$

$$\Gamma_{j}(\nu,\omega)=\Gamma_{jj}(\nu,\omega). \tag{III.23}$$

It is seen from the definition (III.23) and from Eq. (II.18) that the quantity $\Gamma_{jj'}(\nu, \omega)$ has the following properties:

$$\Gamma_{jj'}(\nu, -\omega) = \Gamma_{jj'}(\nu, \omega), \tag{III.24}$$

$$\Gamma^{*}_{jj'}(\nu, \omega) = \Gamma_{j'j}(\nu, \omega) = \Gamma_{jj'}(-\nu, \omega). \tag{III.25}$$

In order to obtain from Eq. (III.23) the damping and the frequency correction, we have to make the substitution $\omega \to \omega + i\delta$ ($\delta > 0$, $\delta \to 0$) and use the identity (see [39])

$$\frac{1}{x \pm i\delta} = P\frac{1}{x} \mp i\pi\delta(x).$$

This change to complex ω is made at the end of this section, where an expression for the complex polarizability will be given.

Equations (III.21) and (III.22) can be solved by assuming that

$$\begin{aligned} G\begin{pmatrix}\nu - \nu\\ j \quad j\end{pmatrix}_\omega &\gg G\begin{pmatrix}\nu - \nu\\ j \quad j'\end{pmatrix}_\omega, \\ \Gamma\begin{pmatrix}\nu - \nu\\ j \quad j\end{pmatrix}_\omega &\gg \Gamma\begin{pmatrix}\nu - \nu\\ j \quad j'\end{pmatrix}_\omega, \end{aligned} \tag{III.26}$$

where $j \neq j'$.

We shall later see under what conditions these inequalities are satisfied.

Taking $j \neq j'$ in Eqs. (III.21) and (III.22), and neglecting in the sums

$$\sum_{j'' \neq j} \Gamma_{jj''}(\nu, \omega)\left\{G\begin{pmatrix}\nu - \nu\\ j'' \quad j\end{pmatrix}_\omega + \Gamma\begin{pmatrix}\nu - \nu\\ j'' \quad j'\end{pmatrix}_\omega\right\}$$

the Green's functions with $j'' \neq j$, we can express the nondiagonal functions (with $j \neq j'$) in terms of the diagonal ones (with $j' = j$). The relations obtained are

$$\begin{aligned} G\begin{pmatrix}\nu - \nu\\ j \quad j'\end{pmatrix}_\omega &= \frac{\Gamma_{jj'}(\nu, \omega)\left(\omega\begin{pmatrix}\nu\\ j\end{pmatrix} - \omega\right)}{\omega^2 - \omega^2\begin{pmatrix}\nu\\ j\end{pmatrix} - 2\omega\begin{pmatrix}\nu\\ j\end{pmatrix}\Gamma_j(\nu, \omega)}\left\{G\begin{pmatrix}\nu - \nu\\ j' \quad j'\end{pmatrix} + \Gamma\begin{pmatrix}\nu - \nu\\ j' \quad j'\end{pmatrix}\right\}_\omega, \\ \Gamma\begin{pmatrix}\nu - \nu\\ j \quad j'\end{pmatrix}_\omega &= \frac{\Gamma_{jj'}(\nu, \omega)\left(\omega\begin{pmatrix}\nu\\ j\end{pmatrix} + \omega\right)}{\omega^2 - \omega^2\begin{pmatrix}\nu\\ j\end{pmatrix} - 2\omega\begin{pmatrix}\nu\\ j\end{pmatrix}\Gamma_j(\nu, \omega)}\left\{G\begin{pmatrix}\nu - \nu\\ j' \quad j'\end{pmatrix} + \Gamma\begin{pmatrix}\nu - \nu\\ j' \quad j'\end{pmatrix}\right\}_\omega. \end{aligned} \tag{III.27}$$

Putting $j = j'$ and substituting Eqs. (III.27) in the sums

$$\sum_{j'' \neq j} \Gamma_{jj''}(\nu, \omega)\left\{G\begin{pmatrix}\nu - \nu\\ j'' \quad j\end{pmatrix} + \Gamma\begin{pmatrix}\nu - \nu\\ j'' \quad j\end{pmatrix}\right\}_\omega$$

we get expressions for the diagonal components of the Green's functions:

$$G\begin{pmatrix}\nu - \nu\\ j \quad j\end{pmatrix}_\omega = -\frac{1}{2\pi}\frac{\omega - \omega\begin{pmatrix}\nu\\ j\end{pmatrix} - \Gamma'_j(\nu, \omega)}{\omega^2 - \omega^2\begin{pmatrix}\nu\\ j\end{pmatrix} - 2\omega\begin{pmatrix}\nu\\ j\end{pmatrix}\Gamma'_j(\nu, \omega)}, \tag{III.28}$$

$$\Gamma\binom{\nu-\nu}{jj}_{\omega}=-\frac{1}{2\pi}\frac{\Gamma_j'(\nu,\omega)}{\omega^2-\omega^2\binom{\nu}{j}-2\omega\binom{\nu}{j}\Gamma_j'(\nu,\omega)}, \tag{III.29}$$

where

$$\Gamma_j'(\nu,\omega)=\Gamma_j(\nu,\omega)+\Phi_j(\nu,\omega), \tag{III.30}$$

$$\Phi_j(\nu,\omega)=\sum_{\tilde{j}\neq j}\frac{2\omega\binom{\nu}{\tilde{j}}\Gamma_{j\tilde{j}}(\nu,\omega)\Gamma_{\tilde{j}j}(\nu,\omega)}{\omega^2-\omega^2\binom{\nu}{\tilde{j}}-2\omega\binom{\nu}{\tilde{j}}\Gamma_{\tilde{j}}(\nu,\omega)}. \tag{III.31}$$

Using Eqs. (III.2), (III.20), (III.24), and (III.25), we now find an expression for the functions $\mathcal{G}\binom{\nu-\nu}{jj'}_{\omega}$, which appear in Eq. (III.1) for the complex polarizability:

$$\mathcal{G}\binom{\nu-\nu}{jj}_{\omega}=\frac{1}{\pi}\frac{\omega\binom{\nu}{j}}{\omega^2-\omega^2\binom{\nu}{j}-2\omega\binom{\nu}{j}\Gamma_j'(\nu,\omega)}, \tag{III.32}$$

$$\mathcal{G}\binom{\nu-\nu}{jj'}_{\omega}=\frac{2}{\pi}\Gamma_{jj'}(\nu,\omega)\,\omega\binom{\nu}{j}\omega\binom{\nu}{j'}\Big\{\Big[\omega^2-\omega^2\binom{\nu}{j}-2\omega\binom{\nu}{j}\Gamma_j(\nu,\omega)\Big]\Big[\omega^2-\omega^2\binom{\nu}{j'}-2\omega\binom{\nu}{j'}\Gamma_{j'}(\nu,\omega)\Big]-$$
$$-4\omega\binom{\nu}{j}\omega\binom{\nu}{j'}\Gamma_{jj'}(\nu,\omega)\Gamma_{j'j}(\nu,\omega)\Big\}^{-1}. \tag{III.33}$$

When two optically active frequencies occur, Eqs. (III.32) and (III.33) are not approximate but exact, in the sense that they follow from an exact solution of Eqs. (III.21) and (III.22).

Let us now consider in which cases the nondiagonal Green's functions $\mathcal{G}\binom{\nu-\nu}{jj'}_{\omega}$ $(j\neq j')$ may be important in calculating the complex polarizability. We shall neglect spatial-dispersion effects, i.e., the wave vector of the light wave in the crystal will be made to tend to zero. The limits of the expressions $\mathcal{G}\binom{\nu-\nu}{jj'}_{\omega}$, $\mathcal{G}\binom{\nu-\nu}{jj'}_{\omega}$, and $\chi_{\alpha\beta}(\nu,\omega)$ are independent of the direction along which the vector ν tends to zero. This is because, from the coefficients $C_{\alpha\beta}\binom{\nu}{kk'}$, which determine the eigenfrequencies and polarization vectors of oscillations having wave vector ν [see (II.9), (II.10)], we have separated the expressions

$$\frac{4\pi}{v_a}\left(\frac{\nu_\alpha\nu_\beta}{\nu^2}\right)\frac{e_k e_{k'}}{(m_k m_{k'})^{1/2}},$$

which take account of the longwave Coulomb interaction and have no unique limit as $\nu\to 0$; cf. the comment on Eq. (II.30), and [11].

In order to decide when the nondiagonal Green's functions $\mathcal{G}\binom{0\ 0}{j\ j'}_{\omega}$ are important, we consider the function $\Gamma_{jj'}(0,\omega)$, which may be written [see (II.14), (II.15), and (II.16)]

$$\Gamma_{jj'}(0,\omega)=\sum_{\substack{kk'\\ \alpha\beta}}w_\alpha\left(k\,\Big|\,{0\atop j}\right)w_\beta^*\left(k'\,\Big|\,{0\atop j'}\right)T_{\alpha\beta}\binom{kk'}{jj';\,\omega}, \tag{III.34}$$

where $T_{\alpha\beta}\left(\begin{smallmatrix}kk'\\jj'\end{smallmatrix}\omega\right)$ is a tensor whose significance is seen from the definition (III.23) of the function $\Gamma_{jj'}(0, \omega)$.

Whether this function is or is not zero depends on the symmetry of the crystal and on the number of ions in the unit cell.

Let us consider, for example, a crystal of the orthorhombic system. Then, only the diagonal components of the tensor $T_{\alpha\alpha}\left(\begin{smallmatrix}kk'\\jj';\ \omega\end{smallmatrix}\right)$ are nonzero, and therefore $\Gamma_{jj'}(0, \omega)$ ($j \neq j'$) will be nonzero if the crystal contains two optical branches having the same polarization. This will evidently occur in crystals containing more than two ions in the unit cell.

Similar considerations apply to cubic crystals.

Let us now consider the value of the nondiagonal Green's functions.

Suppose that a crystal has nonzero $\Gamma_{jj'}(0, \omega)$, ($j \neq j'$). From Eqs. (III.33) and (III.32), it is seen that the inequality $\mathcal{G}\left(\begin{smallmatrix}0 & 0\\j & j\end{smallmatrix}\right)_\omega \gg \mathcal{G}\left(\begin{smallmatrix}0 & 0\\j & j'\end{smallmatrix}\right)_\omega (j \neq j')$ will be satisfied for all frequencies if the differences between the dispersion frequencies in the crystal are much greater than the functions $\Gamma_j(0, \omega)$, $\Gamma_{j'}(0, \omega)$, $\Gamma_{jj'}(0, \omega)$, which determine the widths of the lines and the frequency shifts, i.e., if

$$\omega\begin{pmatrix}0\\j\end{pmatrix} - \omega\begin{pmatrix}0\\j'\end{pmatrix} \gg \Gamma \tag{III.35}$$

for all j and j'. Then, we have also

$$\Gamma_j(0, \omega) \gg \Phi_j(0, \omega). \tag{III.36}$$

If the inequality (III.35) is not satisfied, the approximate method which we have used to solve Eqs. (III.21) and (III.22) is no longer valid, and the equations have to be solved exactly [except in the case of two optically active branches, where Eqs. (III.32) and (III.33) are exact solutions of Eqs. (III.21) and (III.22)]

Let us now consider the simplest case, that of a cubic crystal having two ions in the unit cell. Then, $\Gamma_{jj'}(0, \omega) = 0$ for $j = j'$, and therefore $\mathcal{G}\begin{pmatrix}0 & 0\\j & j'\end{pmatrix} = 0$, $\Phi_j(0, \omega) = 0$.

Using Eqs. (III.1) and (III.32), and taking complex values of ω by means of the substitution $\omega \to \omega + i\delta$ ($\delta > 0$, $\delta \to 0$), we obtain an expression for the complex polarizability:

$$\chi(\omega) = \frac{e^2}{v_a M}\,\frac{1}{\omega_0^2 + 2\omega_0\gamma_1(\omega) - \omega^2 - 2i\omega_0\gamma(\omega)}, \tag{III.37}$$

where M is the reduced mass of the ions, v_a the volume of the unit cell, $\omega_0 = \omega_0\begin{pmatrix}0\\j\end{pmatrix}$,

$$\gamma_1(\omega) = \frac{2}{\hbar^2} P \sum_{\mathbf{y}\tilde{j}\tilde{\tilde{j}}} \left| B\begin{pmatrix}0 & \mathbf{y} & -\mathbf{y}\\ j & \tilde{j} & \tilde{\tilde{j}}\end{pmatrix}\right|^2 \left\{\left(n\begin{pmatrix}\mathbf{y}\\\tilde{j}\end{pmatrix} + n\begin{pmatrix}\mathbf{y}\\\tilde{\tilde{j}}\end{pmatrix} + 1\right)\times\right.$$

$$\times\left(\frac{1}{\omega - \Omega\begin{pmatrix}\mathbf{y}\\\tilde{j}\end{pmatrix} - \Omega\begin{pmatrix}\mathbf{y}\\\tilde{\tilde{j}}\end{pmatrix}} - \frac{1}{\omega + \Omega\begin{pmatrix}\mathbf{y}\\\tilde{j}\end{pmatrix} + \Omega\begin{pmatrix}\mathbf{y}\\\tilde{\tilde{j}}\end{pmatrix}}\right) + \left.\frac{2\left(n\begin{pmatrix}\mathbf{y}\\\tilde{j}\end{pmatrix} - n\begin{pmatrix}\mathbf{y}\\\tilde{\tilde{j}}\end{pmatrix}\right)}{\omega + \Omega\begin{pmatrix}\mathbf{y}\\\tilde{j}\end{pmatrix} - \Omega\begin{pmatrix}\mathbf{y}\\\tilde{\tilde{j}}\end{pmatrix}}\right\} +$$

$$+\frac{3}{\hbar}\sum_{\tilde{j}\,\mathbf{y}} B\begin{pmatrix}0 & 0 & \mathbf{y} & -\mathbf{y}\\ j & j & \tilde{j} & \tilde{j}\end{pmatrix}\left(2n\binom{\mathbf{y}}{\tilde{j}}+1\right)+\frac{6}{\hbar^2}P\sum_{\substack{\mathbf{y}\mathbf{y}'\mathbf{y}''\\ \tilde{j}\tilde{\tilde{j}}\tilde{\tilde{\tilde{j}}}}}\left|B\begin{pmatrix}0 & \mathbf{y} & \mathbf{y}' & \mathbf{y}''\\ j & \tilde{j} & \tilde{\tilde{j}} & \tilde{\tilde{\tilde{j}}}\end{pmatrix}\right|^2\times$$

$$\times\left\{N_1\left(\frac{1}{\omega-\Omega\binom{\mathbf{y}}{\tilde{j}}-\Omega\binom{\mathbf{y}'}{\tilde{\tilde{j}}}-\Omega\binom{\mathbf{y}''}{\tilde{\tilde{\tilde{j}}}}}-\frac{1}{\omega+\Omega\binom{\mathbf{y}}{\tilde{j}}+\Omega\binom{\mathbf{y}'}{\tilde{\tilde{j}}}+\Omega\binom{\mathbf{y}''}{\tilde{\tilde{\tilde{j}}}}}\right)+\right.$$

$$\left.+3N_3\left(\frac{1}{\omega+\Omega\binom{\mathbf{y}}{\tilde{j}}-\Omega\binom{\mathbf{y}'}{\tilde{\tilde{j}}}-\Omega\binom{\mathbf{y}''}{\tilde{\tilde{\tilde{j}}}}}-\frac{1}{\omega-\Omega\binom{\mathbf{y}}{\tilde{j}}+\Omega\binom{\mathbf{y}'}{\tilde{\tilde{j}}}+\Omega\binom{\mathbf{y}''}{\tilde{\tilde{\tilde{j}}}}}\right)\right\}, \qquad \text{(III.38)}$$

$$\gamma(\omega)=\frac{2\pi}{\hbar^2}\sum_{\mathbf{y}\tilde{j}\tilde{\tilde{j}}}\left|B\begin{pmatrix}0 & \mathbf{y} & -\mathbf{y}\\ j & \tilde{j} & \tilde{\tilde{j}}\end{pmatrix}\right|^2\left\{\left(n\binom{\mathbf{y}}{\tilde{j}}+n\binom{\mathbf{y}}{\tilde{\tilde{j}}}+1\right)\left[\delta\left(\omega-\Omega\binom{\mathbf{y}}{\tilde{j}}-\Omega\binom{\mathbf{y}}{\tilde{\tilde{j}}}\right)-\right.\right.$$

$$\left.-\delta\left(\omega+\Omega\binom{\mathbf{y}}{\tilde{j}}+\Omega\binom{\mathbf{y}}{\tilde{\tilde{j}}}\right)\right]+2\left(n\binom{\mathbf{y}}{\tilde{j}}-n\binom{\mathbf{y}}{\tilde{\tilde{j}}}\right)\delta\left(\omega+\Omega\binom{\mathbf{y}}{\tilde{j}}-\right.$$

$$\left.\left.-\Omega\binom{\mathbf{y}}{\tilde{\tilde{j}}}\right)\right\}+\frac{6\pi}{\hbar^2}\sum_{\substack{\mathbf{y}\mathbf{y}'\mathbf{y}''\\ \tilde{j}\tilde{\tilde{j}}\tilde{\tilde{\tilde{j}}}}}\left|B\begin{pmatrix}0 & \mathbf{y} & \mathbf{y}' & \mathbf{y}''\\ j & \tilde{j} & \tilde{\tilde{j}} & \tilde{\tilde{\tilde{j}}}\end{pmatrix}\right|^2\left\{N_1\left[\delta\left(\omega-\Omega\binom{\mathbf{y}}{\tilde{j}}-\Omega\binom{\mathbf{y}'}{\tilde{\tilde{j}}}-\Omega\binom{\mathbf{y}''}{\tilde{\tilde{\tilde{j}}}}\right)-\right.\right.$$

$$\left.-\delta\left(\omega+\Omega\binom{\mathbf{y}}{\tilde{j}}+\Omega\binom{\mathbf{y}'}{\tilde{\tilde{j}}}+\Omega\binom{\mathbf{y}''}{\tilde{\tilde{\tilde{j}}}}\right)\right]+3N_3\left[\delta\left(\omega+\Omega\binom{\mathbf{y}}{\tilde{j}}-\right.\right.$$

$$\left.\left.\left.-\Omega\binom{\mathbf{y}'}{\tilde{\tilde{j}}}-\Omega\binom{\mathbf{y}''}{\tilde{\tilde{\tilde{j}}}}\right)-\delta\left(\omega-\Omega\binom{\mathbf{y}}{\tilde{j}}+\Omega\binom{\mathbf{y}'}{\tilde{\tilde{j}}}+\Omega\binom{\mathbf{y}''}{\tilde{\tilde{\tilde{j}}}}\right)\right]\right\}, \qquad \text{(III.39)}$$

and N_1 and N_3 are defined by Eqs. (III.17).

3. Discussion

In the foregoing section, we have derived expressions for the Green's functions (III.32) and (III.33) in a crystal containing more than one optically active branch. On substituting these expressions in Eq. (III.1), we obtain the complex polarizability of such a crystal.

When there is one optically active branch, the expression for the complex polarizability becomes simpler, and in a cubic crystal of the NaCl type, it becomes Eq. (III.37). This expression is similar in form to the corresponding ones derived by Blackman [7] and by Neuberger and Hatcher [8] using classical methods. Howerver, the functions $\gamma_1(\omega)$ and $\gamma(\omega)$ are different in the classical and quantum cases.

Let us consider further the function $\gamma(\omega)$, which determines the width of the resonance. It consists of two parts, one of which depends on the third-order anharmonic terms in the potential energy, and the other on the fourth-order terms.

At sufficiently high temperatures (the classical case), we can put $n\binom{\mathbf{y}}{\tilde{j}}\approx\frac{kT}{\hbar\omega\binom{\mathbf{y}}{\tilde{j}}}$. Then, the first part is porportional to T, and the second part to T^2, and $\gamma(\omega)$ no longer depends on Planck's constant; see Eqs. (II.41) and (III.17).

The temperature dependence of $\gamma(\omega)$ is different at different frequencies, since the contribution of the cubic terms and of the fourth-order anharmonic terms to $\gamma(\omega)$ varies with the frequency. A numerical calculation of the frequency dependence $\gamma(\omega)$ in the classical case has been made in [8, 21, 22] for NaCl, LiF, and MgO. If only the cubic terms are taken into account, the curve of $\gamma(\omega)$ is, roughly speaking, one with two humps and a deep trough at a frequency close to ω_0. The high-frequency hump is mainly due to absorption processes which are accompanied by the creation of two phonons, and the low-frequency hump to processes in which one phonon is created and another absorbed.

In the region of the trough, i.e., for $\omega \approx \omega_0$, the contributions of the cubic terms and the fourth-order anharmonic terms to $\gamma(\omega)$ are similar in magnitude [21-23], and therefore $\gamma(\omega_0) \propto T^{\alpha}$, where $1 < \alpha < 2$. At frequencies in the region of the humps, $\gamma(\omega) \propto T$. These conclusions are in agreement with experiment [42, 43].

If the frequency ω satisfies the inequalities $3\,\omega_{\max} \gtrsim \omega > \left(\omega\binom{\mathbf{y}}{j} + \omega\binom{\mathbf{y}}{j'}\right)_{\max}$, then the cubic terms are no longer important in $\gamma(\omega)$; in this frequency range, $\gamma(\omega) \propto T^2$.

In calculating the temperature dependence of the optical constants k and n, it must be remembered that $\gamma(\omega)$ and $\chi(\omega)$ may depend on the temperature not only through the quantum numbers but also through the lattice constant, which is a function of temperature. The latter dependence is less important [22], and we shall therefore neglect it.

As already mentioned, a numerical calculation of the frequency dependence $\gamma(\omega)$ and of the optical constants $k(\omega)$ and $n(\omega)$ at sufficiently high temperatures has been made in [8, 21, 22] for NaCl, LiF, and MgO crystals.

In the calculation of $k(\omega)$ and $n(\omega)$ in [22], account was taken not only of the anharmonic terms in the potential energy but also of the deformation of the electron shells of the ions, which leads to the appearance of higher-order electric moments.

A comparison of the theoretical curves with the experimental ones obtained in [42-52] showed that theory and experiment are in good agreement for NaCl and LiF, and agree somewhat less well for MgO.

Let us now consider the behavior of $\gamma(\omega)$ at low temperatures, where quantum effects are important.

From the expression for $\gamma(\omega)$, it is seen to contain terms of two types. One type includes the terms which relate to phonon creation processes, and the other type includes those which contain a contribution from processes in which at least one phonon is absorbed. As the temperature decreases, the terms of the second type diminish much more rapidly than those of the first type, and vanish when T = 0.

As an example, let us consider the part of $\gamma(\omega)$ which depends on the third-order anharmonic terms. The terms of the first and second type in this part are respectively proportional to

$$\left(n\binom{\tilde{\mathbf{y}}}{j} + n\binom{\tilde{\tilde{\mathbf{y}}}}{j} + 1\right), \qquad \left(n\binom{\tilde{\mathbf{y}}}{j} - n\binom{\tilde{\tilde{\mathbf{y}}}}{j}\right).$$

To ascertain the physical significance of these expressions, we write them as

$$\begin{aligned}&\left(n\binom{\tilde{\mathbf{y}}}{j}+1\right)\left(n\binom{\tilde{\tilde{\mathbf{y}}}}{j}+1\right)-n\binom{\tilde{\mathbf{y}}}{j}n\binom{\tilde{\tilde{\mathbf{y}}}}{j},\\ &\left(n\binom{\tilde{\mathbf{y}}}{j}+1\right)n\binom{\tilde{\tilde{\mathbf{y}}}}{j}-n\binom{\tilde{\mathbf{y}}}{j}\left(n\binom{\tilde{\tilde{\mathbf{y}}}}{j}+1\right).\end{aligned} \tag{III.40}$$

This shows that they are proportional to the difference between the probabilities of forward and reverse processes. In the first expression, the forward process is the creation of two phonons and the reverse process is their absorption.

Using Eqs. (III.40) and (III.17), we can also represent $\gamma(\omega)$ as the difference of two expressions, one of which describes the absorption of a photon $\hbar\omega$ and the other its stimulated emission.

If all the numbers $n\left(\begin{smallmatrix} \mathbf{y} \\ j \end{smallmatrix}\right)$ have their thermodynamic equilibrium values, all the differences in Eqs. (III.40) and (III.17) are positive; hence, $\gamma(\omega) > 0$, i.e., the result is absorption of radiation.

When T = 0, the expression which describes the stimulated emission of photons becomes zero.

As already mentioned, the frequency dependence $\gamma(\omega)$ is determined mainly by the third-order anharmonic terms, and has a two-humped form at sufficiently high temperatures (in practice, at room temperature). Since the low-frequency hump is mainly due to processes in which phonons are created and absorbed, it should diminish considerably more quickly than the high-frequency hump as the temperature decreases. There have as yet been few measurements [44] of the temperature dependences of k and n at low temperatures, and so we cannot say that the result is actually in agreement with experiment, but it does agree with the results of [44].

So far, we have assumed that the phonons are in equilibrium. Then $\gamma(\omega) > 0$. It is possible to change the sign of this function at some particular frequency ω, and the most convenient value is the frequency which corresponds to the low-frequency maximum of the function $\gamma(\omega)$. Since the third-order anharmonic terms are predominant at this frequency, in order to change the sign of $\gamma(\omega)$, the expression $n\left(\begin{smallmatrix} \mathbf{y} \\ \tilde{j} \end{smallmatrix}\right) - n\left(\begin{smallmatrix} \mathbf{y} \\ \tilde{\tilde{j}} \end{smallmatrix}\right)$ must be negative in some range of wave vectors $\mathbf{y}$ $\left(\Omega\left(\begin{smallmatrix} \mathbf{y} \\ \tilde{\tilde{j}} \end{smallmatrix}\right) - \Omega\left(\begin{smallmatrix} \mathbf{y} \\ \tilde{j} \end{smallmatrix}\right) = \omega\right)$, and greater than the remaining positive terms in absolute magnitude. Such a change of sign could be brought about by means of systems generating monochromatic phonons. There are hopes that systems of this kind can be constructed; see the experiments on stimulated Raman scattering, and also [53, 54].

To conclude this section, the following are some remarks concerning the expressions for the Green's functions in a crystal having more than one optically active branch.

If the differences between the frequencies of the optically active branches are equal in order of magnitude to the frequencies themselves, and the functions $\Gamma_j(\nu, \omega)$, $\Gamma_{jj'}(\nu, \omega)$ are much less than these differences, we can neglect $4\omega\left(\begin{smallmatrix} \nu \\ j \end{smallmatrix}\right)\omega\left(\begin{smallmatrix} \nu \\ j' \end{smallmatrix}\right)\Gamma_{jj'}(\nu, \omega)\,\Gamma_{j'j}(\nu, \omega)$ in Eq. (III.33) and write

$$\frac{\mathcal{G}\left(\begin{smallmatrix} \nu & -\nu \\ j & j' \end{smallmatrix}\right)_\omega}{\mathcal{G}\left(\begin{smallmatrix} \nu & -\nu \\ j & j \end{smallmatrix}\right)_\omega} = \frac{2\Gamma'_{jj'}(\nu, \omega)\,\omega\left(\begin{smallmatrix} \nu \\ j' \end{smallmatrix}\right)}{\omega^2 - \omega^2\left(\begin{smallmatrix} \nu \\ j' \end{smallmatrix}\right) - 2\omega\left(\begin{smallmatrix} \nu \\ j' \end{smallmatrix}\right)\Gamma_{j'}(\nu, \omega)}. \tag{III.41}$$

Hence, we see that the ratio (III.41) is of the order to Γ/ω_0 for the frequency $\omega = \omega\left(\begin{smallmatrix} \nu \\ j \end{smallmatrix}\right)$, ω_0 denoting a frequency equal in order of magnitude to the optically active frequencies. If $\omega = \omega\left(\begin{smallmatrix} \nu \\ j' \end{smallmatrix}\right)$, the ratio (III.41) is of the order of unity, but at this frequency, the Green's function $\mathcal{G}\left(\begin{smallmatrix} \nu & -\nu \\ j' & j' \end{smallmatrix}\right)_\omega$ is large, and the ratio $\mathcal{G}\left(\begin{smallmatrix} \nu & -\nu \\ j & j' \end{smallmatrix}\right)_\omega \big/ \mathcal{G}\left(\begin{smallmatrix} \nu & -\nu \\ j' & j' \end{smallmatrix}\right)_\omega$ is small, i.e., of the order of Γ/ω_0. Thus, under the conditions discussed above, only the diagonal Green's functions need be taken into account in the first approximation in the expression for the complex polarizability, the contributions from the nondiagonal Green's functions being of the order of Γ/ω_0.

Kashcheev [24] has discussed infrared absorption in a crystal having more than one optically active branch. However, the nondiagonal Green's functions derived by him are inexact (as are the diagonal ones). The inaccuracy lies in neglecting the function $\langle\langle a_{0j}; a_{0j'}\rangle\rangle$ in the calculation of $G^{(2)}_{jj'}$ (in Kashcheev's notation); cf. our Eqs. (III.14) and Eq. (A.4) in [55], from which the expression for $G^{(2)}_{jj'}(\omega)$ was taken. Moreover, there should be a plus sign and not a minus sign between the two fractions in Eq. (37) of [24].

Kashcheev's paper [25] contains the same errors as [24].

CHAPTER IV

Absorption of Microwaves in an Ionic Crystal Lattice Containing Charged Defects

In this chapter, we shall calculate the imaginary part of the permittivity of a lattice containing charged defects. In the Hamiltonian (II.40), terms will be omitted which are of higher than the second degree in the operators b_+ and b_-, since the main component of the absorption, which is due to these terms, has been calculated in the previous chapter.

The imaginary part of the permittivity will be calculated for electric field frequencies much less than ω_0, the frequency at which infrared absorption occurs. This is done, firstly, because simple analytical expressions can be obtained only in this frequency range; secondly, because in this range there exist experimental data with which the theoretical results can be compared.

The lattice in which the defects are situated will be taken to be the NaCl type, and to simplify the calculations we shall assume that the lattice is isotropic for long-wave acoustic and optical vibrations. The polarization vectors and vibration frequencies for such a lattice are calculated in Appendix B.

The results of the calculations in this chapter do not depend on Planck's constant, and can therefore be derived by classical mechanics, but for uniformity's sake we shall use the quantum-mechanical methods discussed in the preceding chapters.

1. Equations for the Retarded Green's Functions

We shall assume for simplicity that the crystal spectrum has only one optically active vibration branch, denoted by the index j. It is seen from Eqs. (II.46)-(II.49) that, in order to find the complex polarizability, we have to calculate the following Green's functions:

$$G\begin{pmatrix}\mathbf{v} & -\mathbf{v}\\ j & j\end{pmatrix},\quad \Gamma\begin{pmatrix}\mathbf{v} & -\mathbf{v}\\ j & j\end{pmatrix},\quad G\begin{pmatrix}\mathbf{y} & -\mathbf{y}\\ \tilde{j} & \tilde{j}\end{pmatrix},\quad \Gamma\begin{pmatrix}\mathbf{y} & -\mathbf{y}\\ \tilde{j} & \tilde{j}\end{pmatrix},$$

$$G_{\substack{l\\k}}\begin{pmatrix}\mathbf{v} & \mathbf{y}\\ j & \tilde{j}\end{pmatrix},\quad \Gamma_{\substack{l\\k}}\begin{pmatrix}\mathbf{v} & \mathbf{y}\\ j & \tilde{j}\end{pmatrix},\quad G_{\substack{l\\k}}\begin{pmatrix}\mathbf{y} & -\mathbf{v}\\ \tilde{j} & j\end{pmatrix},\quad \Gamma_{\substack{l\\k}}\begin{pmatrix}\mathbf{y} & -\mathbf{v}\\ \tilde{j} & j\end{pmatrix}.$$

To do so, as in the previous chapter, we construct sequences of equations for the Green's functions and solve these approximately.

For the functions $G\begin{pmatrix}\mathbf{v} & -\mathbf{v}\\ j & j\end{pmatrix}$, $\Gamma\begin{pmatrix}\mathbf{v} & -\mathbf{v}\\ j & j\end{pmatrix}$, we obtain the equations

$$-i\hbar\,\frac{dG\begin{pmatrix}\mathbf{v} & -\mathbf{v}\\ j & j\end{pmatrix}}{dt} = \hbar\delta(t-\tau) + \left(\hbar\omega\begin{pmatrix}\mathbf{v}\\ j\end{pmatrix} + B\begin{pmatrix}-\mathbf{v} & \mathbf{v}\\ j & j\end{pmatrix}\right)G\begin{pmatrix}\mathbf{v} & -\mathbf{v}\\ j & j\end{pmatrix} + \qquad \text{(IV.1)}$$

$$+ B\begin{pmatrix}-\mathbf{v} & \mathbf{v}\\ j & j\end{pmatrix}\Gamma\begin{pmatrix}\mathbf{v} & -\mathbf{v}\\ j & j\end{pmatrix} + \sum_{\mathbf{y}\tilde{j}}{}' B\begin{pmatrix}-\mathbf{v} & \mathbf{y}\\ j & \tilde{j}\end{pmatrix}\left\{G\begin{pmatrix}\mathbf{y} & -\mathbf{v}\\ \tilde{j} & j\end{pmatrix} + \Gamma\begin{pmatrix}\mathbf{y} & -\mathbf{v}\\ \tilde{j} & j\end{pmatrix}\right\},$$

$$i\hbar\frac{d\Gamma\begin{pmatrix}\mathbf{v}-\mathbf{v}\\ j\ \ j\end{pmatrix}}{dt}=\left(\hbar\omega\begin{pmatrix}\mathbf{v}\\ j\end{pmatrix}+B\begin{pmatrix}-\mathbf{v}\,\mathbf{v}\\ j\,j\end{pmatrix}\right)\Gamma\begin{pmatrix}\mathbf{v}-\mathbf{v}\\ j\ \ j\end{pmatrix}+$$
$$+B\begin{pmatrix}-\mathbf{v}\,\mathbf{v}\\ j\,j\end{pmatrix}G\begin{pmatrix}\mathbf{v}-\mathbf{v}\\ j\ \ j\end{pmatrix}+\sum_{\mathbf{y}\tilde{j}}{}'B\begin{pmatrix}-\mathbf{v}\,\mathbf{y}\\ j\,\tilde{j}\end{pmatrix}\left\{G\begin{pmatrix}\mathbf{y}-\mathbf{v}\\ \tilde{j}\ \ j\end{pmatrix}+\Gamma\begin{pmatrix}\mathbf{y}-\mathbf{v}\\ \tilde{j}\ \ j\end{pmatrix}\right\}. \qquad \text{(IV.2)}$$

Having established the equations for the functions $G\begin{pmatrix}\mathbf{y}-\mathbf{v}\\ \tilde{j}\ \ j\end{pmatrix}$, $\Gamma\begin{pmatrix}\mathbf{y}-\mathbf{v}\\ \tilde{j}\ \ j\end{pmatrix}$, we approximate the right-hand sides in terms of the original functions $G\begin{pmatrix}\mathbf{v}-\mathbf{v}\\ j\ \ j\end{pmatrix}$, $\Gamma\begin{pmatrix}\mathbf{v}-\mathbf{v}\\ j\ \ j\end{pmatrix}$ and thus complete the sequence; doing so at this point corresponds to the Born approximation in scattering theory.

The equations for $G\begin{pmatrix}\mathbf{y}-\mathbf{v}\\ \tilde{j}\ \ j\end{pmatrix}$ and $\Gamma\begin{pmatrix}\mathbf{y}-\mathbf{v}\\ \tilde{j}\ \ j\end{pmatrix}$ are

$$-i\hbar\frac{dG\begin{pmatrix}\mathbf{y}-\mathbf{v}\\ \tilde{j}\ \ j\end{pmatrix}}{dt}=\hbar\Omega\begin{pmatrix}\mathbf{y}\\ \tilde{j}\end{pmatrix}G\begin{pmatrix}\mathbf{y}-\mathbf{v}\\ \tilde{j}\ \ j\end{pmatrix}+B\begin{pmatrix}-\mathbf{y}\,\mathbf{y}\\ \tilde{j}\,\tilde{j}\end{pmatrix}\Gamma\begin{pmatrix}\mathbf{y}-\mathbf{v}\\ \tilde{j}\ \ j\end{pmatrix}+B\begin{pmatrix}-\mathbf{y}\,\mathbf{v}\\ \tilde{j}\,j\end{pmatrix}\left\{G\begin{pmatrix}\mathbf{v}-\mathbf{v}\\ j\ \ j\end{pmatrix}+\Gamma\begin{pmatrix}\mathbf{v}-\mathbf{v}\\ j\ \ j\end{pmatrix}\right\}, \qquad \text{(IV.3)}$$

$$i\hbar\frac{d\Gamma\begin{pmatrix}\mathbf{y}-\mathbf{v}\\ \tilde{j}\ \ j\end{pmatrix}}{dt}=\hbar\Omega\begin{pmatrix}\mathbf{y}\\ \tilde{j}\end{pmatrix}\Gamma\begin{pmatrix}\mathbf{y}-\mathbf{v}\\ \tilde{j}\ \ j\end{pmatrix}+B\begin{pmatrix}-\mathbf{y}\,\mathbf{y}\\ \tilde{j}\,\tilde{j}\end{pmatrix}G\begin{pmatrix}\mathbf{y}-\mathbf{v}\\ \tilde{j}\ \ j\end{pmatrix}+B\begin{pmatrix}-\mathbf{y}\,\mathbf{v}\\ \tilde{j}\,j\end{pmatrix}\left\{G\begin{pmatrix}\mathbf{v}-\mathbf{v}\\ j\ \ j\end{pmatrix}+\Gamma\begin{pmatrix}\mathbf{v}-\mathbf{v}\\ j\ \ j\end{pmatrix}\right\}, \qquad \text{(IV.4)}$$

where

$$\Omega\begin{pmatrix}\mathbf{y}\\ \tilde{j}\end{pmatrix}=\omega\begin{pmatrix}\mathbf{y}\\ \tilde{j}\end{pmatrix}+\frac{B\begin{pmatrix}-\mathbf{y}\,\mathbf{y}\\ \tilde{j}\,\tilde{j}\end{pmatrix}}{\hbar},\quad B\begin{pmatrix}-\mathbf{y}\,\mathbf{y}\\ \tilde{j}\,\tilde{j}\end{pmatrix}$$

being a real quantity which gives a frequency correction of order n/N_0 (n is the number of defects per unit volume of the crystal, N_0 the number of unit cells per unit volume).

The Green's functions $G\begin{pmatrix}\mathbf{y}-\mathbf{y}\\ \tilde{j}\ \ \tilde{j}\end{pmatrix}$, $\Gamma\begin{pmatrix}\mathbf{y}-\mathbf{y}\\ \tilde{j}\ \ \tilde{j}\end{pmatrix}$ can be calculated less accurately than $G\begin{pmatrix}\mathbf{v}-\mathbf{v}\\ j\ \ j\end{pmatrix}$, $\Gamma\begin{pmatrix}\mathbf{v}-\mathbf{v}\\ j\ \ j\end{pmatrix}$, since, in the expression (II.46) for the complex polarizability, they are multiplied by the small quantity $\overline{\delta e_k^l \delta e_{k'}^{l'}} \propto \frac{n}{N_0}$. The sequence of equations for these functions can therefore be completed at the first stage. The equations are

$$-i\hbar\frac{dG\begin{pmatrix}\mathbf{y}-\mathbf{y}\\ \tilde{j}\ \ \tilde{j}\end{pmatrix}}{dt}=\hbar\delta(t-\tau)+\hbar\omega\begin{pmatrix}\mathbf{y}\\ \tilde{j}\end{pmatrix}G\begin{pmatrix}\mathbf{y}-\mathbf{y}\\ \tilde{j}\ \ \tilde{j}\end{pmatrix}, \qquad \text{(IV.5)}$$

$$i\hbar\frac{d\Gamma\begin{pmatrix}\mathbf{y}-\mathbf{y}\\ \tilde{j}\ \ \tilde{j}\end{pmatrix}}{dt}=\hbar\omega\begin{pmatrix}\mathbf{y}\\ \tilde{j}\end{pmatrix}\Gamma\begin{pmatrix}\mathbf{y}-\mathbf{y}\\ \tilde{j}\ \ \tilde{j}\end{pmatrix}. \qquad \text{(IV.6)}$$

Let us now construct the equations for the Green's functions $G_{\substack{l\\k}}\begin{pmatrix}\mathbf{y}\mathbf{y}'\\ jj'\end{pmatrix}$, $\Gamma_{\substack{l\\k}}\begin{pmatrix}\mathbf{y}\mathbf{y}'\\ jj'\end{pmatrix}$. Differentiation of these functions with respect to time gives

$$-i\hbar\frac{dG_{\substack{l\\k}}\begin{pmatrix}\mathbf{y}\mathbf{y}'\\ jj'\end{pmatrix}}{dt}=\hbar\omega\begin{pmatrix}\mathbf{y}\\ j\end{pmatrix}G_{\substack{l\\k}}\begin{pmatrix}\mathbf{y}\mathbf{y}'\\ jj'\end{pmatrix}+\sum_{\mathbf{y}''j''}\left\{\frac{\theta(t-\tau)}{i}\left\langle\delta e_k^l B\begin{pmatrix}-\mathbf{y}\mathbf{y}''\\ jj''\end{pmatrix}\exp\left[2\pi i(\mathbf{y}+\mathbf{y}')\mathbf{x}\begin{pmatrix}l\\k\end{pmatrix}\right]\left[b_+\begin{pmatrix}-\mathbf{y}''\\ j''\end{pmatrix},\ b'_-\begin{pmatrix}\mathbf{y}'\\ j'\end{pmatrix}\right]\right\rangle+\right.$$

$$+\frac{\theta(t-\tau)}{i}\left\langle \delta e^l_k B\begin{pmatrix}-\mathbf{y}\mathbf{y}''\\ jj''\end{pmatrix}\exp\left[2\pi i(\mathbf{y}+\mathbf{y}')\,\mathbf{x}\begin{pmatrix}l\\k\end{pmatrix}\right]\left[b_-\begin{pmatrix}\mathbf{y}''\\ j''\end{pmatrix},\ b'_-\begin{pmatrix}\mathbf{y}'\\ j'\end{pmatrix}\right]\right\rangle\Big\}, \quad \text{(IV.7)}$$

$$i\hbar\frac{d\Gamma_{\substack{l\\k}}\begin{pmatrix}\mathbf{y}\mathbf{y}'\\ jj'\end{pmatrix}}{dt}=\hbar\omega\begin{pmatrix}\mathbf{y}\\ j\end{pmatrix}\Gamma_{\substack{l\\k}}\begin{pmatrix}\mathbf{y}\mathbf{y}'\\ jj'\end{pmatrix}+\sum_{\mathbf{y}''j''}\Big\{\frac{\theta(t-\tau)}{i}\left\langle \delta e^l_k B\begin{pmatrix}-\mathbf{y}\mathbf{y}''\\ jj''\end{pmatrix}\exp\left[2\pi i(\mathbf{y}+\mathbf{y}')\,\mathbf{x}\begin{pmatrix}l\\k\end{pmatrix}\right]\left[b_+\begin{pmatrix}-\mathbf{y}''\\ j''\end{pmatrix},\ b'_-\begin{pmatrix}\mathbf{y}'\\ j'\end{pmatrix}\right]\right\rangle+$$

$$+\frac{\theta(t-\tau)}{i}\left\langle \delta e^l_k B\begin{pmatrix}-\mathbf{y}\mathbf{y}''\\ jj''\end{pmatrix}\exp\left[2\pi i(\mathbf{y}+\mathbf{y}')\,\mathbf{x}\begin{pmatrix}l\\k\end{pmatrix}\right]\left[b_-\begin{pmatrix}\mathbf{y}''\\ j''\end{pmatrix},\ b'_-\begin{pmatrix}\mathbf{y}'\\ j'\end{pmatrix}\right]\right\rangle\Big\}. \quad \text{(IV.8)}$$

The sums on the right-hand sides of these equations can be approximately expressed in terms of the functions $G\begin{pmatrix}-\mathbf{y}'\mathbf{y}'\\ j'j'\end{pmatrix}$, $\Gamma\begin{pmatrix}-\mathbf{y}'\mathbf{y}'\\ j'j'\end{pmatrix}$. To do so, we retain only the terms in which $\mathbf{y}''=-\mathbf{y}'$, $j''=j'$. We could also retain the term in which $\mathbf{y}''=\mathbf{y}$, $j''=j$. This makes a small correction to the frequencies. However, in the functions under consideration, we shall neglect these corrections, thus obtaining the equations

$$-i\hbar\frac{dG_{\substack{l\\k}}\begin{pmatrix}\mathbf{y}\mathbf{y}'\\ jj'\end{pmatrix}}{dt}=\hbar\omega\begin{pmatrix}\mathbf{y}\\ j\end{pmatrix}G_{\substack{l\\k}}\begin{pmatrix}\mathbf{y}\mathbf{y}'\\ jj'\end{pmatrix}+\left\langle \delta e^l_k B\begin{pmatrix}-\mathbf{y}&-\mathbf{y}'\\ j&j'\end{pmatrix}\exp\left[2\pi i(\mathbf{y}+\mathbf{y}')\,\mathbf{x}\begin{pmatrix}l\\k\end{pmatrix}\right]\left\{G\begin{pmatrix}-\mathbf{y}'\mathbf{y}'\\ j'j'\end{pmatrix}+\Gamma\begin{pmatrix}-\mathbf{y}'\mathbf{y}'\\ j'j'\end{pmatrix}\right\}\right\rangle, \quad \text{(IV.9)}$$

$$i\hbar\frac{d\Gamma_{\substack{l\\k}}\begin{pmatrix}\mathbf{y}\mathbf{y}'\\ jj'\end{pmatrix}}{dt}=\hbar\omega\begin{pmatrix}\mathbf{y}\\ j\end{pmatrix}\Gamma_{\substack{l\\k}}\begin{pmatrix}\mathbf{y}\mathbf{y}'\\ jj'\end{pmatrix}+\left\langle \delta e^l_k B\begin{pmatrix}-\mathbf{y}&-\mathbf{y}'\\ j&j'\end{pmatrix}\exp\left[2\pi i(\mathbf{y}+\mathbf{y}')\,\mathbf{x}\begin{pmatrix}l\\k\end{pmatrix}\right]\left\{G\begin{pmatrix}-\mathbf{y}'\mathbf{y}'\\ j'j'\end{pmatrix}+\Gamma\begin{pmatrix}-\mathbf{y}'\mathbf{y}'\\ j'j'\end{pmatrix}\right\}\right\rangle. \quad \text{(IV.10)}$$

2. Solution of the Equations for the Green's Functions. Expression for the Imaginary Part of the Complex Polarizability

To solve Eqs. (IV.1)-(IV.6), (IV.9), and (IV.10), we take Fourier transforms, using Eqs. (III.18) and (III.19).

Let us first consider Eqs. (IV.1)-(IV.4). Using Eqs. (IV.3) and (IV.4), we can express the functions $G\begin{pmatrix}\mathbf{y}&-\mathbf{v}\\ \tilde{j}&j\end{pmatrix}_\omega$ and $\Gamma\begin{pmatrix}\mathbf{y}&-\mathbf{v}\\ \tilde{j}&j\end{pmatrix}_\omega$ in terms of $G\begin{pmatrix}\mathbf{v}&-\mathbf{v}\\ j&j\end{pmatrix}_\omega$ and $\Gamma\begin{pmatrix}\mathbf{v}&-\mathbf{v}\\ j&j\end{pmatrix}_\omega$, neglecting the products $B\begin{pmatrix}-\mathbf{y}\,\mathbf{y}\\ \tilde{j}\tilde{j}\end{pmatrix}\Gamma\begin{pmatrix}\mathbf{y}&-\mathbf{v}\\ \tilde{j}&j\end{pmatrix}_\omega$, $B\begin{pmatrix}-\mathbf{y}\,\mathbf{y}\\ \tilde{j}\tilde{j}\end{pmatrix}G\begin{pmatrix}\mathbf{y}&-\mathbf{v}\\ \tilde{j}&j\end{pmatrix}_\omega$. The resulting expressions are substituted in the Fourier transforms of Eqs. (IV.1) and (IV.2), giving

$$-\omega G\begin{pmatrix}\mathbf{v}&-\mathbf{v}\\ j&j\end{pmatrix}_\omega=\frac{1}{2\pi}+\omega\begin{pmatrix}\mathbf{v}\\ j\end{pmatrix}G\begin{pmatrix}\mathbf{v}&-\mathbf{v}\\ j&j\end{pmatrix}_\omega+\Gamma_j(\mathbf{v},\omega)\left\{G\begin{pmatrix}\mathbf{v}&-\mathbf{v}\\ j&j\end{pmatrix}_\omega+\Gamma\begin{pmatrix}\mathbf{v}&-\mathbf{v}\\ j&j\end{pmatrix}_\omega\right\}, \quad \text{(IV.11)}$$

$$\omega\Gamma\begin{pmatrix}\mathbf{v}&-\mathbf{v}\\ j&j\end{pmatrix}_\omega=\omega\begin{pmatrix}\mathbf{v}\\ j\end{pmatrix}\Gamma\begin{pmatrix}\mathbf{v}&-\mathbf{v}\\ j&j\end{pmatrix}_\omega+\Gamma_j(\mathbf{v},\omega)\left\{G\begin{pmatrix}\mathbf{v}&-\mathbf{v}\\ j&j\end{pmatrix}_\omega+\Gamma\begin{pmatrix}\mathbf{v}&-\mathbf{v}\\ j&j\end{pmatrix}_\omega\right\}, \quad \text{(IV.12)}$$

where

$$\Gamma_j(\mathbf{v},\omega)=\frac{B\begin{pmatrix}\mathbf{v}&-\mathbf{v}\\ j&j\end{pmatrix}}{\hbar}+\frac{1}{\hbar^2}\sum_{\mathbf{y}\tilde{j}}{}'\left|B\begin{pmatrix}\mathbf{v}&-\mathbf{y}\\ j&\tilde{j}\end{pmatrix}\right|^2\left\{\frac{1}{\omega-\Omega\begin{pmatrix}\mathbf{y}\\ \tilde{j}\end{pmatrix}}-\frac{1}{\omega+\Omega\begin{pmatrix}\mathbf{y}\\ \tilde{j}\end{pmatrix}}\right\}. \quad \text{(IV.13)}$$

The prime to the sum signifies that $\binom{y}{\tilde{j}} \neq \binom{-\nu}{j}$. The function $\Gamma_j(\nu, \omega)$ has the properties (III.24) and (III.25). On deriving the functions $G\binom{\nu-\nu}{j\;\;j}_\omega$ and $\Gamma\binom{\nu-\nu}{j\;\;j}_\omega$ from Eqs. (IV.11) and (IV.12) and using Eqs. (II.47) and (III.20), we obtain for $\mathcal{G}\binom{\nu-\nu}{j\;\;j}_\omega$ the expression

$$\mathcal{G}\binom{\nu-\nu}{j\;\;j}_\omega = \frac{1}{\pi}\frac{\omega\binom{\nu}{j}}{\omega^2-\omega^2\binom{\nu}{j}-2\omega\binom{\nu}{j}\Gamma_j(\nu,\omega)}. \tag{IV.14}$$

Substitution of Eq. (IV.14) in Eq. (II.46) gives

$$\chi^{(1)}_{\alpha\beta}(\nu,\omega) = \frac{1}{v_a}\sum_{kk'}\frac{e_k e_{k'}\, w_\alpha\left(k\middle|\begin{smallmatrix}\nu\\ j\end{smallmatrix}\right) w_\beta\left(k'\middle|\begin{smallmatrix}-\nu\\ j\end{smallmatrix}\right)}{\sqrt{m_k m_{k'}}}\frac{1}{\omega^2\binom{\nu}{j}-\omega^2+2\omega\binom{\nu}{j}\Gamma_j(\nu,\omega)}. \tag{IV.15}$$

Let us now consider Eqs. (IV.5) and (IV.6). Taking Fourier transforms and using Eqs. (II.47) and (III.20), we have

$$\mathcal{G}\binom{y-y}{\tilde{j}\;\;\tilde{j}} = -\frac{1}{2\pi}\left\{\frac{1}{\Omega\binom{y}{\tilde{j}}-\omega}+\frac{1}{\Omega\binom{y}{\tilde{j}}+\omega}\right\}. \tag{IV.16}$$

Substitution of this in Eq. (II.46) gives

$$\chi^{(2)}_{\alpha\beta}(\nu,\omega) = \frac{1}{2v_a N}\sum_{\substack{l-l'\,y\\ kk'\tilde{j}}}\frac{w_\alpha\left(k\middle|\begin{smallmatrix}y\\ \tilde{j}\end{smallmatrix}\right) w_\beta\left(k'\middle|\begin{smallmatrix}-y\\ \tilde{j}\end{smallmatrix}\right)}{\sqrt{m_k m_{k'}}\,\omega\binom{y}{\tilde{j}}}\overline{\delta e^l_k \delta e^{l'}_{k'}}\exp\left[2\pi i\,(y-\nu)\,x\binom{ll'}{kk'}\right]\left\{\frac{1}{\Omega\binom{y}{\tilde{j}}-\omega}+\frac{1}{\Omega\binom{y}{\tilde{j}}+\omega}\right\}. \tag{IV.17}$$

We now have only to determine the Green's functions $\mathcal{G}_{\substack{l\\k}}\binom{\nu\;y}{j\;\tilde{j}}_\omega$, $\mathcal{G}_{\substack{l\\k}}\binom{y-\nu}{\tilde{j}\;\;j}_\omega$ and calculate $\chi^{(3)}_{\alpha\beta}(\nu,\omega)$, $\chi^{(4)}_{\alpha\beta}(\nu,\omega)$.

Using the Fourier transforms of Eqs. (IV.9) and (IV.10), and also Eqs. (II.47), (III.20), and (IV.16), we obtain

$$\mathcal{G}_{\substack{l\\k}}\binom{yy'}{jj'}_\omega = \frac{1}{2\pi\hbar}\left\langle \delta e^l_k B\binom{-y-y'}{j\;\;\;j'}\exp\left[2\pi i\,(y+y')\,x\binom{l}{k}\right]\right\rangle \times$$
$$\times\left\{\frac{1}{\omega-\omega\binom{y}{j}}-\frac{1}{\omega+\omega\binom{y}{j}}\right\}\left\{\frac{1}{\omega-\omega\binom{y'}{j'}}-\frac{1}{\omega+\omega\binom{y'}{j'}}\right\}. \tag{IV.18}$$

This function is symmetrical in the pairs of indices $\binom{y}{j}$ and $\binom{y'}{j'}$.

From Eqs. (IV.18), (II.46), and (II.41), we find

$$\chi^{(3)}_{\alpha\beta}(\nu,\omega) = -\frac{1}{4v_a N}\sum_{\substack{l',\,y\\ kk',\,\tilde{j}}}\frac{e_k w_\alpha\left(k\middle|\begin{smallmatrix}\nu\\ j\end{smallmatrix}\right) w_\beta\left(k'\middle|\begin{smallmatrix}y\\ \tilde{j}\end{smallmatrix}\right)}{\sqrt{m_k m_{k'}}\,\omega\binom{\nu}{j}\,\omega\binom{y}{\tilde{j}}}\left\langle \delta e^{l'}_{k'}\, A\binom{-\nu\;-y}{j\;\;\;\tilde{j}}\times\right.$$
$$\left.\times\exp\left[2\pi i\,(\nu+y)\,x\binom{l'}{k'}\right]\right\rangle\left\{\frac{1}{\omega-\omega\binom{\nu}{j}}-\frac{1}{\omega+\omega\binom{\nu}{j}}\right\}\left\{\frac{1}{\omega-\omega\binom{y}{\tilde{j}}}-\frac{1}{\omega+\omega\binom{y}{\tilde{j}}}\right\}, \tag{IV.19}$$

$$\chi^{(4)}_{\alpha\beta}(\nu,\omega) = -\frac{1}{4v_a N}\sum_{\substack{l'\mathrm{y}\\ kk'\tilde{j}}} \frac{e_k w_\alpha\left(k\left|\begin{matrix}\mathrm{y}\\ \tilde{j}\end{matrix}\right.\right) w_\beta\left(k'\left|\begin{matrix}-\nu\\ j\end{matrix}\right.\right)}{\sqrt{m_k m_{k'}}\,\omega\binom{\nu}{j}\omega\binom{\mathrm{y}}{\tilde{j}}} \left\langle \delta e^l_k A\begin{pmatrix}-\mathrm{y} & \nu\\ \tilde{j} & j\end{pmatrix}\times\right.$$

$$\left.\times\exp\left[2\pi i(\mathrm{y}-\nu)\,\mathrm{x}\binom{l}{k}\right]\right\rangle \left\{\frac{1}{\omega-\omega\binom{\nu}{j}}-\frac{1}{\omega+\omega\binom{\nu}{j}}\right\}\left\{\frac{1}{\omega-\omega\binom{\mathrm{y}}{\tilde{j}}}-\frac{1}{\omega+\omega\binom{\mathrm{y}}{\tilde{j}}}\right\}. \qquad \text{(IV.20)}$$

In the braces which contain the frequencies $\omega\binom{\nu}{j}$, we have omitted the terms that cause damping, since we are concerned with frequencies $\omega\ll\omega\binom{\nu}{j}$.

The expressions for $\chi^{1-4}_{\alpha\beta}$ can be combined into a compact form in the following way.

For a cubic crystal with randomly distributed point defects, the nonzero components of the tensors $\chi^{(1-4)}_{\alpha\beta}(\nu,\omega)$ as $\nu\to 0$ are $\chi^{(1-4)}_{xx}=\chi^{(1-4)}_{yy}=\chi^{(1-4)}_{zz}$. If the crystal contains linear defects along the z axis, the tensors $\chi^{(1-4)}_{\alpha\beta}$ for $\nu\to 0$ reduce to $\chi^{(1-4)}_{xx}=\chi^{(1-4)}_{yy}$, $\chi^{(1-4)}_{zz}$. Hence, we shall consider components of the tensors $\chi^{(1-4)}_{\alpha\beta}$ having the form $\chi^{(1-4)}_{\alpha\alpha}$.

In order to derive the imaginary parts of $\chi^{(1-4)}_{\alpha\alpha}$, in which we are principally interested, we must make the substitution $\omega\to\omega+i\delta$ $(\delta>0,\ \delta\to 0)$ in Eqs. (IV.15), (IV.17), (IV.19), and (IV.20). Using the relation

$$\frac{1}{\omega+i\delta\pm\Omega\binom{\mathrm{y}}{j}} = p\frac{1}{\omega\pm\Omega\binom{\mathrm{y}}{j}} \mp i\pi\delta\left(\omega\pm\Omega\binom{\mathrm{y}}{j}\right),$$

Eq. (II.40), and the inequalities

$$\omega\binom{\nu}{j}\gg\omega,\quad \omega^2\binom{\nu}{j}\gg 2\omega\binom{\nu}{j}\Gamma_j(\nu,\omega),$$

we get for the imaginary part of the complete polarizability tensor

$$\operatorname{Im}\chi_{\alpha\alpha}(\nu,\omega) = \frac{\pi}{2v_a}\sum_{\mathrm{y}\tilde{j}}\left\langle\left|\frac{1}{N}\sum_{lk}\frac{w_\alpha\left(k\left|\begin{matrix}\mathrm{y}\\ \tilde{j}\end{matrix}\right.\right)}{\sqrt{m_k}}\delta e^l_k\exp\left[2\pi i(\mathrm{y}-\nu)\,\mathrm{x}\binom{l}{k}\right]-\right.\right.$$

$$\left.\left.-\sum_k\frac{e_k w_\alpha\left(k\left|\begin{matrix}\nu\\ j\end{matrix}\right.\right)}{\sqrt{m_k}}\frac{A\begin{pmatrix}-\nu & \mathrm{y}\\ j & \tilde{j}\end{pmatrix}}{\omega_0^2-\omega^2}\right|^2\right\rangle\frac{1}{\omega\binom{\mathrm{y}}{\tilde{j}}}\delta\left(\omega-\omega\binom{\mathrm{y}}{\tilde{j}}\right), \qquad \text{(IV.21)}$$

where the notation $\omega\binom{\nu}{j}=\omega_0$ is used for the dispersion frequency.

The influence of the defects on the real part of the polarizability will be neglected and assumed to be the same as for an ideal lattice.

As already mentioned, in the derivation of Eq. (IV.21), the scattering of lattice waves by defects is treated in the Born approximation. The validity of this approximation will be discussed later.

In the next two sections we shall calculate $A\begin{pmatrix} \nu & -\mathbf{y} \\ j & \tilde{j} \end{pmatrix}$ for the case where the crystal contains charged defects.

3. Calculation of the Coefficient $\delta\Phi\begin{pmatrix} \nu & -\mathbf{y} \\ j & \tilde{j} \end{pmatrix}$

It is seen from Eqs. (II.15) that the coefficient $A\begin{pmatrix} \nu & -\mathbf{y} \\ j & \tilde{j} \end{pmatrix}$ consists of several terms. In this section, we shall calculate the first term, $\delta\Phi\begin{pmatrix} \nu & -\mathbf{y} \\ j & \tilde{j} \end{pmatrix}$, which does not depend on the displacements of the ions near the defects. In the calculations, we shall make use of the fact that, when ω is in the range $\omega \ll \omega_0$, the lattice vibrations denoted by the indices $\begin{pmatrix} \mathbf{y} \\ \tilde{j} \end{pmatrix}$ are long-wave acoustic vibrations; see Eq. (IV.21).

The change of potential energy of the lattice due to the presence of charged defects is

$$\delta\Phi = \frac{1}{2} \sum_{\binom{l}{k} \neq \binom{l'}{k''}} \frac{e_k\, \delta e_{k'}^{l'} + \delta e_k^l\, e_{k'} + \delta e_k^l\, \delta e_{k'}^{l'}}{\left| \mathbf{x}\binom{l}{k} + \mathbf{u}\binom{l}{k} - \mathbf{x}\binom{l'}{k'} - \mathbf{u}\binom{l'}{k'} \right|}. \tag{IV.22}$$

We shall suppose that the charged defects occupy the same positions as the host lattice ions.

The second derivative of $\delta\Phi$ with respect to the displacements $u_\alpha\binom{l}{k}$, $u_\beta\binom{l'}{k'}$, when $\binom{l}{k} \neq \binom{l'}{k'}$, is

$$\delta\Phi_{\alpha\beta}\binom{ll'}{kk'}_0 = -\left(e_k\, \delta e_{k'}^{l'} + \delta e_k^l\, e_{k'} + \delta e_k^l\, \delta e_{k'}^{l'}\right) D_{\alpha\beta}\binom{ll'}{kk'}, \tag{IV.23}$$

where

$$D_{\alpha\beta}\binom{ll'}{kk'} = \frac{3x_\alpha\binom{ll'}{kk'} x_\beta\binom{ll'}{kk'} - \mathbf{x}^2\binom{ll'}{kk'} \delta_{\alpha\beta}}{\left|\mathbf{x}\binom{ll'}{kk'}\right|^5}. \tag{IV.24}$$

In Eq. (IV.22), the terms such as $e_k \delta e_{k'}^{l'}$ take account of the interaction of the defect with the ions of the host lattice (more precisely, the interaction with them of the increment of charge which results from the presence of the defect) and the terms such as $\delta e_k^l \delta e_{k'}^{l'}$ take account of the interaction between the defects.

In calculating $\delta\Phi\begin{pmatrix} \nu & -\mathbf{y} \\ j & \tilde{j} \end{pmatrix}$, we shall neglect the interaction between the defects; the justification of this for the other terms in $A\begin{pmatrix} \nu & -\mathbf{y} \\ j & \tilde{j} \end{pmatrix}$ will be given in the next section.

The cases of point and linear charged defects will be discussed separately.

For point charged defects, the ratio of the quantities $\delta\Phi_{\alpha\beta}$ which result from the interaction between defects and the interaction between the defect and one of the nearest ions is, in order of magnitude,

$$\frac{\delta\Phi_{\alpha\beta}^{(\delta e^2)}}{\delta\Phi_{\alpha\beta}^{(\delta e \cdot e)}} \approx \frac{\delta e}{e} \frac{n_p}{N_0}, \tag{IV.25}$$

where n_p is the concentration of point defects, N_0 the number of unit cells per unit volume. Since n_p/N_0 is assumed small, we see that the interaction between defects may be neglected.

Next, let us consider linear charged defects, i.e., lines along which charge increments of equal magnitude are situated at uniform distances apart.

If the charge increments δe are on different lines, the ratio (IV.25) becomes

$$\frac{\delta\Phi_{\alpha\beta}^{(\delta e^2)}}{\delta\Phi_{\alpha\beta}^{(e\delta e)}} \approx \frac{\delta e}{e}\frac{n_l^{3/2}}{N_0}, \tag{IV.26}$$

where n_l is the number of charged lines per cm^2.

Since the ratio $n_l^{3/2}/N_0$ is assumed small, the interaction between charge increments δe on different lines may be neglected.

If the charge increments δe are on the same line, and can be at a distance apart which is of the order of the lattice constant, the arguments given above are invalid, but in this case also it can be shown that the interaction is negligible. To do so, we take the first Eq. (A.4), and replace Φ by $\delta\Phi$, $\binom{\mathbf{y}}{j}$ by $\binom{\boldsymbol{\nu}}{j}$, and $\binom{\mathbf{y}'}{j'}$ by $\binom{-\mathbf{y}}{\tilde{j}}$. Since in our case $\delta\Phi_{\alpha\beta}\binom{ll'}{kk'}_0 \propto \dfrac{\delta e^2}{x\binom{ll'}{kk'}^3}$, and the summation in (A.4) is along the line, the main contribution to the sum comes from small $l - l'$ (unlike the case where a three-dimensional summation is necessary and the terms with large $l - l'$ also give a substantial contribution). Since $l - l'$ is small and the magnitude of the wave vector $\mathbf{y}$ is small compared with $1/a$ (where a is the shortest distance between ions), in the second parenthesis in Eq. (A.4) we can expand the exponent in series, obtaining

$$\frac{w_\beta\left(k\middle|\begin{smallmatrix}-\mathbf{y}\\ \tilde{j}\end{smallmatrix}\right)}{\sqrt{m_k}} - \frac{w_\beta\left(k\middle|\begin{smallmatrix}-\mathbf{y}\\ \tilde{j}\end{smallmatrix}\right)}{\sqrt{m_{k'}}}\exp\left[2\pi i\,\mathbf{y}\mathbf{x}\binom{ll'}{kk'}\right] \approx \left(\frac{w_\beta\left(k\middle|\begin{smallmatrix}-\mathbf{y}\\ \tilde{j}\end{smallmatrix}\right)}{\sqrt{m_k}} - \frac{w_\beta\left(k'\middle|\begin{smallmatrix}-\mathbf{y}\\ \tilde{j}\end{smallmatrix}\right)}{\sqrt{m_{k'}}}\right) + \frac{w_\beta\left(k'\middle|\begin{smallmatrix}-\mathbf{y}\\ \tilde{j}\end{smallmatrix}\right)}{\sqrt{m_{k'}}}\,2\pi i\,\mathbf{y}\mathbf{x}\binom{ll'}{kk'}.$$

For long-wave acoustic vibrations, the parenthesis is proportional to $a\mathbf{y}$ if the lattice has no center of symmetry, and to $(a\mathbf{y})^2$ if there is a center of symmetry [see (B.37)]. Thus, for the terms in $\delta\Phi\binom{\boldsymbol{\nu}\ -\mathbf{y}}{j\ \ \tilde{j}}$, which take account of the interaction of charge increments δe on the same line, we find that they are proportional to $a\mathbf{y}$, whereas the leading terms arising from the interaction between the charge increments and the host ions are proportional to unity (see below).

Thus, we can conclude that the interaction between defects is negligible for both point and linear defects. Using Eqs. (A.10) and (IV.23), omitting in the latter the products $\delta e_k^l \delta e_{k'}^{l'}$, we obtain

$$\begin{aligned}
\delta\Phi\binom{\boldsymbol{\nu}\ -\mathbf{y}}{j\ \ \ \tilde{j}} = {}& \frac{1}{N}\sum_{\substack{l'kk'\\ \alpha\beta}} e_k\,\delta e_{k'}^{l'}\frac{w_\alpha\left(k\middle|\begin{smallmatrix}\boldsymbol{\nu}\\ j\end{smallmatrix}\right)}{\sqrt{m_k}}\left\{\sum_{l-l'} D_{\alpha\beta}\binom{ll'}{kk'}\exp\left[2\pi i(\boldsymbol{\nu}-\mathbf{y})\,\mathbf{x}\binom{ll'}{kk'}\right]\right\}\times\\
&\times\frac{w_\beta\left(k\middle|\begin{smallmatrix}-\mathbf{y}\\ \tilde{j}\end{smallmatrix}\right)}{\sqrt{m_k}}\exp\left[2\pi i(\boldsymbol{\nu}-\mathbf{y})\,\mathbf{x}\binom{l'}{k'}\right] - \frac{1}{N}\sum_{\substack{l'kk'\\ \alpha\beta}} e_k\delta e_{k'}^{l'}\frac{w_\alpha\left(k\middle|\begin{smallmatrix}\boldsymbol{\nu}\\ j\end{smallmatrix}\right)}{\sqrt{m_k}}\left\{\sum_{l-l'} D_{\alpha\beta}\binom{ll'}{kk'}\exp\left[2\pi i\,\boldsymbol{\nu}\mathbf{x}\binom{ll'}{kk'}\right]\right\}\frac{w_\beta\left(k'\middle|\begin{smallmatrix}-\mathbf{y}\\ \tilde{j}\end{smallmatrix}\right)}{\sqrt{m_{k'}}}\times\\
&\times\exp\left[2\pi i(\boldsymbol{\nu}-\mathbf{y})\,\mathbf{x}\binom{l'}{k'}\right] + \frac{1}{N}\sum_{\substack{l\,kk'\\ \alpha\beta}} \delta e_k^l\, e_{k'}\frac{w_\alpha\left(k\middle|\begin{smallmatrix}\boldsymbol{\nu}\\ j\end{smallmatrix}\right)}{\sqrt{m_k}}\left\{\sum_{l-l'} D_{\alpha\beta}\binom{ll'}{kk'}\right\}\frac{w_\beta\left(k\middle|\begin{smallmatrix}-\mathbf{y}\\ \tilde{j}\end{smallmatrix}\right)}{\sqrt{m_k}}\exp\left[2\pi i(\boldsymbol{\nu}-\mathbf{y})\,\mathbf{x}\binom{l}{k}\right] -\\
&-\frac{1}{N}\sum_{\substack{l\,kk'\\ \alpha\beta}} \delta e_k^l\, e_{k'}\frac{w_\alpha\left(k\middle|\begin{smallmatrix}\boldsymbol{\nu}\\ j\end{smallmatrix}\right)}{\sqrt{m_k}}\left\{\sum_{l-l'} D_{\alpha\beta}\binom{ll'}{kk'}\exp\left[2\pi i\,\mathbf{y}\mathbf{x}\binom{ll'}{kk'}\right]\right\}\frac{w_\beta\left(k'\middle|\begin{smallmatrix}-\mathbf{y}\\ \tilde{j}\end{smallmatrix}\right)}{\sqrt{m_{k'}}}\exp\left[2\pi i(\boldsymbol{\nu}-\mathbf{y})\,\mathbf{x}\binom{l}{k}\right].
\end{aligned} \tag{IV.27}$$

If the wavelength $\lambda = 1/y$ satisfies the inequalities

$$R \gg \lambda \gg a, \tag{IV.28}$$

where R is the size of the sample, then the dipole sum is given by the expression (see [11, 56, 57])

$$\sum_{l-l'} D_{\alpha\beta}\binom{ll'}{kk'} \exp\left[2\pi i\, \mathbf{y}\mathbf{x}\binom{ll'}{kk'}\right] = \frac{1}{v_a}\left\{f_{\alpha\beta}(kk') - \frac{4\pi y_\alpha y_\beta}{|\mathbf{y}|^2}\right\}. \tag{IV.29}$$

For a cubic crystal, $f_{\alpha\beta}(kk') = \frac{4\pi}{3}\delta_{\alpha\beta}$. If y = 0, then

$$\sum_{l-l'} P_{\alpha\beta}\binom{ll'}{kk'} = \frac{1}{v_a}\{f_{\alpha\beta}(kk') - N_{\alpha\beta}\}, \tag{IV.30}$$

where $N_{\alpha\beta}$ defines the depolarizing field in the crystal, and therefore depends on the shape of the sample.

For long-wave acoustic vibrations, the quantity $\frac{w_\beta\left(k'\middle|\begin{smallmatrix}-\mathbf{y}\\ \tilde{j}\end{smallmatrix}\right)}{\sqrt{m_{k'}}}$ varies only slightly with the ion number k', and we can therefore replace $\frac{w_\beta\left(k'\middle|\begin{smallmatrix}-\mathbf{y}\\ \tilde{j}\end{smallmatrix}\right)}{\sqrt{m_{k'}}}$ by $\frac{w_\beta\left(k\middle|\begin{smallmatrix}-\mathbf{y}\\ \tilde{j}\end{smallmatrix}\right)}{\sqrt{m_k}}$ in the last term in Eq. (IV.27). The error in so doing is proportional to $a\mathbf{y}$, or $(a\mathbf{y})^2$ if the crystal has a center of symmetry [see (B.37)].

After this change, it is seen that the quantities $f_{\alpha\beta}$ (kk') which determine the local fields cancel in the last two terms in Eq. (IV.27), and the remaining expressions in these terms vanish because of the neutrality condition

$$\sum_{k'} e_{k'} = 0. \tag{IV.31}$$

In the second term in Eq. (IV.27), we again replace $\frac{w_\beta\left(k'\middle|\begin{smallmatrix}-\mathbf{y}\\ \tilde{j}\end{smallmatrix}\right)}{\sqrt{m_{k'}}}$ by $\frac{w_\beta\left(k\middle|\begin{smallmatrix}-\mathbf{y}\\ \tilde{j}\end{smallmatrix}\right)}{\sqrt{m_k}}$. If the wavelength of the vibrations $\binom{\mathbf{v}}{j}$ and $\binom{\mathbf{y}}{\tilde{j}}$ is less than the size of the samples, we can obtain from Eq. (IV.29)

$$\delta\Phi\begin{pmatrix}\mathbf{v} & -\mathbf{y}\\ j & \tilde{j}\end{pmatrix} = \frac{4\pi}{Nv_a}\sum_{l'k'}\delta e_{k'}^{l'} \exp\left[2\pi i(\mathbf{v}-\mathbf{y})\,\mathbf{x}\binom{l'}{k'}\right]\left[\sum_{k\alpha\beta}\frac{e_k w_\alpha\left(k\middle|\begin{smallmatrix}\mathbf{v}\\ j\end{smallmatrix}\right)}{\sqrt{m_k}}\left\{\frac{v_\alpha v_\beta}{v^2} - \frac{(\mathbf{y}-\mathbf{v})_\alpha(\mathbf{y}-\mathbf{v})_\beta}{|\mathbf{y}-\mathbf{v}|^2}\right\}\frac{w_\beta\left(k\middle|\begin{smallmatrix}-\mathbf{y}\\ \tilde{j}\end{smallmatrix}\right)}{\sqrt{m_k}}\right]. \tag{IV. 32}$$

The quantity $E'_\beta(\mathbf{v}) = -\frac{4\pi}{v_a}\sum_{k\alpha}\frac{e_k w_\alpha\left(k\middle|\begin{smallmatrix}\mathbf{v}\\ j\end{smallmatrix}\right)}{\sqrt{m_k}}\frac{v_\alpha v_\beta}{v^2}$ is the part of the macroscopic electric field that is due to the lattice itself. Since we have from the start taken account of the interaction between the crystal and this field by means of Eq. (II.30), the field has to be separated from Eq. (IV.32), after which we have

$$\delta\Phi\begin{pmatrix}\mathbf{v}-\mathbf{y}\\ j\ \ \tilde{}\end{pmatrix}=\frac{1}{N}\sum_{l'k'}\delta e_{k'}^{l'}\exp\left[2\pi i(\mathbf{v}-\mathbf{y})\mathbf{x}\begin{pmatrix}l'\\k'\end{pmatrix}\right]\left\{-\frac{4\pi}{v_a}\sum_{k\alpha\beta}\frac{e_k w_\alpha\left(k\middle|\begin{smallmatrix}\mathbf{v}\\ j\end{smallmatrix}\right)}{\sqrt{m_k}}\frac{(\mathbf{y}-\mathbf{v})_\alpha(\mathbf{y}-\mathbf{v})_\beta}{|\mathbf{y}-\mathbf{v}|^2}\frac{w_\beta\left(k\middle|\begin{smallmatrix}-\mathbf{y}\\ \tilde{j}\end{smallmatrix}\right)}{\sqrt{m_k}}\right\}. \quad \text{(IV.33)}$$

Since the vibrations $\begin{pmatrix}\mathbf{v}\\ j\end{pmatrix}$ and $\begin{pmatrix}\mathbf{y}\\ \tilde{j}\end{pmatrix}$ have the same frequency [see Eq. (IV.21)], it follows that $\frac{\nu}{y}=\frac{v}{c}\approx 10^{-5}$, where v is the velocity of sound and c the velocity of light. Thus, ν can be neglected in comparison with y in Eq. (IV.33).

Two important properties of Eq. (IV.33) are that it does not depend on the local field [i.e., on $f_{\alpha\beta}$ (kk')], and that it depends on the direction but not the magnitude of **y**. The latter property is due to the long range of the Coulomb forces.

4. Calculation of $\sum_{j''}\Phi\begin{pmatrix}\mathbf{v}-\mathbf{y}\ \mathbf{y}-\mathbf{v}\\ j\ \ \tilde{j}\ \ \ j''\end{pmatrix}_0 q\begin{pmatrix}\mathbf{y}-\mathbf{v}\\ j''\end{pmatrix}$

Let us now consider the next significant term in $A\begin{pmatrix}\mathbf{v}-\mathbf{y}\\ j\ \ \tilde{j}\end{pmatrix}$. This is

$$\sum_{\mathbf{y}''j''}\Phi\begin{pmatrix}\mathbf{v}-\mathbf{y}\mathbf{y}''\\ j\ \ \tilde{j}\ j''\end{pmatrix}_0 q\begin{pmatrix}\mathbf{y}''\\ j''\end{pmatrix}$$

and is due to the deformation of the lattice near the defects.

Since the coefficient $\Phi\begin{pmatrix}\mathbf{v}-\mathbf{y}\mathbf{y}''\\ j\ \ \tilde{j}\ j''\end{pmatrix}_0$ is unchanged under translation through a lattice vector, we have $\boldsymbol{\nu}-\mathbf{y}+\mathbf{y}''=\mathbf{K}_n$, where $\mathbf{K}_n$ is a reciprocal-lattice vector. Since $\boldsymbol{\nu}$ and **y** are small, and **y''** must lie in the first Brillouin zone, we have $\mathbf{K}_n = 0$, and therefore $\mathbf{y}''=\mathbf{y}-\boldsymbol{\nu}$. The term in question therefore reduces to

$$\sum_{j''}\Phi\begin{pmatrix}\mathbf{v}-\mathbf{y}\ \mathbf{y}-\mathbf{v}\\ j\ \ \tilde{j}\ \ \ j''\end{pmatrix}_0 q\begin{pmatrix}\mathbf{y}-\mathbf{v}\\ j''\end{pmatrix}.$$

We first calculate $q\begin{pmatrix}\mathbf{y}-\mathbf{v}\\ j''\end{pmatrix}$. For this, it is necessary to solve Eq. (II.4). Taking the Fourier transform by means of Eq. (II.7), we obtain from Eq. (II.4)

$$0=\delta\Phi\begin{pmatrix}\mathbf{y}\\ j\end{pmatrix}_0+\omega^2\begin{pmatrix}\mathbf{y}\\ j\end{pmatrix}q\begin{pmatrix}-\mathbf{y}\\ j\end{pmatrix}+\sum_{\mathbf{y}'j'}\delta\Phi\begin{pmatrix}\mathbf{y}\mathbf{y}'\\ jj'\end{pmatrix}_0 q\begin{pmatrix}\mathbf{y}'\\ j'\end{pmatrix}+\frac{1}{2}\sum_{\substack{\mathbf{y}'\mathbf{y}''\\ j'j''}}\left\{\Phi\begin{pmatrix}\mathbf{y}\mathbf{y}'\mathbf{y}''\\ jj'j''\end{pmatrix}_0+\delta\Phi\begin{pmatrix}\mathbf{y}\mathbf{y}'\mathbf{y}''\\ jj'j''\end{pmatrix}_0\right\}q\begin{pmatrix}\mathbf{y}'\\ j'\end{pmatrix}q\begin{pmatrix}\mathbf{y}''\\ j''\end{pmatrix}+\ldots, \quad \text{(IV. 34)}$$

where

$$\delta\Phi\begin{pmatrix}\mathbf{y}\\ j\end{pmatrix}_0=\frac{1}{\sqrt{N}}\sum_{lk\alpha}\frac{\delta\Phi_\alpha\begin{pmatrix}l\\k\end{pmatrix}_0 w_\alpha\left(k\middle|\begin{smallmatrix}\mathbf{y}\\ j\end{smallmatrix}\right)}{\sqrt{m_k}}\exp\left[2\pi i\mathbf{y}\mathbf{x}\begin{pmatrix}l\\k\end{pmatrix}\right]. \quad \text{(IV.35)}$$

For charged defects,

$$\delta\Phi_\alpha\begin{pmatrix}l\\k\end{pmatrix}_0=-\sum_{\substack{l'k'\\ \left(\begin{smallmatrix}l'\\k'\end{smallmatrix}\right)\neq\left(\begin{smallmatrix}l\\k\end{smallmatrix}\right)}}\left(e_k\delta e_{k'}^{l'}+\delta e_k^l e_{k'}+\delta e_k^l\delta e_{k'}^{l'}\right)\frac{x_\alpha\begin{pmatrix}ll'\\kk'\end{pmatrix}}{\left|\mathbf{x}\begin{pmatrix}ll'\\kk'\end{pmatrix}\right|^3}. \quad \text{(IV.36)}$$

Substitution of Eq. (IV.36) in Eq. (IV.35) gives

$$\delta\Phi\binom{\mathbf{y}}{j}_0 = -\frac{1}{\sqrt{N}} \sum_{\binom{l}{k}\neq\binom{l'}{k'}\,\alpha}' (e_k\delta e_{k'}^{l'} + \delta e_k^l e_{k'} + \delta e_k^l \delta e_{k'}^{l'}) \frac{w_\alpha\left(k\middle|\begin{matrix}\mathbf{y}\\ j\end{matrix}\right) x_\alpha\binom{ll'}{kk'}}{\sqrt{m_k}\left|\mathbf{x}\binom{ll'}{kk'}\right|^3} \exp\left[2\pi i\mathbf{y}\mathbf{x}\binom{l}{k}\right]. \quad \text{(IV.37)}$$

We shall consider the case where the lattice and the defect possess central symmetry. Then, the term $\delta e_k^l\, e_{k'}$ gives zero on summation over l'. Thus,

$$\delta\Phi\binom{\mathbf{y}}{j}_0 = -\frac{1}{\sqrt{N}} \sum_{\binom{l}{k}\neq\binom{l'}{k'}\,\alpha}' (e_k\delta e_{k'}^{l'} + \delta e_k^l \delta e_{k'}^{l'}) \frac{w_\alpha\left(k\middle|\begin{matrix}\mathbf{y}\\ j\end{matrix}\right) x_\alpha\binom{ll'}{kk'}}{\sqrt{m_k}\left|\mathbf{x}\binom{ll'}{kk'}\right|^3} \exp\left[2\pi i\mathbf{y}\mathbf{x}\binom{l}{k}\right]. \quad \text{(IV.38)}$$

The first term in the parentheses corresponds to the interaction between the charged defects and the lattice ions, and the second term to the interaction between defects, which we shall neglect. This is allowable both for the interaction of point defects and for that of two charged lines, because of the small density in each case.

For point defects, the ratio of the terms in Eq. (IV.38) which take account of the interaction between charged defects and those which take account of the interaction between a defect and the lattice ions is, in order of magnitude,

$$\frac{\delta\Phi^{(\delta e^2)}}{\delta\Phi^{(e\delta e)}} \approx \frac{\delta e}{e}\left(\frac{n_p}{N_0}\right)^{2/3}. \quad \text{(IV.39)}$$

The corresponding ratio for charge increments δe on different lines is

$$\frac{\delta\Phi^{(\delta e^2)}}{\delta\Phi^{(e\delta e)}} \approx \frac{\delta e}{e}\frac{n_l}{N_0^{2/3}}. \quad \text{(IV.40)}$$

Both these ratios will be assumed small.

The case of interaction between charges δe on the same line will be considered later, and we shall show that this interaction too is negligible.

Let us now calculate the part of $\delta\Phi\binom{\mathbf{y}}{j}_0$, which depends on the interaction between the defects and the host ions. To do so, we must evaluate the sum

$$\sum_{\substack{l-l'\\ \alpha}} \frac{x_\alpha\binom{ll'}{kk'} w_\alpha\left(k\middle|\begin{matrix}\mathbf{y}\\ j\end{matrix}\right)}{\left|\mathbf{x}\binom{ll'}{kk'}\right|^3} \exp\left[2\pi i\mathbf{y}\mathbf{x}\binom{ll'}{kk'}\right] = S(\mathbf{y}). \quad \text{(IV.41)}$$

The summation will be replaced by an integration; the conditions for the validity of this change will be derived below.

Let $\mathbf{y}$ be along the z axis; $\mathbf{w}\left(k\middle|\begin{matrix}\mathbf{y}\\ j\end{matrix}\right)$ at an angle θ_1 to $\mathbf{y}$ and in the xz plane. Using spherical coordinates for $\mathbf{x}\binom{ll'}{kk'}$, we find

$$S(\mathbf{y}) = \frac{1}{v_a}\int_0^{2\pi}\int_0^{\pi}\int_{a'}^{R} W\left(k\left|\begin{smallmatrix}\mathbf{y}\\ j\end{smallmatrix}\right.\right)(\sin\theta_1\sin\theta\cos\varphi + \cos\theta_1\cos\theta)\exp(2\pi iyx\cos\theta)\,dx\sin\theta\,d\theta\,d\varphi. \quad \text{(IV.42)}$$

We have taken the limits of integration over x as a', equal in order of magnitude to the distance between nearest neighbors, and R, the size of the sample. Since we know only the order of magnitude of both limits, the change from summation to integration is justified for values of $|\mathbf{y}|$ such that the result is found to be independent of a' and R.

On carrying out the integration over φ, θ, and x in Eq. (IV.42), we find

$$S(\mathbf{y}) = -\frac{2i\left(\mathbf{w}\left(k\left|\begin{smallmatrix}\mathbf{y}\\ j\end{smallmatrix}\right.\right)\mathbf{y}\right)}{v_a y^2}\left\{\frac{\sin 2\pi yR}{2\pi yR} - \frac{\sin 2\pi ya'}{2\pi ya'}\right\}. \quad \text{(IV.43)}$$

From this, it is seen that, when $(ya') \ll 1$, $(yR) \gg 1$, i.e., for the values with which we are principally concerned, the result is

$$S(\mathbf{y}) = \frac{2i\left(\mathbf{w}\left(k\left|\begin{smallmatrix}\mathbf{y}\\ j\end{smallmatrix}\right.\right)\mathbf{y}\right)}{v_a y^2}. \quad \text{(IV.44)}$$

This expression is independent of the limits of integration a' and R.

Let us now estimate the sum (IV.41) for the case where the charges δe lie on the same line (which we take to be in the z direction), at distances of the order of a' apart, and interact. Summation along the line gives, in order of magnitude,

$$S(\mathbf{y}) = \frac{\pi i y_z w_z\left(k\left|\begin{smallmatrix}\mathbf{y}\\ j\end{smallmatrix}\right.\right)}{a'}. \quad \text{(IV.45)}$$

The ratio of Eqs. (IV.45) and (IV.44) is of the order of $(ya')^2$. Thus, for the important values of y here, the interaction of charges lying on the same line is negligible.

Thus, for wave numbers such that $(ya') \ll 1$, $(yR) \gg 1$, we have

$$\delta\Phi\begin{pmatrix}\mathbf{y}\\ j\end{pmatrix}_0 = -\frac{2i}{v_a\sqrt{N}}\sum_k \frac{e_k\left(\mathbf{w}\left(k\left|\begin{smallmatrix}\mathbf{y}\\ j\end{smallmatrix}\right.\right)\mathbf{y}\right)}{\sqrt{m_k}\,y^2}\sum_{l'k'}\delta e^{l'}_{k'}\exp\left[2\pi i\mathbf{y}\mathbf{x}\begin{pmatrix}l'\\ k'\end{pmatrix}\right]. \quad \text{(IV.46)}$$

In order to obtain $q\begin{pmatrix}\mathbf{y}\\ j\end{pmatrix}$, we must solve Eq. (II.4) or (IV.34). These equations include anharmonic terms which may be significant if the displacements of the ions near a defect are comparable with the shortest distance between ions. If the defect density is not too high, we can neglect in Eq. (II.4) the terms which contain the displacements $v_\alpha\begin{pmatrix}l\\ k\end{pmatrix}$ and coefficients $\delta\Phi_\alpha\ldots\begin{pmatrix}l\ldots\\ k\ldots\end{pmatrix}$ that depend on different defects, and retain only the terms in which the displacements and coefficients $\delta\Phi_\alpha\begin{pmatrix}l\ldots\\ k\ldots\end{pmatrix}$ depend on one defect only. Then, Eq. (II.4) becomes

$$\delta\Phi_\alpha\begin{pmatrix}l\\ k\end{pmatrix}_0 + \sum_{l'k'\beta}\Phi_{\alpha\beta}\begin{pmatrix}ll'\\ kk'\end{pmatrix}_0 v_\beta\begin{pmatrix}l'\\ k'\end{pmatrix} + \sum_{l_0k_0}A_\alpha\left[\mathbf{x}\begin{pmatrix}ll^0\\ kk^0\end{pmatrix}\right] = 0. \quad \text{(IV.47)}$$

Here, the indices $\binom{l^0}{k^0}$ represent the position of the defects, and the function $A_\alpha\left[\mathbf{x}\binom{ll^0}{kk^0}\right]$ decreases rapidly with increasing $\left|\mathbf{x}\binom{ll^0}{kk^0}\right|$.

Taking the Fourier transform, we have

$$\delta\Phi\binom{\mathbf{y}}{j}_0+\omega^2\binom{\mathbf{y}}{j}q\binom{-\mathbf{y}}{j}+A\binom{\mathbf{y}}{j}=0, \qquad \text{(IV.48)}$$

where

$$A\binom{\mathbf{y}}{j}=\frac{1}{\sqrt{N}}\sum_{l^0k^0k\alpha}\frac{w_\alpha\left(k\middle|\begin{matrix}\mathbf{y}\\ j\end{matrix}\right)}{\sqrt{m_k}}\exp\left[2\pi i\mathbf{y}\mathbf{x}\binom{l^0}{k^0}\right]\sum_{l-l_0}A_\alpha\left[\mathbf{x}\binom{ll^0}{kk^0}\right]\exp\left[2\pi i\mathbf{y}\mathbf{x}\binom{ll^0}{kk^0}\right]. \qquad \text{(IV.49)}$$

Since the function $A_\alpha\left[\mathbf{x}\binom{ll^0}{kk^0}\right]$ decreases rapidly with increasing distance from the defect, the exponential in the last sum may be expanded in series.

If the ion configuration around the defect is centrally symmetrical, then

$$\sum_{l-l^0}A_\alpha\left[\mathbf{x}\binom{ll^0}{kk^0}\right]\exp\left[2\pi i\mathbf{x}\binom{ll^0}{kk^0}\mathbf{y}\right]=2\pi i\sum_{\substack{l-l^0\\ \beta}}y_\beta A_\alpha\left[\mathbf{x}\binom{ll^0}{kk^0}\right]x_\beta\binom{ll^0}{kk^0}=\theta\,(kk^0)\,y_\alpha. \qquad \text{(IV.50)}$$

Using this, we obtain for $A\binom{\mathbf{y}}{j}$ the expression

$$A\binom{\mathbf{y}}{j}=\frac{1}{\sqrt{N}}\sum_{l^0k^0k}\frac{\left(\mathbf{w}\left(k\middle|\begin{matrix}\mathbf{y}\\ j\end{matrix}\right)\mathbf{y}\right)}{\sqrt{m_k}}\theta\,(kk^0)\exp\left[2\pi i\mathbf{y}\mathbf{x}\binom{l^0}{k^0}\right]. \qquad \text{(IV.51)}$$

An important point in the present discussion is that this expression is linear in the wave vector and is zero for transverse displacements.

Let us now consider Eq. (IV.48) for longitudinal optical displacements. From Eqs. (IV.46) and (B.38), we find

$$\delta\Phi\binom{\mathbf{y}}{o\parallel}_0=\frac{2e_1}{v_a\sqrt{NM}}\frac{1}{y}\sum_{k'l'}\delta e^{l'}_{k'}\exp\left[2\pi i\mathbf{y}\mathbf{x}\binom{l'}{k'}\right]. \qquad \text{(IV.52)}$$

From the small values of $|\mathbf{y}|$ under consideration here, we can neglect $A\binom{\mathbf{y}}{o\parallel}$ in Eq. (IV.48) in comparison with $\delta\Phi\binom{y}{o\parallel}$, and thus get

$$q\binom{\mathbf{y}}{o\parallel}=-\frac{2e_1}{\omega^2\binom{\mathbf{y}}{o\parallel}v_a\sqrt{NM}}\frac{1}{y}\sum_{k'l'}\delta e^{l'}_{k'}\exp\left[-2\pi i\mathbf{y}\mathbf{x}\binom{l'}{k'}\right]. \qquad \text{(IV.53)}$$

For longitudinal acoustic displacements, using Eqs. (B.37) and (IV.46), we get

$$\delta\Phi\binom{\mathbf{y}}{a\parallel}_0=-\frac{6}{5}\frac{e_1(m_2-m_1)}{v_a\sqrt{N}(m_1+m_2)^{3/2}M}\frac{\beta}{\omega^2\binom{\mathbf{y}}{o\parallel}}4\pi^2a^2y\sum_{k'l'}\delta e^{l'}_{k'}\exp\left[2\pi i\mathbf{y}\mathbf{x}\binom{l'}{k'}\right]. \qquad \text{(IV.54)}$$

From Eqs. (IV.48), (B.37), and (IV.54), it follows that

$$q\begin{pmatrix} \mathbf{y} \\ a \parallel \end{pmatrix} \propto \frac{1}{y}. \tag{IV.55}$$

A more accurate calculation of this quantity is possible if the displacements of the ions near the defect are not too large. Then, the quantity $A\begin{pmatrix} \mathbf{y} \\ a \parallel \end{pmatrix}$ in Eq. (IV.48) can be neglected in comparison with $\delta\Phi\begin{pmatrix} \mathbf{y} \\ a \parallel \end{pmatrix}$, giving

$$q\begin{pmatrix} \mathbf{y} \\ a \parallel \end{pmatrix} = \frac{e_1}{\omega^2\begin{pmatrix} \delta \\ o \parallel \end{pmatrix} v_a \sqrt{NM}} \frac{(m_2 - m_1)}{\sqrt{m_1 m_2}} \frac{1}{y} \sum_{l'k'} \delta e_{k'}^{l'} \exp\left[-2\pi i y \mathbf{x}\begin{pmatrix} l' \\ k' \end{pmatrix}\right]. \tag{IV.56}$$

Using the amplitude expressions (IV.53) and (IV.56), and also Eq. (II.7), we can find the distribution of ion displacements when defects are present. When the lattice contains a single charged defect at the point $\begin{pmatrix} l^0 \\ k^0 \end{pmatrix}$, we get

$$v_\alpha\begin{pmatrix} l \\ k \end{pmatrix} = \frac{e_k \delta e_{k^0}^{l^0}}{2M\omega^2\begin{pmatrix} 0 \\ o \parallel \end{pmatrix}} \frac{\left(\mathbf{x}\begin{pmatrix} ll^0 \\ kk^0 \end{pmatrix} \mathbf{e}_\alpha\right)}{\left|\mathbf{x}\begin{pmatrix} ll^0 \\ kk^0 \end{pmatrix}\right|^3}, \qquad v_\alpha\begin{pmatrix} l^0 \\ k^0 \end{pmatrix} = 0, \tag{IV.57}$$

where $\mathbf{e}_\alpha$ is a unit vector in the direction α, and $\omega^2\begin{pmatrix} 0 \\ o \parallel \end{pmatrix}$ has the form (B.38). Since the expressions (IV.53) and (IV.56) are exact when $|\mathbf{y}|$ is sufficiently small, the first expression (IV.57) is valid at a large distance from the defect, i.e., when $\left|\mathbf{x}\begin{pmatrix} ll^0 \\ kk^0 \end{pmatrix}\right| \gg a$.

It is fairly difficult to determine the displacements of the ions near the charged defect, but the problem becomes simpler if we assume that, in Eq. (II.4), only those coefficients $\Phi_{\alpha\beta}\begin{pmatrix} ll' \\ kk' \end{pmatrix}$ are nonzero which take account of the interaction between nearest neighbors [see (B. 30)].

By solving Eq. (II.4) for the displacements of the ions along the (1, 0, 0) axis, omitting $\delta\Phi_{\alpha\beta}\begin{pmatrix} ll' \\ kk' \end{pmatrix}_0$ and the anharmonic terms, we find

$$\left(1 + \frac{1}{m}\right) v_x(n) - v_x(n-1) = \frac{v_x(n+m)}{m} - \frac{A}{m} \sum_{k=1}^{m-1} \frac{(m-k)(-1)^{n+k}}{(n+k)^2}, \tag{IV.58}$$

where $A = \frac{e\delta e_0^0}{a^2\beta}$. In deriving this relation, it is assumed that the ions situated on the (1, 0, 0) axis are numbered $n = 1, 2, \ldots$, and have charges $e_n = e(-1)^n$, while the defect is numbered $n = 0$ and has charge δe_0^0.

Taking the limit $m = \infty$ in Eq. (IV.58) for $n = 1$, and $v_x(0) = 0$, we find the displacements of the ions nearest the defect to be

$$v_x(1) = \frac{\pi^2}{12} \frac{e_1 \delta e_0^0}{a^2\beta}. \tag{IV.59}$$

The allowance for the Coulomb interaction between the ions can obviously only decrease this quantity [see the expression (B.38) for the longitudinal optical frequencies], and so the displacements in a real crystal are such that

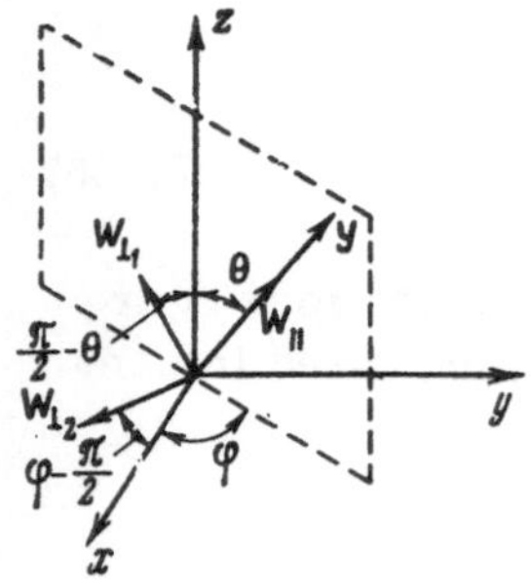

Fig. 1

$$\frac{v(1)}{a} < \frac{\pi^2}{12}\frac{e_1\delta e_0^0}{a^2\beta}. \quad \text{(IV.60)}$$

We can make an estimate of the right-hand side of this inequality for NaCl. Taking $|\delta e_0^0| = |e|, \beta = \frac{\lambda_{+-}}{\rho^2}e^{-\frac{a}{\rho}}$, and using the numerical values of λ_{+-}, ρ, and a as given in [11], we find

$$\frac{v(1)}{a} < 0.52.$$

This inequality shows that there are cases where the anharmonic terms in Eqs. (II.4) and (IV.34) may be neglected even if $\delta e \sim e$.

Let us now consider the calculation of the expression

$$\sum_{j''} \Phi\begin{pmatrix} \nu - y & y - \nu \\ j \; \tilde{j} & j'' \end{pmatrix}_0 q\begin{pmatrix} y - \nu \\ j'' \end{pmatrix}.$$

From Eqs. (A.9), (B.37), and (B.38),

$$\Phi\begin{pmatrix} \nu - y & y \\ j \; \tilde{j} & o \end{pmatrix} \propto y, \quad \Phi\begin{pmatrix} \nu - y & y \\ j \; \tilde{j} & a \end{pmatrix} \propto y^3.$$

Since

$$q\begin{pmatrix} y \\ o\,\| \end{pmatrix} \propto \frac{1}{y}, \quad q\begin{pmatrix} y \\ a\,\| \end{pmatrix} \propto \frac{1}{y}, \quad q\begin{pmatrix} y \\ a\perp \end{pmatrix} = q\begin{pmatrix} y \\ o\perp \end{pmatrix} = 0,$$

it is clear that, at microwave frequencies, only the coefficient $\Phi\begin{pmatrix} \nu - y & y \\ j \; \tilde{j} & o\,\| \end{pmatrix}_0$ is important. This is useful, since, for the long-wave optical vibrations, the quantity $q\begin{pmatrix} y \\ o\,\| \end{pmatrix}$ can be calculated with high accuracy, even when the displacements of the ions near the defect are large.

We can now simplify Eq. (A.9) by taking into account only the interaction with nearest neighbors; this is allowable, since $\Phi_{\alpha\beta\gamma}\begin{pmatrix} l & 0 & 0 \\ 1 & 2 & 2 \end{pmatrix}$ decrease rapidly with increasing distance. The result is

$$\Phi\begin{pmatrix} \nu - y & y \\ j \; \tilde{j} & j'' \end{pmatrix} = \frac{4\pi i\sigma a}{\sqrt{Nm_1m_2}} \sum_\alpha y_\alpha \left(\frac{w_\alpha\left(1 \middle| \begin{matrix} \nu \\ j \end{matrix}\right)}{\sqrt{m_1}} - \frac{w_\alpha\left(2 \middle| \begin{matrix} \nu \\ j \end{matrix}\right)}{\sqrt{m_2}} \right) \times$$

$$\times \left(w_\alpha\left(1 \middle| \begin{matrix} -y \\ \tilde{j} \end{matrix}\right) w_\alpha\left(2 \middle| \begin{matrix} y \\ j'' \end{matrix}\right) - w_\alpha\left(2 \middle| \begin{matrix} -y \\ \tilde{j} \end{matrix}\right) w_\alpha\left(1 \middle| \begin{matrix} y \\ j'' \end{matrix}\right) \right), \quad \text{(IV.61)}$$

where

$$\sigma = \left\{ \Phi_{\alpha\alpha\alpha}\begin{pmatrix} l & 0 & 0 \\ 1 & 2 & 2 \end{pmatrix} \right\}_{x_\alpha\left(\begin{smallmatrix} l & 0 \\ 1 & 2 \end{smallmatrix}\right) = a}$$

The components of the vectors $w_\alpha\left(k \middle| \begin{matrix} y \\ j \end{matrix}\right), y_\alpha$ can be written in the spherical coordinates shown in Fig. 1. The result is

$$\mathbf{y} = y\{\sin\theta\cos\varphi,\ \sin\theta\sin\varphi,\ \cos\theta\},$$

$$\mathbf{w}\left(k\,\Big|\,{\mathbf{y} \atop \|}\right) = \left(\mathbf{w}\left(k\,\Big|\,{\mathbf{y} \atop \|}\right)\mathbf{e}_{\|}(\mathbf{y})\right)\{\sin\theta\cos\varphi,\ \sin\theta\sin\varphi,\ \cos\theta\},$$

$$\mathbf{w}\left(k\,\Big|\,{\mathbf{y} \atop \perp_1}\right) = \left(\mathbf{w}\left(k\,\Big|\,{\mathbf{y} \atop \perp_1}\right)\mathbf{e}_{\perp_1}(\mathbf{y})\right)\left\{\sin\left(\frac{\pi}{2}-\theta\right)\cos(\pi+\varphi),\right. \tag{IV.62}$$

$$\left.\sin\left(\frac{\pi}{2}-\theta\right)\sin(\pi+\varphi),\ \cos\left(\frac{\pi}{2}-\theta\right)\right\} = \left(\mathbf{w}\left(k\,\Big|\,{\mathbf{y} \atop \perp_1}\right)\mathbf{e}_{\perp_1}(\mathbf{y})\right)\{-\cos\theta\cos\varphi,\ -\cos\theta\sin\varphi,\ \sin\theta\},$$

$$\mathbf{w}\left(k\,\Big|\,{\mathbf{y} \atop \perp_2}\right) = \left(\mathbf{w}\left(k\,\Big|\,{\mathbf{y} \atop \perp_2}\right)\mathbf{e}_{\perp_2}(y)\right)\{\sin\theta\sin\varphi,\ -\sin\theta\cos\varphi,\ 0\}\frac{1}{|\sin\theta|}.$$

The polarization vectors are orthonormal.

For point defects, we are interested in the polarizability components $\chi_{xx} = \chi_{yy} = \chi_{zz}$; for lines of charges in the z direction, the relevant components are χ_{zz}, $\chi_{xx} = \chi_{yy}$.

Thus, in order to find all the components of the tensor $\chi_{\alpha\beta}$ in the cases concerned, we have to determine the components χ_{zz} and χ_{xx}. Accordingly, the vector $w_\alpha\left(k\,\Big|\,{\mathbf{v} \atop j}\right)$, which is parallel to the electric field, is in the z direction or the x direction.

Substituting in Eq. (IV.61) the expressions in Eqs. (IV.62), (B.37), and (B.38), we have for the case where $\mathbf{w}\left(k|_j\right)$ is parallel to the z axis

$$\Phi\begin{pmatrix}\mathbf{v} & -\mathbf{y} & \mathbf{y}\\ j & a\| & o\|\end{pmatrix} = -\frac{4\pi i\sigma a}{\sqrt{MN}}\left(\frac{w_z\left(1\,\Big|\,{\nu \atop j}\right)}{\sqrt{m_1}} - \frac{w_z\left(2\,\Big|\,{\nu \atop j}\right)}{\sqrt{m_2}}\right)y\cos^3\theta\frac{1}{\sqrt{m_1+m_2}},$$

$$\Phi\begin{pmatrix}\mathbf{v} & -\mathbf{y} & \mathbf{y}\\ j & a\perp_1 & o\|\end{pmatrix} = -\frac{4\pi i\sigma a}{\sqrt{MN}}\left(\frac{w_z\left(1\,\Big|\,{\nu \atop j}\right)}{\sqrt{m_1}} - \frac{w_z\left(2\,\Big|\,{\mathbf{v} \atop j}\right)}{\sqrt{m_2}}\right)y\cos^2\theta\sin\theta\frac{1}{\sqrt{m_1+m_2}}, \tag{IV.63}$$

$$\Phi\begin{pmatrix}\mathbf{v} & -\mathbf{y} & \mathbf{y}\\ j & a_{\perp_2} & o\|\end{pmatrix} = 0.$$

When $\mathbf{w}\left(k\,\Big|\,{\mathbf{v} \atop j}\right)$ is parallel to the x axis,

$$\Phi\begin{pmatrix}\mathbf{v} & -\mathbf{y} & \mathbf{y}\\ j & a\| & o\|\end{pmatrix} = -\frac{4\pi i\sigma a}{\sqrt{MN}}\left(\frac{w_x\left(1\,\Big|\,{\mathbf{v} \atop j}\right)}{\sqrt{m_1}} - \frac{w_x\left(2\,\Big|\,{\mathbf{v} \atop j}\right)}{\sqrt{m_2}}\right)y\sin^3\theta\cos^3\varphi\frac{1}{\sqrt{m_1+m_2}},$$

$$\Phi\begin{pmatrix}\mathbf{v} & -\mathbf{y} & \mathbf{y}\\ j & a\perp_1 & o\|\end{pmatrix} = \frac{4\pi i\sigma a}{\sqrt{MN}}\left(\frac{w_x\left(1\,\Big|\,{\mathbf{v} \atop j}\right)}{\sqrt{m_1}} - \frac{w_x\left(2\,\Big|\,{\mathbf{v} \atop j}\right)}{\sqrt{m_2}}\right)y\sin^2\theta\cos\theta\cos^3\varphi\frac{1}{\sqrt{m_1+m_2}}, \tag{IV.64}$$

$$\Phi\begin{pmatrix}\mathbf{v} & -\mathbf{y} & \mathbf{y}\\ j & a\perp_2 & o\|\end{pmatrix} = -\frac{4\pi i\sigma a}{\sqrt{MN}}\left(\frac{w_x\left(1\,\Big|\,{\mathbf{v} \atop j}\right)}{\sqrt{m_1}} - \frac{w_x\left(2\,\Big|\,{\mathbf{v} \atop j}\right)}{\sqrt{m_2}}\right)y\sin^2\theta\cos^2\varphi\sin\varphi\frac{1}{\sqrt{m_1+m_2}}.$$

Substituting in Eqs. (IV.63) and (IV.64) the expressions for $w_x\left(k\,\Big|\,{\mathbf{v} \atop j}\right)$ given by Eqs. (B.38), and using Eq. (IV.53), we obtain

$$\Phi\begin{pmatrix}\mathbf{v} & -\mathbf{y} & \mathbf{y}\\ j & \widetilde{j} & o\|\end{pmatrix}_0 q\begin{pmatrix}\mathbf{y}\\ o\|\end{pmatrix} = -\frac{8\pi e_1\sigma a}{M\sqrt{M(m_1+m_2)}\,\omega^2\begin{pmatrix}0\\ o\|\end{pmatrix}}\frac{1}{v_a N}A(\alpha,\widetilde{j})\sum_{l'k'}\delta e_{k'}^{l'}\exp\left[-2\pi i\mathbf{y}\mathbf{x}\begin{pmatrix}l'\\ k'\end{pmatrix}\right], \tag{IV.65}$$

where $A_{\alpha\widetilde{j}}$ is given by Table 1. Using Eqs. (B.37) and (B.38), we can write the coefficient

TABLE 1

α	$\tilde{j}$	$A(\alpha, \tilde{j})$	$B(\alpha, \tilde{j})$	$C(\alpha, \tilde{j})$
z	$\parallel$	$\cos^3\theta$	$\cos\theta$	$\cos\theta$
z	$\perp_1$	$\cos^2\theta\sin\theta$	0	$\sin\theta$
z	$\perp_2$	0	0	0
x	$\parallel$	$\sin^3\theta\cos^3\varphi$	$\sin\theta\sin\varphi$	$\sin\theta\cos\varphi$
x	$\perp_1$	$-\sin^2\theta\cos\theta\cos^3\varphi$	0	$-\cos\theta\cos\varphi$
x	$\perp_2$	$\sin^2\theta\cos^2\varphi\sin\varphi$	0	$\sin\varphi$

$\delta\Phi\begin{pmatrix}\nu & -\mathbf{y}\\ j & \tilde{j}\end{pmatrix}$ [see Eq. (IV.33)] as

$$\delta\Phi\begin{pmatrix}\nu & -\mathbf{y}\\ j & \tilde{j}\end{pmatrix} = -\frac{4\pi e_1}{\sqrt{M(m_1+m_2)}}\frac{1}{v_a N}B(\alpha,\tilde{j})\sum_{l'k'}\delta e_{k'}^{l'}\exp\left[-2\pi i\mathbf{y}\mathbf{x}\begin{pmatrix}l'\\ k'\end{pmatrix}\right], \tag{IV.66}$$

where $B(\alpha, \tilde{j})$ is given by Table 1.

With the notation

$$\delta = \frac{2a\sigma}{M\omega^2\begin{pmatrix}0\\ o\,\parallel\end{pmatrix}} \tag{IV.67}$$

and combining Eqs. (IV.66) and (IV.65), we get the following expression for the coefficient $A\begin{pmatrix}\nu & -\mathbf{y}\\ j & \tilde{j}\end{pmatrix}$:

$$A\begin{pmatrix}\nu & -\mathbf{y}\\ j & \tilde{j}\end{pmatrix} = -\frac{4\pi e_1}{\sqrt{M(m_1+m_2)}}\frac{1}{v_a N}\{B(\alpha,\tilde{j}) + A(\alpha,\tilde{j})\,\delta\}\sum_{l'k'}\delta e_{k'}^{l'}\exp\left[-2\pi i\mathbf{y}\mathbf{x}\begin{pmatrix}l'\\ k'\end{pmatrix}\right]. \tag{IV.68}$$

In addition to the terms given here, the expression for $A\begin{pmatrix}\nu & -\mathbf{y}\\ j & \tilde{j}\end{pmatrix}$ [see Eq. (II.15)] also includes terms of the form

$$\sum_{\mathbf{y}''j''}\delta\Phi\begin{pmatrix}\nu & -\mathbf{y} & \mathbf{y}''\\ j & \tilde{j} & j''\end{pmatrix}_0 q\begin{pmatrix}\mathbf{y}''\\ j''\end{pmatrix} + {}^1/_2\sum_{\substack{\mathbf{y}''\mathbf{y}'''\\ j''j'''}}\Phi\begin{pmatrix}\nu & -\mathbf{y} & \mathbf{y}'' & \mathbf{y}'''\\ j & \tilde{j} & j'' & j'''\end{pmatrix}_0 q\begin{pmatrix}\mathbf{y}''\\ j''\end{pmatrix}q\begin{pmatrix}\mathbf{y}'''\\ j'''\end{pmatrix} + \cdots.$$

These can be estimated in the same way as the terms denoted by $A\begin{pmatrix}\mathbf{y}\\ j\end{pmatrix}$ in Eq. (IV.48). They are proportional to $|\mathbf{y}|$ and are negligible when the latter is small.

5. Expressions for the Imaginary Part of the Permittivity in the Case of Point and Linear Charged Defects

Using Eqs. (IV.68) and (B.37), we can write the expressions (IV.21) for the imaginary part of the complex polarizability in the form

$$\mathrm{Im}\,\chi_{\alpha\alpha}(\omega) = \frac{\pi}{2v_a N(m_1+m_2)}\sum_{\mathbf{y}\tilde{j}}\frac{1}{\omega\begin{pmatrix}\mathbf{y}\\ \tilde{j}\end{pmatrix}}\delta\left(\omega-\omega\begin{pmatrix}\mathbf{y}\\ \tilde{j}\end{pmatrix}\right)\times$$

$$\times\sum_{\substack{l-l'\\ kk'}}\overline{\delta e_k^l\delta e_{k'}^{l'}}\exp\left[2\pi i\mathbf{y}\mathbf{x}\begin{pmatrix}ll'\\ kk'\end{pmatrix}\right]|C(\alpha,\tilde{j}) + \Delta\varepsilon\{B(\alpha,\tilde{j}) + A(\alpha,\tilde{j})\,\delta\}|^2, \tag{IV.69}$$

where $\Delta\varepsilon = \frac{4\pi e^2 N_0}{M(\omega_0^2 - \omega^2)}$, and $C(\alpha, \tilde{j})$ is given in Table 1.

Let us first calculate the imaginary part of the permittivity $\varepsilon''(\omega) = \varepsilon''_{zz}(\omega) = 4\pi \operatorname{Im}\chi_{zz}(\omega)$ for the case of point charged defects. To do so, we must find $\overline{\delta e_k^l \, \delta e_{k'}^{l'}}$. We shall assume that the crystal contains N_p defects having a charge difference δe_1 and an equal number having an opposite charge difference δe_2, the defects being distributed randomly. Then,

$$\overline{\delta e_k^l \delta e_{k'}^{l'}} = \delta_{ll'} \delta_{kk'} \overline{\delta e_k^2} + \overline{\delta e_k^l}\, \overline{\delta e_{k'}^{l'}} \qquad \text{(IV.70)}$$

where

$$\overline{\delta e_k^l} = \frac{N_p}{N} \delta e_k = \frac{n_p}{N_0} \delta e_k, \quad \overline{\delta e^2}_k = \frac{N_p}{N} \delta e_k^2 = \frac{n_p}{N_0} \delta e_k^2,$$

$\overline{\delta e_1} + \overline{\delta e_2} = 0$, and n_p is the number of point defects having one particular charge sign.

From Table 1 and the expressions for the acoustic vibration spectrum

$$\omega^2\binom{y}{\parallel} = (2\pi)^2 v_\parallel^2 y^2, \qquad \omega^2\binom{y}{\perp} = (2\pi)^2 v_\perp^2 y^2, \qquad \text{(IV.71)}$$

where $v_\parallel$ and $v_\perp$ are the longitudinal and transverse phase velocities of acoustic waves, and changing from summation over y to integration, we find

$$\varepsilon''(\omega)^p = \frac{2 n_p \delta e^2}{\rho} \omega \left\{ \frac{1}{v_\parallel^3} [1/3 + 2\Delta\varepsilon(1/3 + 1/5\,\delta) + \Delta\varepsilon^2 (1/3 + 2/5\,\delta + 1/7\,\delta^2)] + \frac{2}{v_\perp^3} [1/3 + 2/15\,\Delta\varepsilon\delta + 1/35\,\Delta\varepsilon^2\delta^2] \right\}, \qquad \text{(IV. 72)}$$

where $\rho = (m_1 + m_2)/v_a$ is the density of the crystal.

Let us next consider the case of linear charges. We shall assume that the crystal contains N_l lines in the z direction, and N_z ions of type 1 on each line, at a distance d apart, whose charge is changed by δe_1. The total charge on the lines is therefore $N_l N_z \delta e_1$.

We shall also assume that the crystal is neutralized by N_p point defects having an opposite charge difference δe_2, arranged randomly. Hence, $N_l N_z \delta e_1 + N_p \delta e_2 = 0$. We shall further assume that the lines are also arranged randomly but always in the z direction. This is important in relation to the anisotropy of absorption.

The mean $\overline{\delta e_2^l \delta e_2^{l'}}$ is given by an expression similar to Eq. (IV.70). For $\overline{\delta e_1^l \, \delta e_1^{l'}}$, the result is

$$\overline{\delta e_1^l \delta e_1^{l'}} = \overline{\delta e_1^2} g_{ll'} + \overline{\delta e_1^l}\, \overline{\delta e_1^{l'}}, \qquad \text{(IV.73)}$$

where $g_{ll'}$ is a function which is zero unless l and l' refer to the position of charge increments δe_1 on the same line, when it is unity:

$$\overline{\delta e_1^l} = \frac{N_l N_z}{N} \delta e_1 = \frac{n_l N_z}{N_0} \delta e_1.$$

If the charged lines traverse the whole crystal, then $N_z \propto 1/d$, and n_l may conveniently be expressed in terms of the number of lines per cm^2. Then, $\overline{\delta e_1^l} = \frac{n_l}{N_0 d}\,\delta e_1$.

From Eq. (IV.73), we easily find

$$\sum_{l-l'} \overline{\delta e_1^l \delta e_1^{l'}} \exp\left[2\pi i y x\binom{ll'}{kk'}\right] = \overline{\delta e_1^2}\pi\Phi(y_z) + \delta_{y,0}N^2\overline{\delta e_1^l}\,\overline{\delta e_1^{l'}}, \qquad \text{(IV.74)}$$

where

$$\Phi(y_z) = \frac{1}{\pi}\,\frac{\sin \pi y_z N_z d}{\pi y_z d}.$$

If $\pi y N_z d \ll 1$, i.e., if the acoustic wavelength is much greater than the length of the line ($\pi L \ll \lambda$), then $\Phi(y_z) = N_z/\pi$.

If $\pi y N_z d \gg 1$ ($\pi L \gg \lambda$), then $\Phi(y_z) = \frac{1}{\pi d}\,\delta(y_z)$.

Using Eqs. (IV.69)-(IV.71) and (IV.74), we have for the case where $\pi y N_z d \gg 1$,

$$\varepsilon_{xx}(\omega)^l = \varepsilon''(\omega)^p + \frac{\pi n_l N_z \delta e_1^2}{2\rho d}\left\{\frac{1}{v_\parallel^2}\left[1 + 2\Delta\varepsilon\left(1 + {}^3/_4\delta\right) + \Delta\varepsilon^2\left(1 + {}^3/_2\delta + {}^5/_8\delta^2\right)\right] + \frac{1}{v_\perp^2}\left[1 + {}^1/_2\Delta\varepsilon\delta + {}^1/_8\Delta\varepsilon^2\delta^2\right]\right\}, \qquad \text{(IV.75)}$$

$$\varepsilon''_{xx}(\omega)^l = \varepsilon''_{yy}(\omega)^l,$$

$$\varepsilon''_{zz}(\omega)^l = \varepsilon''(\omega)^p + \frac{\pi n_l N_z \delta e_1^2}{\rho d v_\perp^2}.$$

In order to derive $\varepsilon''(\omega)^p$ in these expressions, we must make the substitution $2n_p\delta e^2 \to n_p\delta e_2^2$ in Eq. (IV.72).

If $\pi y N_z d \ll 1$, then

$$\varepsilon''_{xx}(\omega)^l = \varepsilon''_{yy}(\omega)^l = \varepsilon''_{zz}(\omega)^l = \varepsilon''(\omega)^p, \qquad \text{(IV.76)}$$

where $\varepsilon''(\omega)^p$ is obtained from Eq. (IV.72) by the substitution $2n_p\delta e^2 \to n_p\delta e^2 + n_l N_z^2 \delta e_1^2$.

6. Discussion of the Results. Comparison of Theory and Experiment

In the previous section, we derived expressions for the imaginary part of the permittivity of an ionic crystal containing point and line charged defects. In deriving these expressions, the Born approximation was used. It is shown in Appendix C that this approximation gives the correct result for the imaginary part of the permittivity if the following inequalities hold:

$$\frac{C_k^{(0)}(0\parallel)}{\tilde{\omega}^2}\left(\frac{\omega}{\tilde{\omega}}\right)^2 \ll 1, \quad \frac{C_k^{(0)}(0\perp_1)}{\tilde{\omega}^2}\left(\frac{\omega}{\tilde{\omega}}\right)^2 \ll 1, \qquad \text{(IV.77)}$$

where $\tilde{\omega}$ is of the order of the mean frequency in the lattice spectrum, and $C_k^{(0)}(0_\parallel)$, $C_k^{(0)}(0_{\perp 1})$ are given by Eqs. (C.38) and (C.39).

The inequalities (IV.77) may not be satisfied when the charged defect has resonance frequencies in the range $\omega \ll \omega_0$ (or $\omega \ll \tilde{\omega}$), and $C_k^{(0)}(0_\parallel)$, $C_k^{(0)}(0_{\perp 1})$ may then be large.

To find the resonance frequencies of real charged defects, we must take into account not only the electrostatic forces between the defect and the remaining ions but also the short-range repulsion of the electron shells and the mass difference between the defect ion and the lattice ions.

Thus, the calculation of the resonance frequencies of charged defects is a complicated problem, and it can be solved only by applying numerical methods. If there are resonance frequencies in the range $\omega \ll \omega_0$, it seems to be easier to determine them experimentally.

Let us now examine more closely Eqs. (IV.72) and (IV.75)-(IV.77).

If we put $\Delta\varepsilon = 0$ in Eq. (IV.72), the result is Lysenko's equation (see [35]). The physical significance of this equation has been discussed in Chapter I. Since $\Delta\varepsilon \approx 5$ for ionic crystals of the NaCl type, and is several hundreds for crystals such as $SrTiO_3$ and $BaTiO_3$, the allowance for it in Eqs. (IV.72) and (IV.75)-(IV.77) is extremely important.

In order to see the significance of $\Delta\varepsilon$, we take $\delta = 0$ in Eqs. (IV.72) and (IV.75)-(IV.77), i.e., we neglect the deformation of the lattice near the defect. Then, the equations take the simpler forms

$$\varepsilon''(\omega)^p = \frac{2n_p\delta e^2}{\rho}\,\omega\left\{{}^1/_3\,\frac{(\Delta\varepsilon+1)^2}{v_{\|}^3} + {}^2/_3\,\frac{1}{v_{\perp}^3}\right\} \qquad \text{(IV.78)}$$

for point defects and

$$\varepsilon''_{xx}(\omega)^l = \varepsilon''(\omega)^p + \frac{\pi n_l N_z \delta e_1^2}{2\rho d}\left\{\frac{1}{v_{\|}^2}(\Delta\varepsilon+1)^2 + \frac{1}{v_{\perp}^2}\right\},$$
$$\varepsilon''_{xx}(\omega)^l = \varepsilon''_{yy}(\omega)^l, \qquad \text{(IV.79)}$$

$$\varepsilon''_{zz}(\omega)^l = \varepsilon''(\omega)^p + \frac{\pi n_l N_z \delta e_1^2}{\rho d}\,\frac{1}{v_{\perp}^2} \qquad \text{(IV.80)}$$

for linear defects in the case where the inequality $\pi y N_z d \gg 1$ holds.

It is seen from Eqs. (IV.78)-(IV.80) that $\Delta\varepsilon$ occurs in the terms which contain $v_{\|}$, the velocity of longitudinal acoustic waves.

The quantity $\Delta\varepsilon$ arises from the interaction of the dipoles formed when the defect ion and the lattice ions are displaced by the transverse optical wave with wave vector $\boldsymbol{\nu}$ and the acoustic wave with wave vector $\mathbf{y}$. This interaction is zero except for longitudinal acoustic waves. If the deformation of the lattice near the defect is taken into account, it is possible to have an interaction between the optical wave $\binom{\mathbf{v}}{j}$ and the transverse acoustic waves, and Eqs. (IV.72) and (IV.75)-(IV.77) therefore include terms of the form $\Delta\varepsilon^2\delta^2/v_{\perp}^2$, which contain the velocity of these waves as well as $\Delta\varepsilon$. The interaction between the optical wave and the acoustic wave increases with the magnitude of the dipoles formed by the passage of the optical wave, i.e., with decreasing frequency ω_0. Thus, we arrive at the following physical picture of the absorption of microwaves in the presence of defects.

The electric field of the wave generates in the crystal transverse optical vibrations having wave vector $\boldsymbol{\nu}$ and frequency ω. Since $\omega \ll \omega_0$, these vibrations are subject to little or no direct damping. They create defects which emit acoustic waves and thereby take energy from the optical vibrations.

Among the various kinds of lattice defects, the charged defects provide the most effective mechanism for absorption of electromagnetic waves by the lattice in the range $\omega \ll \omega_0$. This is due to the long range of the Coulomb interaction. To illustrate, let us consider a defect consisting of a force constant changed by Δk (where k is the proportionality factor between the force and the relative displacement of two adjacent ions). Clearly, this defect will appear only when the relative displacements of the ions are large. However, for long acoustic waves the displacements are negligible, and so, for such defects, $\varepsilon''(\omega)$ is small and is given in order of magnitude by

$$\varepsilon''(\omega) \sim \left(\frac{\omega}{\omega_0}\right)^3. \tag{IV.81}$$

If the coupling constants change by the same amount on either side of the ion concerned (the defect is centrally symmetrical), $\varepsilon''(\omega)$ has the even smaller value

$$\varepsilon''(\omega) \sim \left(\frac{\omega}{\omega_0}\right)^5. \tag{IV.82}$$

The isotopic defects also give the form (IV.82) for $\varepsilon''(\omega)$ (see [26]). The estimates (IV.81) and (IV.82) assume that the defects considered have no resonance at frequencies $\omega \ll \omega_0$.

Charged defects have different properties. Since their interaction with distant ions is just as strong as with neighboring ions, and since the relative displacements of the defect and the distant ions may be large even for long-wave acoustic vibrations, the interaction of the optical wave and the acoustic wave does not depend on the magnitude of the wave vector of the latter when this is small [see Eq. (IV.33)]. Thus, the frequency dependence of ε'' for charged defects when $\omega \ll \omega_0$ is determined only by the density of states, which is proportional to ω/ω_0.

Equations (IV.72) and (IV.75)-(IV.77) have been derived on the assumption that the crystal and the charged defect in it have central symmetry. The assumption regarding the crystal is essential, since crystals with no center of symmetry may exhibit further effects due to their piezoelectric properties.

The assumption regarding the defect is not essential. Equations (IV.72) and (IV.75)-(IV.77) are valid also for charged defects with no center of symmetry (provided that there are not so many defects as to have a large effect on the properties of the lattice), for example, defects consisting of interstitial ions in the NaCl-type lattice. The correctness of this statement for the terms in Eqs. (IV.72) and (IV.75)-(IV.77) which originate from the coefficient $\delta\Phi\begin{pmatrix}\nu & -y\\ j & \tilde{j}\end{pmatrix}$ is easily proved by considering the expression (IV.33) for this coefficient, which is seen to be independent of the local field in the crystal and to depend only on the macroscopic field:

$$-\frac{4\pi}{v_a}\sum_{k\alpha}\frac{e_k w_\alpha\left(k\left|\begin{matrix}\nu\\ j\end{matrix}\right.\right)}{\sqrt{m_k}}\frac{y_\alpha y_\beta}{|\mathbf{y}|^2}.$$

The fact that $\varepsilon''(\omega)$ is independent of the position of the charged defects (and therefore of the symmetry of the defect) is due ultimately to the long range of the interaction between the defects and the lattice ions, and to the long wavelength of the acoustic vibrations emitted.

In deriving Eqs. (IV.72), (IV.75), and (IV.76), it has also been assumed that the crystal contains two ions per unit cell, but these equations are written in terms of quantities such as ρ, $\Delta\varepsilon = \varepsilon' - 1$, $v_\parallel$, and $v_\perp$, which are meaningful for a crystal having any number of ions in the unit cell.

Thus, from the foregoing discussion, we should expect that Eqs. (IV.72), (IV.75), and (IV.76) have a wider range of applicability than has been postulated in their derivation. We shall therefore apply these equations to account for the microwave dielectric losses dependent on defect content in crystals less simple than NaCl.

Rupprecht and Bell [58] have studied dielectric losses at microwave frequencies in cubic $SrTiO_3$, and determined the dependence of these losses on frequency, temperature, and numbers of various defects in the crystal.

The results of this work can be represented by the following relations for $\tan\delta = \varepsilon''/\varepsilon'$ and the real part ε' of the permittivity:

$$(T - T_c)\tan\delta = \alpha + \beta T + \gamma T^2, \tag{IV.83}$$

$$\varepsilon' = \frac{C}{T - T_c}, \tag{IV.84}$$

where T_c = 37°K is the Curie point, $C = 8.25 \times 10^4$ °K and the coefficients β and γ depend only slightly on the number of defects, being determined mainly by the anharmonicity of the lattice vibrations.

The orders of magnitude of these coefficients and the temperature dependence of the terms βT and γT^2 can be explained by using Eqs. (III.37) and (III.39). The coefficient α depends considerably on the defect concentration, and tends to zero for pure crystals. All three coefficients α, β, and γ are proportional to the frequency in the range from 3 to 36 Gc.

To explain the frequency dependence of the coefficients β and γ, numerical calculations are necessary, using Eqs. (III.37) and (III.39) and the exact phonon spectrum of the crystal.

The situation is simpler for the coefficient α. Its frequency dependence can be explained by assuming that the crystal contains point charged defects.

To test the theory, let us determine by means of Eq. (IV.72) the density of charged defects, using the experimental values of α from [58], and compare the result with the impurity concentration used in the experiment. For an $SrTiO_3$ single crystal containing 0.1% Gd^{3+} (corresponding to $n_p = 1.7 \times 10^{19}\,cm^{-3}$), $\alpha = 0.033$. Assuming that $\Delta\varepsilon \approx \varepsilon' \approx C/(T - T_e)$, we find from Eq. (IV.72) $n_p \approx 5 \times 10^{19}$ cm^{-3} ($\rho = 5$ g/cm^3, $v_\parallel = 8 \times 10^5$ cm/sec, $|\delta e| \approx e$, $\omega = 2\pi \times 22 \times 10^9$ cps). We have estimated the value of δ as zero, since its exact value for $SrTiO_3$ is unknown. Thus, there is agreement in order of magnitude.

The losses in an $SrTiO_3$ sample having oxygen vacancies have also been studied in [58]. Such defects were not found to influence $\tan\delta$ at concentrations up to 0.03%. It appears that the majority of these defects carry no additional charge, i.e., $\delta e = 0$.

We can also attempt to elucidate the nature of the dielectric losses observed in [59, 60] in the range $10^5 - 10^8$ cps, which depend only slightly on the frequency, assuming that the crystal contains linear charged defects.

Using the results of [60], we can determine the concentration of charged lines necessary to account for the frequency-independent losses. For $SrTiO_3$ at $\omega \approx 2\pi \times 10^5 - 2\pi \times 10^6$, $\tan\delta \approx 10^{-3}$. Putting $\Delta\varepsilon \approx \varepsilon' = 180$, $d = 4 \times 10^{-8}$, $N_z = 1/d$, we find from Eq. (IV.75) $n_l = 10^9$ cm^{-2} for $\delta = 0$. If $\delta = -6$ (as in NaCl), then $n_l = 3 \times 10^8$ cm^{-2}.

If it is assumed that the charged lines are formed because of the presence of dislocations, their concentration is found to have reasonable values.

Conclusions

This work may be regarded as consisting of two parts.

The first part deals with the quantum theory of infrared absorption and dispersion in ideal ionic crystals.

In the second part, we consider the problem of microwave absorption in ionic crystals containing charged defects, both points and lines.

In the first part, expressions are derived for the complex permittivity of an ideal ionic crystal, with allowance for anharmonic terms of the third and fourth order in the lattice Hamiltonian, and using the double-time retarded Green's function method.

The resulting expressions are analyzed as functions of frequency and temperature, with particular attention to low temperatures, where quantum effects are important.

The case where the lattice spectrum contains more than one optically active branch is discussed, as is the mutual interaction of these branches.

The expressions obtained are compared with the experimental data, and it is shown that, at high temperatures, the experimental and theoretical results are in good agreement. For example, the theory predicts that the frequency dependence of the damping function $\gamma(\omega)$ in NaCl-type crystals at high temperatures should have a two-humped form with a trough at the dispersion frequency ω_0, and that $\gamma(\omega) \propto T$ in the region of the humps but $\gamma(\omega) \propto T^{\alpha}$ in that of the trough, where $1 < \alpha < 2$. These theoretical conclusions are confirmed by experiment.

There is very little experimental information on the low-temperature region, where quantum effects are important, and hence we cannot decide whether theory and experiment do or do not agree in this temperature range.

In the second part of the work, expressions are derived for the imaginary part of the permittivity of an ionic lattice containing point and linear charged defects, these expressions being valid at frequencies $\omega \ll \omega_0$ (where ω_0 is the dispersion frequency).

The methods used are the same as in the first part.

The calculation of the scattering of lattice waves by charged defects is made in the Born approximation, and the limits of validity of this approximation are established.

The resulting expressions are analyzed as functions of frequency and of defect concentration. The influence of various types of lattice defects (isotopic, charged, and due to change of short-range forces between ions) on electromagnetic wave absorption in the frequency range $\omega \ll \omega_0$ is compared. It is found that charged defects provide the most effective absorption mechanism in this frequency range.

The expressions obtained are compared with the experimental results of Rupprecht and Bell [58], who measured the dielectric losses in cubic $SrTiO_3$ in the frequency range from 3 to 36 Gc, and found that $\varepsilon''(\omega) \propto \omega$ in this range. It is shown that the frequency dependence, the temperature dependence, and the order of magnitude of the part of the dielectric losses which depends on the impurity ion concentration can all be satisfactorily explained by assuming that the crystal contains point charged defects.

An attempt is also made to explain the nature of the dielectric losses observed in [59, 60] in the frequency range $10^6 - 10^8$ cps, which depend only slightly on the frequency. This is done on the basis of the assumption that the crystal contains linear charged defects.

I am deeply grateful to V. A. Chuenkov for valuable advice and for discussing the work.

APPENDIX A

The Form of the Coefficients $\Phi\binom{\mathbf{y}\mathbf{y}'\ldots\mathbf{y}^{(n)}}{jj'\ldots j^{(n)}}_0$ and $\delta\Phi\binom{\mathbf{y}\mathbf{y}'\ldots\mathbf{y}^{(n)}}{jj'\ldots j^{(n)}}_0$ for Forces between Pairs of Ions

In chapter II, expressions have been given for $\Phi\binom{\mathbf{y}\mathbf{y}'\ldots\mathbf{y}^{(n)}}{jj'\ldots j^{(n)}}_0$ and $\delta\Phi\binom{\mathbf{y}\mathbf{y}'\ldots\mathbf{y}^{(n)}}{jj'\ldots j^{(n)}}_0$ for the case where no assumption is made regarding the forces between the ions. Let us now assume that these forces act between pairs of ions. Then, the potential energy of the displaced ions is

$$\Phi = {}^1\!/_2 \sum_{\binom{l}{k}\neq\binom{l'}{k'}} \tilde{\Phi}\Big[\Big(\mathbf{x}\binom{l}{k}+\mathbf{u}\binom{l}{k}-\mathbf{x}\binom{l'}{k'}-\mathbf{u}\binom{l'}{k'}\Big)_x^2,$$

$$\Big(\mathbf{x}\binom{l}{k}+\mathbf{u}\binom{l}{k}-\mathbf{x}\binom{l'}{k'}-\mathbf{u}\binom{l'}{k'}\Big)_y^2,\ \Big(\mathbf{x}\binom{l}{k}+\mathbf{u}\binom{l}{k}-\mathbf{x}\binom{l'}{k'}-\mathbf{u}\binom{l'}{k'}\Big)_z^2\Big]. \qquad (A.1)$$

Expanding $\tilde{\Phi}$ in series of powers of the differences of displacements, we have

$$\Phi = {}^1\!/_2 \sum_{\binom{l}{k}\neq\binom{l'}{k'}} \Big\{\tilde{\Phi}\Big[\mathbf{x}\binom{ll'}{kk'}_x^2;\ \mathbf{x}\binom{ll'}{kk'}_y^2;\ \mathbf{x}\binom{ll'}{kk'}_z^2\Big] + \sum_{\alpha}\tilde{\Phi}_\alpha\binom{l}{k}_0\Big(\mathbf{u}\binom{l}{k}-\mathbf{u}\binom{l'}{k'}\Big)_\alpha -$$

$$- {}^1\!/_2\sum_{\alpha\beta}\tilde{\Phi}_{\alpha\beta}\binom{ll'}{kk'}_0\Big(\mathbf{u}\binom{l}{k}-\mathbf{u}\binom{l'}{k'}\Big)_\alpha\Big(\mathbf{u}\binom{l}{k}-\mathbf{u}\binom{l'}{k'}\Big)_\beta -$$

$$-\frac{1}{3!}\sum_{\alpha\beta\gamma}\tilde{\Phi}_{\alpha\beta\gamma}\binom{ll\,l'}{kkk'}_0\Big(\mathbf{u}\binom{l}{k}-\mathbf{u}\binom{l'}{k'}\Big)_\alpha\Big(\mathbf{u}\binom{l}{k}-\mathbf{u}\binom{l'}{k'}\Big)_\beta\Big(\mathbf{u}\binom{l}{k}-\mathbf{u}\binom{l'}{k'}\Big)_\gamma -$$

$$-\frac{1}{4!}\sum_{\alpha\beta\gamma\delta}\tilde{\Phi}_{\alpha\beta\gamma\delta}\binom{lll\,l'}{kkkk'}_0\Big(\mathbf{u}\binom{l}{k}-\mathbf{u}\binom{l'}{k'}\Big)_\alpha\Big(\mathbf{u}\binom{l}{k}-\mathbf{u}\binom{l'}{k'}\Big)_\beta\Big(\mathbf{u}\binom{l}{k}-\mathbf{u}\binom{l'}{k'}\Big)_\gamma\Big(\mathbf{u}\binom{l}{k}-\mathbf{u}\binom{l'}{k'}\Big)_\delta\Big\}. \qquad (A.2)$$

The negative signs here arise because of the single differentiation with respect to $u_\alpha\binom{l'}{k'}$, which has a minus sign in front of it.

By direct differentiation of Φ with respect to the displacements, it can be shown that the derivatives of Φ are the same as those of $\tilde{\Phi}$, i.e.,

$$\Phi_{\alpha\beta\ldots\gamma}\binom{ll\ldots l'}{kk\ldots k'}_0 = \tilde{\Phi}_{\alpha\beta\ldots\gamma}\binom{ll\ldots l'}{kk\ldots k'}_0. \qquad (A.3)$$

Substituting Eq. (II.6) in Eq. (A.2) and using Eq. (A.3), we get the following expressions for the coefficients concerned:

$$\Phi\binom{\mathbf{y}\mathbf{y}'}{jj'}_0 = -{}^1\!/_2\frac{1}{N}\sum_{\substack{\binom{l}{k}\neq\binom{l'}{k'}\\ \alpha\beta}}\Phi_{\alpha\beta}\binom{ll'}{kk'}_0\exp\Big[2\pi i(\mathbf{y}+\mathbf{y}')\mathbf{x}\binom{l}{k}\Big]\times$$

$$\times\left(\frac{w_\alpha\left(k\middle|\begin{smallmatrix}\mathbf{y}\\ j\end{smallmatrix}\right)}{\sqrt{m_k}}-\frac{w_\alpha\left(k'\middle|\begin{smallmatrix}\mathbf{y}\\ j\end{smallmatrix}\right)}{\sqrt{m_{k'}}}\exp\Big[-2\pi i\mathbf{y}\mathbf{x}\binom{ll'}{kk'}\Big]\right)\left(\frac{w_\beta\left(k\middle|\begin{smallmatrix}\mathbf{y}'\\ j'\end{smallmatrix}\right)}{\sqrt{m_k}}-\frac{w_\beta\left(k'\middle|\begin{smallmatrix}\mathbf{y}'\\ j'\end{smallmatrix}\right)}{\sqrt{m_{k'}}}\exp\Big[-2\pi i\mathbf{y}'\mathbf{x}\binom{ll'}{kk'}\Big]\right),$$

$$\Phi\binom{\mathbf{y}\mathbf{y}'\mathbf{y}''}{jj'j''}_0 = -{}^1\!/_2\frac{1}{N^{3/2}}\sum_{\substack{\binom{l}{k}\neq\binom{l'}{k'}\\ \alpha\beta\gamma}}\Phi_{\alpha\beta\gamma}\binom{ll\,l'}{kkk'}_0\exp\Big[2\pi i(\mathbf{y}+\mathbf{y}'+\mathbf{y}'')\mathbf{x}\binom{l}{k}\Big]\times$$

$$\times\left(\frac{w_\alpha\left(k\middle|\begin{smallmatrix}\mathbf{y}\\ j\end{smallmatrix}\right)}{\sqrt{m_k}}-\frac{w_\alpha\left(k'\middle|\begin{smallmatrix}\mathbf{y}\\ j\end{smallmatrix}\right)}{\sqrt{m_{k'}}}\exp\Big[-2\pi i\mathbf{y}\mathbf{x}\binom{ll'}{kk'}\Big]\right)\times$$

$$\times\left(\frac{w_\beta\left(k\left|\begin{smallmatrix}\mathbf{y}'\\ j'\end{smallmatrix}\right.\right)}{\sqrt{m_k}}-\frac{w_\beta\left(k'\left|\begin{smallmatrix}\mathbf{y}'\\ j'\end{smallmatrix}\right.\right)}{\sqrt{m_{k'}}}\exp\left[-2\pi i\mathbf{y}'\mathbf{x}\begin{pmatrix}ll'\\ kk'\end{pmatrix}\right]\right)\left(\frac{w_\gamma\left(k\left|\begin{smallmatrix}\mathbf{y}''\\ j''\end{smallmatrix}\right.\right)}{\sqrt{m_k}}-\frac{w_\gamma\left(k'\left|\begin{smallmatrix}\mathbf{y}''\\ j''\end{smallmatrix}\right.\right)}{\sqrt{m_{k'}}}\exp\left[-2\pi i\mathbf{y}'\mathbf{x}\begin{pmatrix}ll'\\ kk'\end{pmatrix}\right]\right),$$

$$\Phi\begin{pmatrix}\mathbf{y}\mathbf{y}'\mathbf{y}''\mathbf{y}'''\\ jj'j''j'''\end{pmatrix}_0=-{}^1\!/_2\frac{1}{N^2}\sum_{\substack{\binom{l}{k}\neq\binom{l'}{k'}\\ \alpha\beta\gamma\delta}}\Phi_{\alpha\beta\gamma\delta}\begin{pmatrix}lll'\\ kkk'\end{pmatrix}_0\exp\left[2\pi i(\mathbf{y}+\mathbf{y}'+\mathbf{y}''+\mathbf{y}''')\mathbf{x}\binom{l}{k}\right]\times$$

$$\times\left(\frac{w_\alpha\left(k\left|\begin{smallmatrix}\mathbf{y}\\ j\end{smallmatrix}\right.\right)}{\sqrt{m_k}}-\frac{w_\alpha\left(k'\left|\begin{smallmatrix}\mathbf{y}\\ j\end{smallmatrix}\right.\right)}{\sqrt{m_{k'}}}\exp\left[-2\pi i\mathbf{y}\mathbf{x}\begin{pmatrix}ll'\\ kk'\end{pmatrix}\right]\right)\left(\frac{w_\beta\left(k\left|\begin{smallmatrix}\mathbf{y}'\\ j'\end{smallmatrix}\right.\right)}{\sqrt{m_k}}-\frac{w_\beta\left(k'\left|\begin{smallmatrix}\mathbf{y}'\\ j'\end{smallmatrix}\right.\right)}{\sqrt{m_{k'}}}\exp\left[-2\pi i\mathbf{y}'\mathbf{x}\begin{pmatrix}ll'\\ kk'\end{pmatrix}\right]\right)\times$$

$$\times\left(\frac{w_\gamma\left(k\left|\begin{smallmatrix}\mathbf{y}''\\ j''\end{smallmatrix}\right.\right)}{\sqrt{m_k}}-\frac{w_\gamma\left(k'\left|\begin{smallmatrix}\mathbf{y}''\\ j''\end{smallmatrix}\right.\right)}{\sqrt{m_{k'}}}\exp\left[-2\pi i\mathbf{y}''\mathbf{x}\begin{pmatrix}ll'\\ kk'\end{pmatrix}\right]\right)\left(\frac{w_\delta\left(k\left|\begin{smallmatrix}\mathbf{y}'''\\ j'''\end{smallmatrix}\right.\right)}{\sqrt{m_k}}-\frac{w_\delta\left(k'\left|\begin{smallmatrix}\mathbf{y}'''\\ j'''\end{smallmatrix}\right.\right)}{\sqrt{m_{k'}}}\exp\left[-2\pi i\mathbf{y}'''\mathbf{x}\begin{pmatrix}ll'\\ kk'\end{pmatrix}\right]\right). \quad (A.4)$$

If the potential energy is unchanged by translation through a lattice vector, then

$$\Phi_{\alpha\beta\gamma}\begin{pmatrix}ll'\\ kkk'\end{pmatrix}_0=\Phi_{\alpha\beta\gamma}\begin{pmatrix}l-l' & l-l' & 0\\ k & k & k'\end{pmatrix}_0,$$

$$\Phi_{\alpha\beta\gamma}\begin{pmatrix}lll'\\ kkkk'\end{pmatrix}_0=\Phi_{\alpha\beta\gamma\delta}\begin{pmatrix}l-l' & l-l' & l-l' & 0\\ k & k & k & k'\end{pmatrix}_0. \quad (A.5)$$

In this case, the coefficients $\Phi\begin{pmatrix}\mathbf{y}\mathbf{y}'\mathbf{y}''\\ jj'j''\end{pmatrix}_0$, $\Phi\begin{pmatrix}\mathbf{y}\mathbf{y}'\mathbf{y}''\mathbf{y}'''\\ jj'j''j'''\end{pmatrix}_0$ become

$$\Phi\begin{pmatrix}\mathbf{y}\mathbf{y}'\mathbf{y}''\\ jj'j''\end{pmatrix}_0=-\frac{1}{2}\frac{1}{N^{1/2}}\Delta(\mathbf{y}+\mathbf{y}'+\mathbf{y}'')\sum_{\substack{\binom{l}{k}\neq\binom{0}{k'}\\ \alpha\beta\gamma}}\Phi_{\alpha\beta\gamma}\begin{pmatrix}ll0\\ kkk'\end{pmatrix}\exp\left[2\pi i(\mathbf{y}+\mathbf{y}'+\mathbf{y}'')\mathbf{x}\binom{0}{k}\right]\left(\frac{w_\alpha\left(k\left|\begin{smallmatrix}\mathbf{y}\\ j\end{smallmatrix}\right.\right)}{\sqrt{m_k}}-\frac{w_\alpha\left(k'\left|\begin{smallmatrix}\mathbf{y}\\ j\end{smallmatrix}\right.\right)}{\sqrt{m_{k'}}}\times\right.$$

$$\left.\times\exp\left[-2\pi i\mathbf{y}\mathbf{x}\binom{l}{kk}\right]\right)\left(\frac{w_\beta\left(k\left|\begin{smallmatrix}\mathbf{y}'\\ j'\end{smallmatrix}\right.\right)}{\sqrt{m_k}}-\frac{w_\beta\left(k'\left|\begin{smallmatrix}\mathbf{y}'\\ j'\end{smallmatrix}\right.\right)}{\sqrt{m_{k'}}}\exp\left[-2\pi i\mathbf{y}'\mathbf{x}\binom{l}{kk'}\right]\right)\left(\frac{w_\gamma\left(k\left|\begin{smallmatrix}\mathbf{y}''\\ j''\end{smallmatrix}\right.\right)}{\sqrt{m_k}}-\frac{w_\gamma\left(k'\left|\begin{smallmatrix}\mathbf{y}''\\ j''\end{smallmatrix}\right.\right)}{\sqrt{m_{k'}}}\exp\left[2\pi i\mathbf{y}''\mathbf{x}\binom{l}{kk'}\right]\right),$$

$$\Phi\begin{pmatrix}\mathbf{y}\mathbf{y}'\mathbf{y}''\mathbf{y}'''\\ jj'j''j'''\end{pmatrix}_0=-\frac{1}{2}\frac{1}{N}\Delta(\mathbf{y}+\mathbf{y}'+\mathbf{y}''+\mathbf{y}''')\sum_{\substack{\binom{l}{k}\neq\binom{0}{k'}\\ \alpha\beta\gamma\delta}}\Phi_{\alpha\beta\gamma\delta}\begin{pmatrix}ll l0\\ kkkk'\end{pmatrix}_0\times$$

$$\times\exp\left[2\pi i(\mathbf{y}+\mathbf{y}'+\mathbf{y}''+\mathbf{y}''')\mathbf{x}\binom{0}{k}\right]\left(\frac{w_\alpha\left(k\left|\begin{smallmatrix}\mathbf{y}\\ j\end{smallmatrix}\right.\right)}{\sqrt{m_k}}-\frac{w_\alpha\left(k'\left|\begin{smallmatrix}\mathbf{y}\\ j\end{smallmatrix}\right.\right)}{\sqrt{m_{k'}}}\times\right.$$

$$\left.\times\exp\left[-2\pi i\mathbf{y}\mathbf{x}\binom{l}{kk'}\right]\right)\left(\frac{w_\beta\left(k\left|\begin{smallmatrix}\mathbf{y}'\\ j'\end{smallmatrix}\right.\right)}{\sqrt{m_k}}-\frac{w_\beta\left(k'\left|\begin{smallmatrix}\mathbf{y}'\\ j'\end{smallmatrix}\right.\right)}{\sqrt{m_{k'}}}\exp\left[-2\pi i\mathbf{y}'\mathbf{x}\binom{l}{kk'}\right]\right)\left(\frac{w_\gamma\left(k\left|\begin{smallmatrix}\mathbf{y}''\\ j''\end{smallmatrix}\right.\right)}{\sqrt{m_k}}-\frac{w_\gamma\left(k'\left|\begin{smallmatrix}\mathbf{y}''\\ j''\end{smallmatrix}\right.\right)}{\sqrt{m_{k'}}}\times\right.$$

$$\left.\times\exp\left[-2\pi i\mathbf{y}''\mathbf{x}\binom{l}{kk'}\right]\right)\left(\frac{w_\delta\left(k\left|\begin{smallmatrix}\mathbf{y}'''\\ j'''\end{smallmatrix}\right.\right)}{\sqrt{m_k}}-\frac{w_\delta\left(k'\left|\begin{smallmatrix}\mathbf{y}'''\\ j'''\end{smallmatrix}\right.\right)}{\sqrt{m_{k'}}}\exp\left[-2\pi i\mathbf{y}'''\mathbf{x}\binom{l}{kk'}\right]\right). \quad (A.6)$$

The most important of the coefficients $\Phi\begin{pmatrix}\mathbf{y}\mathbf{y}'\mathbf{y}''\\ jj'j''\end{pmatrix}_0$ in our case is $\Phi\begin{pmatrix}\boldsymbol{\nu} & -\mathbf{y}\mathbf{y}'\\ j & j'j''\end{pmatrix}_0$, where ν is the small wave vector of the radiation and $\mathbf{y}$ is the wave vector of the long-wave acoustic phonon. Since we are considering wave vectors in the range $-\frac{\mathrm{K}}{2} < \mathrm{y} \leqslant \frac{\mathrm{K}}{2}$, the equation $\nu - \mathbf{y} + \mathbf{y}' = 0$ must be satisfied, so that we can in practive take $\mathbf{y} = \mathbf{y}'$.

If the crystal has a center of symmetry, the expression for $\Phi\begin{pmatrix}\nu & -\mathrm{y} & \mathrm{y} \\ j & j' & j''\end{pmatrix}_0$ becomes

$$\Phi\begin{pmatrix}\nu & -\mathrm{y} & \mathrm{y} \\ j & j' & j''\end{pmatrix}_0 = \frac{i}{2N^{1/2}} \sum_{\binom{l}{k} \neq \binom{0}{k'}} \frac{\Phi_{\alpha\beta\gamma}\begin{pmatrix} l & l & 0 \\ k & k & k'\end{pmatrix}_0}{\sqrt{m_k m_{k'}}} \sin 2\pi \mathrm{y}\mathrm{x}\binom{l}{kk'} \left(\frac{w_\alpha\left(k \middle| \begin{smallmatrix}\nu \\ j\end{smallmatrix}\right)}{\sqrt{m_k}} - \frac{w_\alpha\left(k' \middle| \begin{smallmatrix}\nu \\ j\end{smallmatrix}\right)}{\sqrt{m_{k'}}} \right) \times$$

$$\times \left(w_\beta\left(k' \middle| \begin{smallmatrix}-\mathrm{y} \\ j'\end{smallmatrix}\right) w_\gamma\left(k \middle| \begin{smallmatrix}\mathrm{y} \\ j''\end{smallmatrix}\right) - w_\beta\left(k \middle| \begin{smallmatrix}-\mathrm{y} \\ j'\end{smallmatrix}\right) w_\gamma\left(k' \middle| \begin{smallmatrix}\mathrm{y} \\ j''\end{smallmatrix}\right) \right). \tag{A.7}$$

The products of the form $w_\beta\left(k \middle| \begin{smallmatrix}-\mathrm{y} \\ j'\end{smallmatrix}\right) w_\gamma\left(k \middle| \begin{smallmatrix}\mathrm{y} \\ j''\end{smallmatrix}\right)$ cancel from Eq. (A.7), since, when there is central symmetry

$$\sum_{l \neq 0} \Phi_{\alpha\beta\gamma}\begin{pmatrix} l & l & 0 \\ k & k & k'\end{pmatrix}_0 = 0.$$

If the crystal contains two ions per unit cell, then, by using the relations

$$\Phi_{\alpha\beta\gamma}\begin{pmatrix} -l & -l & 0 \\ 2 & 2 & 1\end{pmatrix}_0 = \Phi_{\alpha\beta\gamma}\begin{pmatrix} 0 & 0 & l \\ 2 & 2 & 1\end{pmatrix}_0 = \Phi_{\alpha\beta\gamma}\begin{pmatrix} l & 0 & 0 \\ 1 & 2 & 2\end{pmatrix}_0 = -\Phi_{\alpha\beta\gamma}\begin{pmatrix} l & l & 0 \\ 1 & 1 & 2\end{pmatrix}_0, \tag{A.8}$$

we can put Eq. (A.7) in the form

$$\Phi\begin{pmatrix}\nu & -\mathrm{y} & \mathrm{y} \\ j & j' & j''\end{pmatrix}_0 = \frac{i}{N^{1/2}} \sum_{\substack{l \neq 0 \\ \alpha\beta\gamma}} \frac{\Phi_{\alpha\beta\gamma}\begin{pmatrix} l & 0 & 0 \\ 1 & 2 & 2\end{pmatrix}}{\sqrt{m_1 m_2}} \sin 2\pi \mathrm{y}\mathrm{x}\binom{l}{12} \left(\frac{w_\alpha\left(1 \middle| \begin{smallmatrix}\nu \\ j\end{smallmatrix}\right)}{\sqrt{m_1}} - \frac{w_\alpha\left(2 \middle| \begin{smallmatrix}\nu \\ j\end{smallmatrix}\right)}{\sqrt{m_2}} \right) \times$$

$$\times \left(w_\beta\left(1 \middle| \begin{smallmatrix}-\mathrm{y} \\ j'\end{smallmatrix}\right) w_\gamma\left(2 \middle| \begin{smallmatrix}\mathrm{y} \\ j''\end{smallmatrix}\right) - w_\beta\left(2 \middle| \begin{smallmatrix}-\mathrm{y} \\ j'\end{smallmatrix}\right) w_\gamma\left(1 \middle| \begin{smallmatrix}\mathrm{y} \\ j''\end{smallmatrix}\right) \right). \tag{A.9}$$

It will be more convenient to write $\Phi\begin{pmatrix}\mathrm{y} & \mathrm{y}' \\ j & j'\end{pmatrix}_0$ in a somewhat different form. Multiplying the expressions in the parentheses in Eq. (A.4) and changing the variables of summation, we get

$$\Phi\begin{pmatrix}\mathrm{y} & \mathrm{y}' \\ j & j'\end{pmatrix}_0 = -\frac{1}{N} \sum_{\substack{\binom{l}{k} \neq \binom{l'}{k'} \\ \alpha\beta}} \Phi_{\alpha\beta}\begin{pmatrix} l & l' \\ k & k'\end{pmatrix}_0 \frac{w_\alpha\left(k \middle| \begin{smallmatrix}\mathrm{y} \\ j\end{smallmatrix}\right)}{\sqrt{m_k}} \frac{w_\beta\left(k' \middle| \begin{smallmatrix}\mathrm{y}' \\ j'\end{smallmatrix}\right)}{\sqrt{m_k}} \exp\left[2\pi i(\mathrm{y} + \mathrm{y}')\mathrm{x}\binom{l}{k}\right] +$$

$$+\frac{1}{N} \sum_{\substack{\binom{l}{k} \neq \binom{l'}{k'} \\ \alpha\beta}} \Phi_{\alpha\beta}\begin{pmatrix} l & l' \\ k & k'\end{pmatrix}_0 \frac{w_\alpha\left(k \middle| \begin{smallmatrix}\mathrm{y} \\ j\end{smallmatrix}\right)}{\sqrt{m_k}} \frac{w_\beta\left(k' \middle| \begin{smallmatrix}\mathrm{y}' \\ j'\end{smallmatrix}\right)}{\sqrt{m_{k'}}} \exp\left[2\pi i\left(\mathrm{y}\mathrm{x}\binom{l}{k} + \mathrm{y}'\mathrm{x}\binom{l'}{k'}\right)\right]. \tag{A.10}$$

APPENDIX B

Calculation of the Polarization Vectors $\mathrm{w}_\alpha\left(\mathrm{k} \middle| \begin{smallmatrix}\mathrm{y} \\ \mathrm{j}\end{smallmatrix}\right)$ and the Frequencies $\omega\binom{\mathrm{y}}{\mathrm{j}}$ for small $|\mathrm{y}|$ in the Isotropic Case

We shall solve Eq. (II.9) by expressing $C_{\alpha\beta}\binom{\mathrm{y}}{kk'}$, $w_\alpha\left(k \middle| \begin{smallmatrix}\mathrm{y} \\ j\end{smallmatrix}\right)$, and $\omega\binom{\mathrm{y}}{j}$ as expansions in powers of y (see [11])

$$C_{\alpha\beta}\left(\begin{smallmatrix}\mathbf{y}\\ kk'\end{smallmatrix}\right) = C^{(0)}_{\alpha\beta}\left(\begin{smallmatrix}\mathbf{y}\\ kk'\end{smallmatrix}\right) + i\sum_{\gamma}\bar{C}^{(1)}_{\alpha\beta,\gamma}(kk')\,y_\gamma + {}^1/_2\sum_{\gamma\lambda}\bar{C}^{(2)}_{\alpha\beta,\gamma\lambda}(kk')\,y_\gamma y_\lambda + \cdots,$$
$$w_\alpha\left(k\left|\begin{smallmatrix}\mathbf{y}\\ j\end{smallmatrix}\right.\right) = w^{(0)}_\alpha\left(k\left|\begin{smallmatrix}\mathbf{y}\\ j\end{smallmatrix}\right.\right) + iw^{(1)}_\alpha\left(k\left|\begin{smallmatrix}\mathbf{y}\\ j\end{smallmatrix}\right.\right) + {}^1/_2 w^{(2)}_\alpha\left(k\left|\begin{smallmatrix}\mathbf{y}\\ j\end{smallmatrix}\right.\right) + \cdots,$$
$$\omega\left(\begin{smallmatrix}\mathbf{y}\\ j\end{smallmatrix}\right) = \omega^{(0)}\left(\begin{smallmatrix}\mathbf{y}\\ j\end{smallmatrix}\right) + \omega^{(1)}\left(\begin{smallmatrix}\mathbf{y}\\ j\end{smallmatrix}\right) + \omega^{(2)}\left(\begin{smallmatrix}\mathbf{y}\\ j\end{smallmatrix}\right) + \cdots . \tag{B.1}$$

The expressions for the expansion coefficients $\bar{C}^{(0)}_{\alpha\beta}(kk')$, $\bar{C}^{(1)}_{\alpha\beta,\gamma}(kk')$, and $\bar{C}^{(2)}_{\alpha\beta,\gamma\lambda}(kk')$ in an ionic crystal are very complicated. They are given in [11]. The coefficient $C^{(0)}_{\alpha\beta}\left(\begin{smallmatrix}\mathbf{y}\\ kk'\end{smallmatrix}\right)$ may be written as

$$C^{(0)}_{\alpha\beta}\left(\begin{smallmatrix}\mathbf{y}\\ kk'\end{smallmatrix}\right) = \bar{C}^{(0)}_{\alpha\beta}(kk') + \frac{4\pi e_k e_{k'}}{v_a\sqrt{m_k m_{k'}}}\,\frac{y_\alpha y_\beta}{|\mathbf{y}|^2}. \tag{B.2}$$

The bar in the coefficients $\bar{C}^{(0)}_{\alpha\beta}(kk')$, $\bar{C}^{(1)}_{\alpha\beta,\gamma}(kk')$, and $\bar{C}^{(2)}_{\alpha\beta,\gamma\lambda}(kk')$ signifies that the long-range part of the Coulomb interaction has been separated from these expressions.

Let us first consider acoustic vibrations. Then $\omega^{(0)}\left(\begin{smallmatrix}\mathbf{y}\\ j\end{smallmatrix}\right)=0$. We shall also assume that the crystal has central symmetry; then $\bar{C}^{(1)}_{\alpha\beta\gamma}(kk') = 0$, and also $w^{(1)}_\alpha\left(k\left|\begin{smallmatrix}\mathbf{y}\\ j\end{smallmatrix}\right.\right)=0$ (see [11] for further details).

Substituting Eq. (B.1) in Eq. (II.9) and collecting terms of various orders, we obtain the following equations for $w^{(0)}_\alpha\left(k\left|\begin{smallmatrix}\mathbf{y}\\ j\end{smallmatrix}\right.\right)$ and $w^{(2)}_\alpha\left(k\left|\begin{smallmatrix}\mathbf{y}\\ j\end{smallmatrix}\right.\right)$:

$$\sum_{k'\beta} C^{(0)}_{\alpha\beta}\left(\begin{smallmatrix}\mathbf{y}\\ kk'\end{smallmatrix}\right) w^{(0)}_\beta\left(k'\left|\begin{smallmatrix}\mathbf{y}\\ j\end{smallmatrix}\right.\right) = 0, \tag{B.3}$$

$$\sum_{k'\beta} C^{(0)}_{\alpha\beta}\left(\begin{smallmatrix}\mathbf{y}\\ kk'\end{smallmatrix}\right) w^{(2)}_\beta\left(k'\left|\begin{smallmatrix}\mathbf{y}\\ j\end{smallmatrix}\right.\right) = -\sum_{k'\beta\gamma\lambda}\bar{C}^{(2)}_{\alpha\beta,\gamma\lambda}(kk')\,y_\gamma y_\lambda w^{(0)}_\beta\left(k'\left|\begin{smallmatrix}\mathbf{y}\\ j\end{smallmatrix}\right.\right) + 2\left[\omega^{(1)}\left(\begin{smallmatrix}\mathbf{y}\\ j\end{smallmatrix}\right)\right]^2 w^{(0)}_\alpha\left(k\left|\begin{smallmatrix}\mathbf{y}\\ j\end{smallmatrix}\right.\right). \tag{B.4}$$

From the conditions (see [11])

$$\sum_{k'\beta} C^{(0)}_{\alpha\beta}\left(\begin{smallmatrix}\mathbf{y}\\ kk'\end{smallmatrix}\right)\sqrt{m_{k'}} = \sum_{k'\beta}\bar{C}^{(0)}_{\alpha\beta}(kk')\sqrt{m_{k'}} = 0, \tag{B.5}$$

it follows that the solution of Eqs. (B.3) is

$$w^{(0)}_\alpha\left(k\left|\begin{smallmatrix}\mathbf{y}\\ j\end{smallmatrix}\right.\right) = \sqrt{m_k}\,u_\alpha(j), \tag{B.6}$$

where $u_\alpha(j)$ is a certain vector.

Equations (B.4) are in general too complicated. In order to simplify them, we average over directions in the crystal, i.e., make an anisotropic crystal isotropic. We can derive expressions relating the coefficients in Eqs. (B.4) for an isotropic crystal to those for an anisotropic crystal. We shall use the following properties of the coefficients $\bar{C}^{(2)}_{\alpha\beta,\gamma\lambda}(kk')$ (see [11]):

$$\bar{C}^{(2)}_{\alpha\beta,\gamma\lambda}(kk') = \bar{C}^{(2)}_{\alpha\beta,\lambda\gamma}(kk'), \tag{B.7}$$

$$\overline{C}^{(2)}_{\alpha\beta,\gamma\lambda}(kk') = \overline{C}^{(2)}_{\beta\alpha,\gamma\lambda}(k'k). \tag{B.8}$$

For forces between pairs of ions, this tensor is also symmetrical in the indices α and β, i.e.,

$$\overline{C}^{(2)}_{\alpha\beta,\gamma\lambda}(kk') = \overline{C}^{(2)}_{\beta\alpha,\gamma\lambda}(kk'). \tag{B.9}$$

The general form of a tensor dependent on four indices in the isotropic case is

$$\overline{C}^{(2)}_{\alpha\beta,\gamma\lambda}(kk') = a(kk')\,\delta_{\alpha\beta}\delta_{\gamma\lambda} + b(kk')\,\delta_{\alpha\lambda}\delta_{\beta\gamma} + d(kk')\,\delta_{\alpha\gamma}\delta_{\beta\lambda}. \tag{B.10}$$

From Eq. (B.7), it follows that b(kk') = d(kk'), and Eq. (B.10) becomes

$$\overline{C}^{(2)}_{\alpha\beta,\gamma\lambda}(kk') = a(kk')\,\delta_{\alpha\beta}\delta_{\gamma\lambda} + b(kk')\,\{\delta_{\alpha\gamma}\delta_{\beta\lambda} + \delta_{\alpha\lambda}\delta_{\beta\gamma}\}. \tag{B.11}$$

The form of this tensor shows that it is symmetrical in α and β, and therefore must also be symmetrical in γ and λ; see Eq. (B.8).

Expressions for a (kk') and b (kk') can be obtained by means of various contractions of the tensor. They are

$$\begin{aligned} a(kk') &= {}^{1}/_{15}\,(2\overline{C}^{(2)}_{\alpha\alpha,\gamma\gamma}(kk') - \overline{C}^{(2)}_{\alpha\beta,\alpha\beta}(kk')),\\ b(kk') &= {}^{1}/_{10}\,(\overline{C}^{(2)}_{\alpha\beta,\alpha\beta}(kk') - {}^{1}/_{3}\,\overline{C}^{(2)}_{\alpha\alpha,\gamma\gamma}(kk')). \end{aligned} \tag{B.12}$$

In these equations, summation over repeated indices is implied.

The tensor $\overline{C}^{(0)}_{\alpha\beta}(kk')$ in the isotropic case is

$$\overline{C}^{(0)}_{\alpha\beta}(kk') = C(kk')\,\delta_{\alpha\beta}, \tag{B.13}$$

where

$$C(kk') = {}^{1}/_{3}\,\overline{C}^{(0)}_{\alpha\alpha}(kk'). \tag{B.14}$$

Thus we have the relations (B.11)-(B.14) between the coefficients for an anisotropic crystal and those for a crystal which has been averaged over directions.

Substitution of Eqs. (B.11) and (B.13) in Eq. (B.4) gives

$$\begin{aligned} &\sum_{k'} C(kk')\, w^{(2)}_{\alpha}\left(k'\,\middle|\,\begin{matrix}\mathbf{y}\\ j\end{matrix}\right) + \sum_{k'\beta}\left(\frac{4\pi}{v_a}\frac{e_k e_{k'}}{\sqrt{m_k m_{k'}}}\frac{y_\alpha y_\beta}{y^2}\right) w^{(2)}_{\beta}\left(k'\,\middle|\,\begin{matrix}\mathbf{y}\\ j\end{matrix}\right) = \\ &= -\sum_{k'} a(kk')\, y^2 w^{(0)}_{\alpha}\left(k'\,\middle|\,\begin{matrix}\mathbf{y}\\ j\end{matrix}\right) - 2\sum_{k'} b(kk')\, y_\alpha\left(\mathbf{y}\mathbf{w}^{(0)}\left(k'\,\middle|\,\begin{matrix}\mathbf{y}\\ j\end{matrix}\right)\right) + 2\left[\omega^{(1)}\begin{pmatrix}\mathbf{y}\\ j\end{pmatrix}\right]^2 w^{(0)}_{\alpha}\left(k\,\middle|\,\begin{matrix}\mathbf{y}\\ j\end{matrix}\right). \end{aligned} \tag{B.15}$$

For longitudinal vibrations, and with the assumption that the direction α is parallel to $\mathbf{y}$, Eq. (B.15) becomes

$$\sum_{k'} C(kk')^l w^{(2)}_{\alpha}\left(k'\,\middle|\,\begin{matrix}\mathbf{y}\\ j\end{matrix}\right) = -\sum_{k'} C^{(2)}(kk')^l y^2\, w^{(0)}_{\alpha}\left(k'\,\middle|\,\begin{matrix}\mathbf{y}\\ j\end{matrix}\right) + 2\left[\omega^{(1)}\begin{pmatrix}\mathbf{y}\\ j\end{pmatrix}\right]^2 w^{(0)}_{\alpha}\left(k\,\middle|\,\begin{matrix}\mathbf{y}\\ l\end{matrix}\right), \tag{B.16}$$

where

$$C\,(kk')^l = C\,(kk') + \frac{4\pi e_k e_{k'}}{v_a \sqrt{m_k m_{k'}}}, \tag{B.17}$$

$$C^{(2)}\,(kk')^l = a\,(kk') + 2b\,(kk'). \tag{B.18}$$

For transverse vibrations,

$$\sum_{k'} C\,(kk')^t w_\alpha^{(2)}\left(k' \,\middle|\, \begin{matrix} \mathrm{y} \\ j \end{matrix}\right) = -\sum_{k'} C^{(2)}\,(kk')^t y^2 w_\alpha^{(0)}\left(k' \,\middle|\, \begin{matrix} \mathrm{y} \\ j \end{matrix}\right) + 2\left[\omega^{(1)}\begin{pmatrix} \mathrm{y} \\ j \end{pmatrix}\right]^2 w_\alpha^{(0)}\left(k \,\middle|\, \begin{matrix} \mathrm{y} \\ j \end{matrix}\right), \tag{B.19}$$

where

$$C\,(kk')^t = C\,(kk'), \tag{B.20}$$

$$C^{(2)}\,(kk')^t = a\,(kk'). \tag{B.21}$$

Let us now determine from Eqs. (B.16) and (B.19) the quantities $w_\alpha^{(2)}\left(k \,\middle|\, \begin{matrix} \mathrm{y} \\ j \end{matrix}\right)$, $\omega^{(1)}\begin{pmatrix} \mathrm{y} \\ j \end{pmatrix}$ for a crystal containing two ions per unit cell. Since the coefficients $C_{\alpha\beta}^{(0)}\begin{pmatrix} \mathrm{y} \\ kk' \end{pmatrix}$ (and therefore the coefficients $C\,(kk')^l$, $C\,(kk')^t$) are linearly related [see Eq. (B.5)], the determinants of Eqs. (B.16) and (B.19) are zero. The condition for a set of inhomogeneous equations with zero determinant to be soluble is that the solutions of the homogeneous equations should be orthogonal to the inhomogeneous part. This condition yields an expression for $\omega^{(1)}\begin{pmatrix} \mathrm{y} \\ l,t \end{pmatrix}^2$. If we multiply Eqs. (B.16) and (B.19) by $\sqrt{m_k}$, sum over k, and cancel the vector $u_\alpha(j)$, we get

$$\left[\omega^{(1)}\begin{pmatrix} \mathrm{y} \\ l,t \end{pmatrix}\right]^2 = \frac{1}{2\sum_k m_k} \sum_{kk'} \sqrt{m_k m_{k'}} \cdot C^{(2)}\,(kk')^{l,t} y^2. \tag{B.22}$$

To determine $w_\alpha^{(2)}\left(k \,\middle|\, \begin{matrix} \mathrm{y} \\ j \end{matrix}\right)$ from Eqs. (B.16) and (B.19), we can use one equation from each set, for example those with k = 2. This is possible because the equations with k = 1 and 2 are linearly related. We can also take $w_\alpha^{(2)}\left(1 \,\middle|\, \begin{matrix} \mathrm{y} \\ j \end{matrix}\right) = 0$ by adding a solution of the homogeneous equations to that of the inhomogeneous equations.

Using Eqs. (B.16), (B.19), and (B.22), we find

$$w_\alpha^{(2)}\left(2 \,\middle|\, \begin{matrix} \mathrm{y} \\ l,t \end{matrix}\right) = \sqrt{m_2}\, y^2 f^{l,t} u_\alpha\,(j), \tag{B.23}$$

where

$$f^{l,t} = \frac{1}{C\,(2,2)^{l,t}} \left\{ \frac{1}{m_1 + m_2} \sum_{kk'} \sqrt{m_k m_{k'}}\, C^{(2)}\,(kk')^{l,t} - \frac{1}{\sqrt{m_2}} \sum_{k'} C^{(2)}\,(2,k')^{l,t} \sqrt{m_{k'}} \right\}. \tag{B.24}$$

Thus, the solutions are

$$\begin{aligned} w_\alpha\left(1 \,\middle|\, \begin{matrix} \mathrm{y} \\ j \end{matrix}\right) &= \sqrt{m_1}\, u_\alpha\,(j), \\ w_\alpha\left(2 \,\middle|\, \begin{matrix} \mathrm{y} \\ j \end{matrix}\right) &= \sqrt{m_2}\, u_\alpha\,(j)\,(1 + {}^1/_2 f^j y^2). \end{aligned} \tag{B.25}$$

If $u_\alpha(j)$ is regarded as an arbitrary function, which is equivalent to multiplying the solutions (B.25) by an arbitrary factor, these solutions can be normalized. The normalization conditions (II.13) give

$$w_\alpha\left(1\left|\begin{smallmatrix} y \\ l,t \end{smallmatrix}\right.\right)=\sqrt{\frac{m_1}{m_1+m_2}}\left(1-\frac{m_2}{2(m_1+m_2)}f^{l,t}y^2\right),$$

$$w_\alpha\left(2\left|\begin{smallmatrix} y \\ l,t \end{smallmatrix}\right.\right)=\sqrt{\frac{m_1}{m_1+m_2}}\left(1+\frac{m_1}{2(m_1+m_2)}f^{l,t}y^2\right). \qquad \text{(B.26)}$$

Let us now calculate $f^{l,t}$. Carrying out the summation over k and k' in (B.24), we have

$$f^{l,t}=\frac{1}{C(2,2)^{l,t}}\left\{\frac{m_1}{m_1+m_2}\left(C^{(2)}(1,1)^{l,t}-C^{(2)}(2,2)^{l,t}\right)+\sqrt{\frac{m_1}{m_2}}\frac{m_2-m_1}{m_1+m_2}C^{(2)}(1,2)^{l,t}\right\}, \qquad \text{(B.27)}$$

where

$$C(2,2)^l={}^1/_3\,\overline{C}^{(0)}_{\alpha\alpha}(2,2)+\frac{4\pi e^2}{v_a m_2},$$

$$C^{(2)}(kk')^l={}^1/_{15}\left(\overline{C}^{(2)}_{\alpha\alpha,\gamma\gamma}(kk')+2\overline{C}^{(2)}_{\alpha\beta,\alpha\beta}(kk')\right),$$

$$C(2,2)^t={}^1/_3\,\overline{C}^{(0)}_{\alpha\alpha}(2,2), \qquad \text{(B.28)}$$

$$C^{(2)}(kk')^t={}^1/_{15}\left(2\overline{C}^{(2)}_{\alpha\alpha,\gamma\gamma}(kk')-\overline{C}^{(2)}_{\alpha\beta,\alpha\beta}(kk')\right).$$

Using Eqs. (31.22) in [11], we obtain

$$\overline{C}^{(0)}_{\alpha\alpha}(kk')=\frac{1}{(m_k m_{k'})^{1/2}}\sum_l \Phi^N_{\alpha\alpha}\binom{l}{kk'}-\frac{4\pi e_k e_{k'}}{v_a(m_k m_{k'})^{1/2}},$$

$$\overline{C}^{(2)}_{\alpha\alpha,\beta\beta}(kk')=-\frac{4\pi^2}{(m_k m_{k'})^{1/2}}\sum_l \Phi^N_{\alpha\alpha}\binom{l}{kk'}x_\beta\binom{l}{kk'}x_\beta\binom{l}{kk'}, \qquad \text{(B.29)}$$

$$\overline{C}^{(2)}_{\alpha\beta,\alpha\beta}(kk')=-\frac{4\pi^2}{(m_k m_{k'})^{1/2}}\sum_l \Phi^N_{\alpha\beta}\binom{l}{kk'}x_\alpha\binom{l}{kk'}x_\beta\binom{l}{kk'},$$

where Φ^N denotes the non-Coulomb part of the potential energy. Summation over the repeated indices α and β is implied.

The coefficients (B.28) can be evaluated, assuming that in Φ^N the interaction between nearest neighbors is predominant. We put

$$\Phi^N_{xx}\binom{a_x}{12}=\Phi^N_{yy}\binom{a_y}{12}=\Phi^N_{zz}\binom{a_z}{12}=-\beta,$$

$$\Phi^N_{xx}\binom{0}{kk}=\Phi^N_{yy}\binom{0}{kk}=\Phi^N_{zz}\binom{0}{kk}=2\beta. \qquad \text{(B.30)}$$

Then,

$$C(2,2)^t=\frac{2\beta}{m_2}-\frac{4\pi}{3v_a}\frac{e^2}{m_2}, \qquad C(2,2)^l=\frac{2\beta}{m_2}+\frac{8\pi}{3v_a}\frac{e^2}{m_2},$$

$$C(1,2)^t=-\frac{2\beta}{\sqrt{m_1m_2}}+\frac{4\pi}{3v_a}\frac{e^2}{\sqrt{m_1m_2}}, \qquad C(1,2)^l=-\frac{2\beta}{\sqrt{m_1m_2}}-\frac{8\pi}{3v_a}\frac{e^2}{\sqrt{m_1m_2}}, \qquad \text{(B.31)}$$

$$C^{(2)}(1,2)^t=\frac{4\pi^2}{\sqrt{m_1m_2}}\beta\,{}^2/_5\,a^2, \qquad C^{(2)}(1,2)^l=\frac{4\pi^2}{\sqrt{m_1m_2}}\beta\,{}^6/_5\,a^2,$$

$$C^{(2)}(kk)^t=0, \qquad C^{(2)}(kk)^l=0.$$

Substitution of Eq. (B.31) in Eq. (B.27) gives

$$
\begin{aligned}
f^l &= \frac{m_2 - m_1}{m_1 + m_2} \frac{6\beta}{5\left(2\beta + \frac{8\pi e^2}{3v_a}\right)} 4\pi^2 a^2, \\
f^t &= \frac{m_2 - m_1}{m_1 + m_2} \frac{2\beta}{5\left(2\beta - \frac{4\pi e^2}{3v_a}\right)} 4\pi^2 a^2.
\end{aligned}
\tag{B.32}
$$

We can also derive expressions for the acoustic frequencies when $|\mathbf{y}|$ is small. Using Eqs. (B.22) and (B.31), we find

$$
\begin{aligned}
\omega\begin{pmatrix} \mathbf{y} \\ a\parallel \end{pmatrix}^2 &= 4\pi^2 v_\parallel^2 y^2, \quad v_\parallel^2 = \frac{6\beta a^2}{5(m_1 + m_2)}, \\
\omega\begin{pmatrix} \mathbf{y} \\ a\perp \end{pmatrix}^2 &= 4\pi^2 v_\perp^2 y^2, \quad v_\perp^2 = \frac{2\beta a^2}{5(m_1 + m_2)}.
\end{aligned}
\tag{B.33}
$$

Let us now calculate $\omega\begin{pmatrix} \mathbf{y} \\ j \end{pmatrix}$ and $w_\alpha\left(k \middle| \begin{smallmatrix} \mathbf{y} \\ j \end{smallmatrix}\right)$ for small $|\mathbf{y}|$ at optical frequencies. In this case, $\omega^{(0)}\begin{pmatrix} \mathbf{y} \\ j \end{pmatrix} \neq 0,$ and Eqs. (II.9) for the zero-order quantities become

$$
\left[\omega^{(0)}\begin{pmatrix} \mathbf{y} \\ l,t \end{pmatrix}\right]^2 w_\alpha^{(0)}\left(k \middle| \begin{matrix} \mathbf{y} \\ j \end{matrix}\right) = \sum_{k'} C\,(kk')^{l,t} w_\alpha^{(0)}\left(k' \middle| \begin{matrix} \mathbf{y} \\ j \end{matrix}\right).
\tag{B.34}
$$

The solutions of these equations are

$$
\begin{aligned}
w_\alpha^{(0)}\left(1 \middle| \begin{matrix} \mathbf{y} \\ j \end{matrix}\right) &= \sqrt{\frac{m_2}{m_1 + m_2}}, \\
w_\alpha^{(0)}\left(2 \middle| \begin{matrix} \mathbf{y} \\ j \end{matrix}\right) &= -\sqrt{\frac{m_2}{m_1 + m_2}}.
\end{aligned}
\tag{B.35}
$$

The transverse and longitudinal optical frequencies are

$$
\begin{aligned}
\omega^{(0)}\begin{pmatrix} \mathbf{y} \\ o\perp \end{pmatrix}^2 &= \left(2\beta - \frac{4\pi}{3v_a} e^2\right)\frac{1}{M}, \\
\omega^{(0)}\begin{pmatrix} \mathbf{y} \\ o\parallel \end{pmatrix}^2 &= \left(2\beta + \frac{8\pi}{3v_a} e^2\right)\frac{1}{M},
\end{aligned}
\tag{B.36}
$$

where M is the reduced mass.

The relations (B.26) and (B.35) determine the polarization vectors to within a factor having modulus unity. The condition $w_\alpha^*\left(k \middle| \begin{smallmatrix} \mathbf{y} \\ j \end{smallmatrix}\right) = w_\alpha\left(k \middle| \begin{smallmatrix} -\mathbf{y} \\ j \end{smallmatrix}\right)$ given in Chapter II shows that the polarization vectors may be taken to be real, and then $w_\alpha\left(k \middle| \begin{smallmatrix} \mathbf{y} \\ j \end{smallmatrix}\right)$ is unchanged when the sign of **y** is reversed; or they may be taken to be imaginary, and then they change sign when **y** is inverted.

We shall take the polarization vectors to be imaginary, and define unit vectors $\mathbf{e}_\parallel(\mathbf{y})$, $\mathbf{e}_\perp(\mathbf{y})$ such that $\mathbf{e}_\parallel(-\mathbf{y}) = -\mathbf{e}_\parallel(\mathbf{y})$, $\mathbf{e}_\perp(-\mathbf{y}) = -\mathbf{e}_\perp(\mathbf{y})$. These vectors are respectively parallel and perpendicular to **y**. They can be used in expressions for the polarization vectors, as follows.

For long-wave acoustic vibrations,

$$
\begin{aligned}
\mathbf{w}\left(1 \middle| \begin{matrix} \mathbf{y} \\ \parallel,\perp \end{matrix}\right) &= i\sqrt{\frac{m_1}{m_1 + m_2}}\left(1 - \frac{m_2}{2(m_1 + m_2)} f^{\parallel,\perp} y^2\right)\begin{Bmatrix} \mathbf{e}_\parallel(\mathbf{y}) \\ \mathbf{e}_\perp(\mathbf{y}) \end{Bmatrix}, \\
\mathbf{w}\left(2 \middle| \begin{matrix} \mathbf{y} \\ \parallel,\perp \end{matrix}\right) &= i\sqrt{\frac{m_2}{m_1 + m_2}}\left(1 + \frac{m_1}{2(m_1 + m_2)} f^{\parallel,\perp} y^2\right)\begin{Bmatrix} \mathbf{e}_\parallel(\mathbf{y}) \\ \mathbf{e}_\perp(\mathbf{y}) \end{Bmatrix},
\end{aligned}
$$

where

$$f^{\parallel}=\frac{m_2-m_1}{M}\frac{v_{\parallel}^2}{\omega^2\begin{pmatrix}\mathbf{y}\\ o\parallel\end{pmatrix}}4\pi^2,$$

$$f^{\perp}=\frac{m_2-m_1}{M}\frac{v_{\perp}^2}{\omega^2\begin{pmatrix}\mathbf{y}\\ o\perp\end{pmatrix}}4\pi^2,$$

$$v_{\parallel}^2=\frac{6\beta a^2}{5(m_1+m_2)},$$

$$v_{\perp}^2=\frac{2\beta a^2}{5(m_1+m_2)}. \tag{B.37}$$

For long-wave optical vibrations,

$$\mathbf{w}\left(1\left|\begin{matrix}\mathbf{y}\\ \parallel,\perp\end{matrix}\right.\right)=i\sqrt{\frac{m_2}{m_1+m_2}}\begin{Bmatrix}\mathbf{e}_{\parallel}(\mathbf{y})\\ \mathbf{e}_{\perp}(\mathbf{y})\end{Bmatrix},$$

$$\mathbf{w}\left(2\left|\begin{matrix}\mathbf{y}\\ \parallel,\perp\end{matrix}\right.\right)=-i\sqrt{\frac{m_1}{m_1+m_2}}\begin{Bmatrix}\mathbf{e}_{\parallel}(\mathbf{y})\\ \mathbf{e}_{\perp}(\mathbf{y})\end{Bmatrix},$$

$$\omega^2\begin{pmatrix}\mathbf{y}\\ o\parallel\end{pmatrix}=\frac{2\beta}{M}+\frac{8\pi e^2}{3v_aM},$$

$$\omega^2\begin{pmatrix}\mathbf{y}\\ o\perp\end{pmatrix}=\frac{2\beta}{M}-\frac{4\pi e^2}{3v_aM}, \tag{B.38}$$

$$\sum_k\frac{e_k\left(\mathbf{w}\left(k\left|\begin{matrix}\mathbf{y}\\ \parallel,\perp\end{matrix}\right.\right)\mathbf{e}_{\parallel,\perp}\right)}{\sqrt{m_k}}=i\frac{e_1}{\sqrt{M}}.$$

APPENDIX C

Applicability of the Born Approximation in Calculating the Scattering of Lattice Waves by Charged Defects

In order to assess the limits of applicability of the Born approximation in calculating the scattering of lattice waves by charged defects, we must find expressions which take account of the scattering in higher-order approximations of perturbation theory, and estimate their magnitudes. This can be done by means of the methods used in Chapter IV, but here, for clarity, diagram methods will be applied. We begin from the classical equations of motion of ions in a lattice containing defects. A Fourier transformation with respect to time brings these equations to the form

$$-m_k\omega^2u_\alpha\begin{pmatrix}l\\ k\end{pmatrix}+\sum_{l'k'\beta}\left[\Phi_{\alpha\beta}\begin{pmatrix}ll'\\ kk'\end{pmatrix}_0+\delta\tilde{\Phi}_{\alpha\beta}\begin{pmatrix}ll'\\ kk'\end{pmatrix}_0\right]u_\beta\begin{pmatrix}l'\\ k'\end{pmatrix}=0, \tag{C.1}$$

where

$$\delta\tilde{\Phi}_{\alpha\beta}\begin{pmatrix}ll'\\ kk'\end{pmatrix}_0=\delta\Phi_{\alpha\beta}\begin{pmatrix}ll'\\ kk'\end{pmatrix}_0+\sum_{l''k''\gamma}\Phi_{\alpha\beta\gamma}\begin{pmatrix}l\,l'\,l''\\ kk'k''\end{pmatrix}_0v_\gamma\begin{pmatrix}l''\\ k''\end{pmatrix}. \tag{C.2}$$

The free Green's function $g_{\alpha\beta}\begin{pmatrix}ll''\\ kk''\end{pmatrix}$ is defined by means of the equation

$$-m_k\omega^2g_{\alpha\gamma}\begin{pmatrix}ll''\\ kk''\end{pmatrix}+\sum_{l'k'\beta}\Phi_{\alpha\beta}\begin{pmatrix}ll'\\ kk'\end{pmatrix}_0g_{\beta\gamma}\begin{pmatrix}l'l''\\ k'k''\end{pmatrix}=\delta_{\alpha\gamma}\delta_{kk''}\delta_{ll''}. \tag{C.3}$$

Eq. (C.3) gives

$$g_{\alpha\gamma}\begin{pmatrix} ll'' \\ kk'' \end{pmatrix} = \sum_{\mathbf{y}j} \frac{\chi_\alpha\left(\begin{smallmatrix} l \\ k \end{smallmatrix}\middle|\begin{smallmatrix} \mathbf{y} \\ j \end{smallmatrix}\right)\chi^{*}_\gamma\left(\begin{smallmatrix} l'' \\ k'' \end{smallmatrix}\middle|\begin{smallmatrix} \mathbf{y} \\ j \end{smallmatrix}\right)}{\omega_j^2(\mathbf{y}) - \omega^2}, \tag{C.4}$$

where

$$\chi_\alpha\left(\begin{smallmatrix} l \\ k \end{smallmatrix}\middle|\begin{smallmatrix} \mathbf{y} \\ j \end{smallmatrix}\right) = \frac{1}{\sqrt{Nm_k}}\, w_\alpha\left(k\middle|\begin{smallmatrix} \mathbf{y} \\ j \end{smallmatrix}\right)\exp\left[2\pi i\mathbf{y}\mathbf{x}\begin{pmatrix} l \\ k \end{pmatrix}\right]. \tag{C.5}$$

By means of the equations of motion (II.9), it is easy to verify that the function (C.4) satisfies Eq. (C.3).

Let us now consider the equation for the Green's function $\mathcal{G}_{\alpha\gamma}\begin{pmatrix} ll'' \\ kk'' \end{pmatrix}$, which takes account of the influence of the defects:

$$-\omega^2 m_k \mathcal{G}_{\alpha\gamma}\begin{pmatrix} ll'' \\ kk'' \end{pmatrix} + \sum_{l'k'\beta}\left[\Phi_{\alpha\beta}\begin{pmatrix} ll' \\ kk' \end{pmatrix}_0 + \delta\widetilde{\Phi}_{\alpha\beta}\begin{pmatrix} ll' \\ kk' \end{pmatrix}_0\right]\mathcal{G}_{\beta\gamma}\begin{pmatrix} l'l'' \\ k'k'' \end{pmatrix} = \delta_{\alpha\gamma}\delta_{kk''}\delta_{ll''}. \tag{C.6}$$

Using Eq. (C.3) and the property of the free Green's function expressed by the equation

$$g_{\sigma\beta}\begin{pmatrix} ll' \\ kk' \end{pmatrix}^{*} = g_{\beta\sigma}\begin{pmatrix} l'l \\ k'k \end{pmatrix}, \tag{C.7}$$

we can write Eq. (C.6) as

$$\mathcal{G}_{\alpha\beta}\begin{pmatrix} ll'' \\ kk'' \end{pmatrix} = g_{\alpha\beta}\begin{pmatrix} ll'' \\ kk'' \end{pmatrix} - \sum_{\substack{l'k'\gamma \\ \tilde{l}\tilde{k}\delta}} g_{\alpha\gamma}\begin{pmatrix} ll' \\ kk' \end{pmatrix}\delta\widetilde{\Phi}_{\gamma\delta}\begin{pmatrix} l'\tilde{l} \\ k'\tilde{k} \end{pmatrix}_0 \mathcal{G}_{\delta\beta}\begin{pmatrix} \tilde{l}l'' \\ \tilde{k}k'' \end{pmatrix}. \tag{C.8}$$

In this equation, we take Fourier components by means of the relations

$$\mathcal{G}\begin{pmatrix} \mathbf{y}'\mathbf{y}'' \\ j'j'' \end{pmatrix} = \sum_{\substack{lk\alpha \\ l''k''\gamma}} \sqrt{\frac{m_k}{N}}\, w^{*}_\alpha\left(k\middle|\begin{smallmatrix} \mathbf{y}' \\ j' \end{smallmatrix}\right)\exp\left[-2\pi i\mathbf{y}\mathbf{x}\begin{pmatrix} l \\ k \end{pmatrix}\right]\mathcal{G}_{\alpha\gamma}\begin{pmatrix} ll'' \\ kk'' \end{pmatrix}\sqrt{\frac{m_{k''}}{N}}\, w_\gamma\left(k''\middle|\begin{smallmatrix} \mathbf{y}'' \\ j'' \end{smallmatrix}\right)\exp\left[2\pi i\mathbf{y}''\mathbf{x}\begin{pmatrix} l'' \\ k'' \end{pmatrix}\right], \tag{C.9}$$

$$\delta\widetilde{\Phi}\begin{pmatrix} \mathbf{y}\mathbf{y}' \\ jj' \end{pmatrix} = \frac{1}{N}\sum_{\substack{lk\alpha \\ l'k'\beta}} \frac{\delta\widetilde{\Phi}_{\alpha\beta}\begin{pmatrix} ll' \\ kk' \end{pmatrix}_0}{\sqrt{m_k m_{k'}}}\, w_\alpha\left(k\middle|\begin{smallmatrix} \mathbf{y} \\ j \end{smallmatrix}\right) w_\beta\left(k'\middle|\begin{smallmatrix} \mathbf{y}' \\ j' \end{smallmatrix}\right)\exp\left[2\pi i\left(\mathbf{y}\mathbf{x}\begin{pmatrix} l \\ k \end{pmatrix} + \mathbf{y}'\mathbf{x}\begin{pmatrix} l' \\ k' \end{pmatrix}\right)\right]. \tag{C.10}$$

The result is

$$\mathcal{G}\begin{pmatrix} \mathbf{y}\mathbf{y}'' \\ jj'' \end{pmatrix} = g\begin{pmatrix} \mathbf{y}\mathbf{y}'' \\ jj'' \end{pmatrix} - \sum_{\substack{\mathbf{y}'\tilde{\mathbf{y}} \\ j'\tilde{j}}} g\begin{pmatrix} \mathbf{y}\mathbf{y}' \\ jj' \end{pmatrix}\delta\widetilde{\Phi}\begin{pmatrix} -\mathbf{y}'\tilde{\mathbf{y}} \\ j'\tilde{j} \end{pmatrix}\mathcal{G}\begin{pmatrix} \tilde{\mathbf{y}}\mathbf{y}'' \\ \tilde{j}j'' \end{pmatrix}, \tag{C.11}$$

where

$$g\begin{pmatrix} \mathbf{y}\mathbf{y}'' \\ jj'' \end{pmatrix} = \delta_{jj''}\delta_{\mathbf{y}\mathbf{y}'}\frac{1}{\omega_j^2(\mathbf{y}) - \omega^2} = \delta_{jj''}\delta_{\mathbf{y}\mathbf{y}''}g_j(\mathbf{y}). \tag{C.12}$$

For the case of charged defects, $\delta\tilde{\Phi}\binom{\mathbf{y}\mathbf{y}'}{jj'}$ has the form (if the interaction between defects is neglected)

$$\delta\tilde{\Phi}\binom{\mathbf{y}\mathbf{y}'}{jj'} = -\frac{1}{N}\sum_{lk}\delta e_k^l \exp\left[2\pi i(\mathbf{y}+\mathbf{y}')\mathbf{x}\binom{l}{k}\right]U_k\binom{\mathbf{y}\mathbf{y}'}{jj'}. \quad \text{(C.13)}$$

An expression for the function $U_k\binom{\mathbf{y}\mathbf{y}'}{jj'}$ will be given later.

Let us first take the case of randomly distributed point charged defects.

Iteration of Eq. (C.11), using Eq. (C.13) and the relationships

$$\overline{\delta e_k^l \delta e_{k'}^{l'}} = \overline{\delta e_k^2}\,\delta_{ll'}\,\delta_{kk'} + \overline{\delta e_k}\,\overline{\delta e_{k'}},$$

$$\overline{\delta e_k^l\,\delta e_{k'}^{l'}\delta e_{k''}^{l''}} = \overline{\delta e_k^3}\,\delta_{ll'}\,\delta_{l'l''}\delta_{kk'}\delta_{k'k''} + \overline{\delta e_k}\,\overline{\delta e_{k'}^2}\,\delta_{l'l''}\delta_{k'k''} + \overline{\delta e_{k'}}\,\overline{\delta e_k^2}\,\delta_{ll''}\delta_{kk''} + \overline{\delta e_{k''}}\,\overline{\delta e_k^2}\,\delta_{ll'}\,\delta_{kk'} + \overline{\delta e_k}\,\overline{\delta e_{k'}}\,\overline{\delta e_{k''}}, \quad \text{(C.14)}$$

to average the resulting expressions over the distribution of defects, and with the notation

$$\overline{\mathcal{G}\binom{\mathbf{y}\mathbf{y}''}{jj''}} = \mathcal{G}_{jj''}(\mathbf{y})\,\delta_{\mathbf{y},\,\mathbf{y}''}, \quad \text{(C.15)}$$

gives

$$\mathcal{G}_{jj'}(\mathbf{y}) = g_j(\mathbf{y})\,\delta_{jj'} + \sum_{k_1}\overline{\delta e_{k_1}}\,g_j(\mathbf{y})\,U_{k_1}\binom{-\mathbf{y}\mathbf{y}}{jj'}g_{j'}(\mathbf{y}) + \quad \text{(a), (b)}$$

$$+ v_a\sum_{k_1j_1}\overline{\delta e_{k_1}^2}\int g_j(\mathbf{y})\,U_k\binom{-\mathbf{y}\mathbf{y}_1}{jj_1}g_{j_1}(\mathbf{y}_1)\,U_{k_1}\binom{-\mathbf{y}_1\mathbf{y}}{j_1j'}g_{j'}(\mathbf{y})\,d\mathbf{y}_1 + \quad \text{(c)}$$

$$+ \sum_{j_1k_1k_2}\overline{\delta e_{k_1}}\,\overline{\delta e_{k_2}}\,g_j(\mathbf{y})\,U_{k_1}\binom{-\mathbf{y}\mathbf{y}}{jj_1}g_{j_1}(\mathbf{y})\,U_{k_2}\binom{-\mathbf{y}\mathbf{y}}{j_1j'}g_{j'}(\mathbf{y}) + \quad \text{(d)}$$

$$+ v_a^2\sum_{j_1j_2k_1}\overline{\delta e_{k_1}^3}\iint g_j(\mathbf{y})\,U_{k_1}\binom{-\mathbf{y}\mathbf{y}_1}{jj_1}g_{j_1}(\mathbf{y}_1)\,U_{k_1}\binom{-\mathbf{y}_1\mathbf{y}_2}{j_1j_2}g_{j_2}(\mathbf{y}_2)\,U_{k_1}\binom{-\mathbf{y}_2\mathbf{y}}{j_2j'}g_{j'}(\mathbf{y})\,d\mathbf{y}_1d\mathbf{y}_2 + \ldots. \quad \text{(e) (C.16)}$$

A diagram can be constructed for each term (a)-(e) in Eq. (C.16). Let the functions $g_j(\mathbf{y})$ be denoted by a horizontal line, and the expression $\delta e_{k_1}U_{k_1}\binom{-\mathbf{y}\mathbf{y}_2}{jj_2}$ by a wavy line joining the point k_1 to two horizontal lines corresponding to the functions $g_j(\mathbf{y})$ and $g_{j_2}(\mathbf{y}_2)$.

Summation (or integration) with respect to the indices $\binom{\mathbf{y}}{j}$ of an interior horizontal line is necessary, taking account of the law of conservation of momentum.

Figure 2 shows the diagrams which correspond to the terms (a)-(e) in Eq. (C.16). The structure of each diagram may be called either reducible or irreducible. A diagram is said to be irreducible if it cannot be divided into two parts joined by one horizontal line. Diagrams (a), (b), (c), and (e) are irreducible; diagram (d) is reducible.

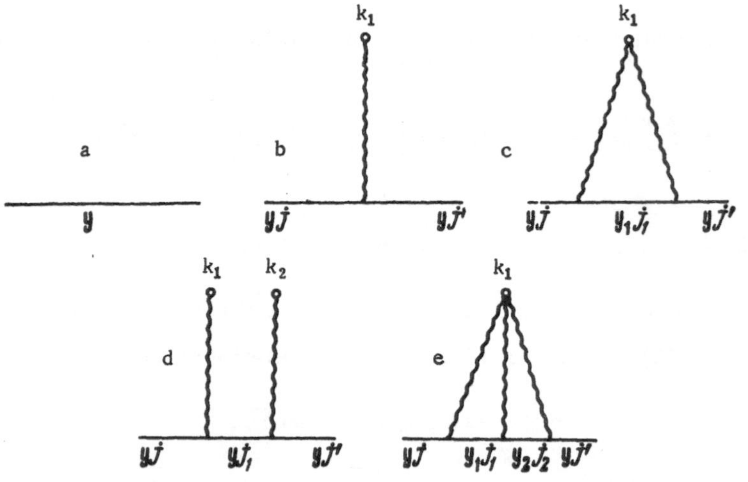

Fig. 2

It is known that the function $\mathcal{G}_{jj'}(y)$ can be expressed in terms of the quantity $\sigma_{jj_1}(y)$, which is the sum of expressions corresponding to irreducible diagrams only (see, e.g., [61]). This can be done by means of the equation

$$\mathcal{G}_{jj'}(y) = g_j(y)\,\delta_{jj'} + \sum_{j_1} g_j(y)\,\sigma_{jj_1}(y)\,\mathcal{G}_{j_1j'}(y). \tag{C.17}$$

In $\sigma_{jj_1}(y)$, we shall take account of irreducible diagrams that are proportional to n_p/N_0, but neglect those which are proportional to higher powers of this small quantity. This means that diagrams (b), (c), and (e) will be included, while diagrams such as that shown in Fig. 3 will be omitted.

Thus, we obtain for $\sigma_{jj'}(y)$ the expression

$$\sigma_{jj'}(y) = \sum_k \Big\{ \overline{\delta e_k}\, U_k\left(-\begin{matrix} yy \\ jj' \end{matrix}\right) + \overline{\delta e_k^2} \sum_{j_1} \int U_k\left(-\begin{matrix} yy_1 \\ jj_1 \end{matrix}\right) g_{j_1}(y_1)\, U_k\left(-\begin{matrix} y_1y \\ j_1j' \end{matrix}\right) dy_1 +$$
$$+ \overline{\delta e_k^3}\, v_a^2 \sum_{j_1 j_2} \iint U_k\left(-\begin{matrix} yy_1 \\ jj_1 \end{matrix}\right) g_{j_1}(y_1)\, U_k\left(-\begin{matrix} y_1y_2 \\ j_1j_2 \end{matrix}\right) g_{j_2}(y_2)\, U_k\left(-\begin{matrix} y_2y \\ j_2j' \end{matrix}\right) dy_1 dy_2 + \ldots \Big\}. \tag{C.18}$$

The equation (C.17) for the diagonal function $\mathcal{G}_{jj}(y)$ can be written

$$\mathcal{G}_{jj}(y) = g_j(y) + g_j(y)\,\sigma_{jj}(y)\,\mathcal{G}_{jj}(y) + \sum_{j_1 \neq j} g_j(y)\,\sigma_{jj_1}(y)\,\mathcal{G}_{j_1j}(y). \tag{C.19}$$

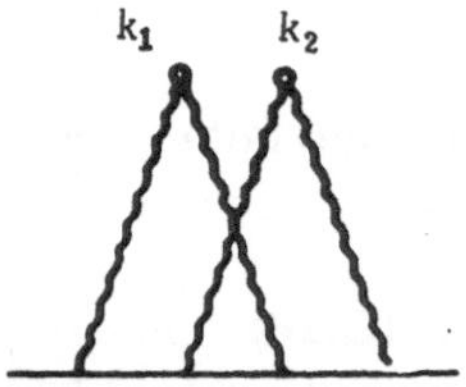

Fig. 3

Since $\mathcal{G}_{j_1j}(y) \sim \frac{n_p}{N_0}\,(j_1 \neq j)$, the third term on the right of this equation is of the order of $(n_p/N_0)^2$, and is negligible in comparison with the first two. Then, using Eq. (C.12), we have

$$\mathcal{G}_{jj}(y) = \frac{1}{\omega_j^2(y) - \omega^2 - \sigma_{jj}(y)}. \tag{C.20}$$

where

$$\sigma_{jj}(\mathbf{y})=\frac{n_p}{N_0}\sum_k\Big\{\delta e_k U_k\begin{pmatrix}-\mathbf{y}\mathbf{y}\\ jj\end{pmatrix}+\delta e_k^2 v_a\sum_{j_1}\int U_k\begin{pmatrix}-\mathbf{y}\mathbf{y}_1\\ jj_1\end{pmatrix}g_{j_1}(\mathbf{y}_1)\,U_k\begin{pmatrix}-\mathbf{y}_1\mathbf{y}\\ j_1j\end{pmatrix}d\mathbf{y}_1+$$

$$+\delta e_k^3 v_a^2\sum_{j_1j_2}\iint U_k\begin{pmatrix}-\mathbf{y}\mathbf{y}_1\\ jj_1\end{pmatrix}g_{j_1}(\mathbf{y}_1)\,U_k\begin{pmatrix}-\mathbf{y}_1\mathbf{y}_2\\ j_1j_2\end{pmatrix}g_{j_2}(\mathbf{y}_2)\,U_k\begin{pmatrix}-\mathbf{y}_2\mathbf{y}\\ j_2j\end{pmatrix}d\mathbf{y}_1d\mathbf{y}_2+\ldots\Big\}. \tag{C.21}$$

The latter quantity may also be written as

$$\sigma_{jj}(\mathbf{y})=\frac{n_p}{N_0}\sum_k\Big\{\delta e_k U_k\begin{pmatrix}-\mathbf{y}\mathbf{y}\\ jj\end{pmatrix}+\delta e_k v_a\sum_{j'}\int U_k\begin{pmatrix}-\mathbf{y}\mathbf{y}'\\ jj'\end{pmatrix}g_{j'}(\mathbf{y}')\,V_k\begin{pmatrix}-\mathbf{y}'\mathbf{y}\\ j'j\end{pmatrix}d\mathbf{y}'\Big\}, \tag{C.22}$$

where $V_k\begin{pmatrix}-\mathbf{y}'\mathbf{y}\\ j'j\end{pmatrix}$ satisfies the equation

$$V_k\begin{pmatrix}-\mathbf{y}'\mathbf{y}\\ j'j\end{pmatrix}=\delta e_k U_k\begin{pmatrix}-\mathbf{y}'\mathbf{y}\\ j'j\end{pmatrix}+\delta e_k v_a\sum_{j''}\int U_k\begin{pmatrix}-\mathbf{y}'\mathbf{y}''\\ j'j''\end{pmatrix}g_{j''}(\mathbf{y}'')\,V_k\begin{pmatrix}-\mathbf{y}''\mathbf{y}\\ j''j\end{pmatrix}d\mathbf{y}''. \tag{C.23}$$

For linear charged defects,

$$\sigma_{jj}(\mathbf{y})=\frac{n_l N_z}{N_0}\Big\{\delta e_1 U_1\begin{pmatrix}-\mathbf{y}'\mathbf{y}\\ jj\end{pmatrix}+\delta e_1 v_a\pi\sum_{j'}\int U_1\begin{pmatrix}-\mathbf{y}\mathbf{y}'\\ jj'\end{pmatrix}g_{j'}(\mathbf{y}')\,V_1\begin{pmatrix}-\mathbf{y}'\mathbf{y}\\ j'j\end{pmatrix}\Phi(y_z-y_z')\,d\mathbf{y}'\Big\}, \tag{C.24}$$

where $V_1\begin{pmatrix}-\mathbf{y}'\mathbf{y}\\ j'j\end{pmatrix}$ satisfies the equation

$$V_1\begin{pmatrix}-\mathbf{y}'\mathbf{y}\\ j'j\end{pmatrix}=\delta e_1 U_1\begin{pmatrix}-\mathbf{y}'\mathbf{y}\\ j'j\end{pmatrix}+\delta e_1 v_a\pi\sum_{j''}\int U_1\begin{pmatrix}-\mathbf{y}'\mathbf{y}''\\ j'j''\end{pmatrix}g_{j''}(\mathbf{y}'')\,V_1\begin{pmatrix}-\mathbf{y}''\mathbf{y}\\ j''j\end{pmatrix}\Phi(y_z-y_z'')\,d\mathbf{y}'', \tag{C.25}$$

and the function $\Phi(y_z)$ is given by Eq. (IV.74).

If we multiply Eq. (C.20) by $-\omega_j(\mathbf{y})/\pi$ and take $\mathbf{y}=\boldsymbol{\nu}$, $j=j_{\perp 0}$, an expression is obtained for the Green's function $\mathcal{G}\begin{pmatrix}\boldsymbol{\nu}\,-\boldsymbol{\nu}\\ j\;\;j\end{pmatrix}_\omega$ that is most important in our case [see Eq. (IV.14)]. The interaction between lattice waves and charged defects is taken into account more fully in Eq. (C.20) than in Eq. (IV.14). Let us now estimate the expansion parameters in Eqs. (C.21)-(C.24). To do so, we must use a specific expression for $U_k\begin{pmatrix}\mathbf{y}\mathbf{y}'\\ jj'\end{pmatrix}$. By means of Eqs. (IV.27), (A.6), (IV.34), and (IV.43), we find

$$U_k\begin{pmatrix}\mathbf{y}\mathbf{y}'\\ jj'\end{pmatrix}=U_k^{(1)}\begin{pmatrix}\mathbf{y}\mathbf{y}'\\ jj'\end{pmatrix}+U_k^{(2)}\begin{pmatrix}\mathbf{y}\mathbf{y}'\\ jj'\end{pmatrix}, \tag{C.26}$$

where

$$U_k^{(1)}\begin{pmatrix}\mathbf{y}\mathbf{y}'\\ jj'\end{pmatrix}=\frac{4\pi}{3v_a}\sum_{k'}e_{k'}\Bigg\{\frac{w_\alpha\left(k'\middle|\begin{matrix}\mathbf{y}\\ j\end{matrix}\right)w_\beta\left(k\middle|\begin{matrix}\mathbf{y}'\\ j'\end{matrix}\right)}{\sqrt{m_k m_{k'}}}f(|\mathbf{y}|)\left(\delta_{\alpha\beta}-\frac{3y_\alpha y_\beta}{|\mathbf{y}|^2}\right)+$$

$$+\frac{w_\alpha\left(k\middle|\begin{matrix}\mathbf{y}\\ j\end{matrix}\right)w_\beta\left(k'\middle|\begin{matrix}\mathbf{y}'\\ j'\end{matrix}\right)}{\sqrt{m_k m_{k'}}}f(|\mathbf{y}'|)\left(\delta_{\alpha\beta}-\frac{3y'_\alpha y'_\beta}{|\mathbf{y}'|^2}\right)-\frac{w_\alpha\left(k'\middle|\begin{matrix}\mathbf{y}\\ j\end{matrix}\right)w_\beta\left(k'\middle|\begin{matrix}\mathbf{y}'\\ j'\end{matrix}\right)}{m_{k'}}f(|\mathbf{y}+\mathbf{y}'|)\left(\delta_{\alpha\beta}-\frac{3(\mathbf{y}+\mathbf{y}')_\alpha(\mathbf{y}+\mathbf{y}')_\beta}{|\mathbf{y}+\mathbf{y}'|^2}\right)\Bigg\}, \tag{C.27}$$

and the function $f(|\mathbf{y}|)$ is

$$f(|\mathbf{y}|)=\frac{3\sin 2\pi ya'}{(2\pi ya')^3}-\frac{3\cos 2\pi ya'}{(2\pi ya')^2}. \tag{C.28}$$

When $2\pi ya'$ (a' being of the order of the dimensions of the unit cell), this function is almost unity.

The second part of the expression (C.26) is given by

$$U_k^{(2)}\binom{\mathbf{y}\mathbf{y}'}{jj'}=-\sum_{j''}\Gamma\binom{\mathbf{y}\mathbf{y}'\ -\mathbf{y}-\mathbf{y}'}{jj'\quad j''}\zeta\binom{-\mathbf{y}-\mathbf{y}'}{j''}, \tag{C.29}$$

$$\Gamma\binom{\mathbf{y}\mathbf{y}'\ -\mathbf{y}-\mathbf{y}'}{jj'\quad j''}=\sum_{\alpha l}\frac{\Phi_{\alpha\alpha\alpha}\binom{l\ 00}{1\ 22}}{\sqrt{m_1m_2}}\left[\left\{\frac{w_\alpha\left(1\left|\begin{matrix}\mathbf{y}\\ j\end{matrix}\right.\right)w_\alpha\left(2\left|\begin{matrix}\mathbf{y}'\\ j'\end{matrix}\right.\right)w_\alpha\left(1\left|\begin{matrix}-\mathbf{y}-\mathbf{y}'\\ j''\end{matrix}\right.\right)}{\sqrt{m_1}}+\right.\right.$$

$$\left.+\frac{w_\alpha\left(2\left|\begin{matrix}\mathbf{y}\\ j\end{matrix}\right.\right)w_\alpha\left(1\left|\begin{matrix}\mathbf{y}'\\ j'\end{matrix}\right.\right)w_\alpha\left(2\left|\begin{matrix}-\mathbf{y}-\mathbf{y}'\\ j''\end{matrix}\right.\right)}{\sqrt{m_2}}\right\}\sin 2\pi\mathbf{y}'\mathbf{x}\binom{l}{12}+\left\{\frac{w_\alpha\left(2\left|\begin{matrix}\mathbf{y}\\ j\end{matrix}\right.\right)w_\alpha\left(1\left|\begin{matrix}\mathbf{y}'\\ j'\end{matrix}\right.\right)w_\alpha\left(1\left|\begin{matrix}-\mathbf{y}-\mathbf{y}'\\ j''\end{matrix}\right.\right)}{\sqrt{m_1}}+\right.$$

$$\left.+\frac{w_\alpha\left(1\left|\begin{matrix}\mathbf{y}\\ j\end{matrix}\right.\right)w_\alpha\left(2\left|\begin{matrix}\mathbf{y}'\\ j'\end{matrix}\right.\right)w_\alpha\left(2\left|\begin{matrix}-\mathbf{y}-\mathbf{y}'\\ j''\end{matrix}\right.\right)}{\sqrt{m_2}}\right\}\sin 2\pi\mathbf{y}\mathbf{x}\binom{l}{12}-\left\{\frac{w_\alpha\left(1\left|\begin{matrix}\mathbf{y}\\ j\end{matrix}\right.\right)w_\alpha\left(1\left|\begin{matrix}\mathbf{y}'\\ j'\end{matrix}\right.\right)w_\alpha\left(2\left|\begin{matrix}-\mathbf{y}-\mathbf{y}'\\ j''\end{matrix}\right.\right)}{\sqrt{m_1}}+\right.$$

$$\left.\left.+\frac{w_\alpha\left(2\left|\begin{matrix}\mathbf{y}\\ j\end{matrix}\right.\right)w_\alpha\left(2\left|\begin{matrix}\mathbf{y}'\\ j'\end{matrix}\right.\right)w_\alpha\left(1\left|\begin{matrix}-\mathbf{y}-\mathbf{y}'\\ j''\end{matrix}\right.\right)}{\sqrt{m_2}}\right\}\sin 2\pi(\mathbf{y}+\mathbf{y}')\mathbf{x}\binom{l}{12}\right], \tag{C.30}$$

$$\zeta\binom{-\mathbf{y}-\mathbf{y}'}{j''}=\frac{2}{v_a}\sum_k\frac{e_k}{\sqrt{m_k}}\frac{\left(\mathbf{w}\left(k\left|\begin{matrix}\mathbf{y}+\mathbf{y}'\\ j''\end{matrix}\right.\right)(\mathbf{y}+\mathbf{y}')\right)}{|\mathbf{y}+\mathbf{y}'|^2\,\omega^2\binom{\mathbf{y}+\mathbf{y}'}{j''}}\frac{\sin 2\pi|\mathbf{y}+\mathbf{y}'|a'}{2\pi|\mathbf{y}+\mathbf{y}'|a'} \tag{C.31}$$

In Eq. (C.26), $U_k^{(1)}\binom{\mathbf{y}\mathbf{y}'}{jj'}$ is derived from the first term of Eq. (C.2), and $U_k^{(2)}\binom{\mathbf{y}\mathbf{y}'}{jj'}$ from the second term.

Unlike the corresponding expressions (IV.33) and (IV.65), the expressions (C.27) and (C.29)-(C.31) for $U_k^{(1)}\binom{\mathbf{y}\mathbf{y}'}{jj'}$, $U_k^{(2)}\binom{\mathbf{y}\mathbf{y}'}{jj'}$ are approximately correct for any values of $|\mathbf{y}|$, $|\mathbf{y}'|$, and j, j'.

The following are some noteworthy properties of $U_k^{(1)}\binom{\mathbf{y}\mathbf{y}'}{jj'}$, $U_k^{(2)}\binom{\mathbf{y}\mathbf{y}'}{jj'}$. Their definitions show that they are real and are unchanged when the pairs of indices $\binom{\mathbf{y}}{j}$ and $\binom{\mathbf{y}'}{j'}$ are transposed. If j' pertains to the acoustic branch, and the inequality $y \gg y'$ is satisfied, we can use Eq. (B.37) to show that these quantities are proportional to the small ratio y'/y, i.e.,

$$U_k^{(1)}\binom{\mathbf{y}\mathbf{y}'}{jj'}\sim\frac{y'}{y},\quad U_k^{(2)}\binom{\mathbf{y}\mathbf{y}'}{jj'}\sim\frac{y'}{y}. \tag{C.32}$$

A rough estimate of the ratio of successive terms in the series (C.21) for point defects (with $\omega = 0$ also) gives

$$\alpha=\widetilde{\Delta\varepsilon}\left(\frac{\delta e_k}{e_k}\right) \tag{C.33}$$

if $U_k\binom{yy'}{jj'}$ is replaced by $U_k^{(1)}\binom{yy'}{jj'}$, and

$$\beta = \widetilde{\Delta\varepsilon}\left(\frac{\delta e_k}{e_k}\right)\delta, \tag{C.34}$$

if $U_k\binom{yy'}{jj'}$ is replaced by $U_k^{(2)}\binom{yy'}{jj'}$. Here, $\widetilde{\Delta\varepsilon} = \frac{4\pi N_0 e^2}{M\tilde{\omega}^2}$, where $\tilde{\omega}^2$ is of the order of the mean square frequency in the lattice spectrum, and δ is given by Eq. (IV.67).

For linear defects, the parameters α and β have the same order of magnitude as for point defects. A numerical estimate of these parameters shows that they are not small when $\delta e_k \sim e_k$. For example, in NaCl, they are $\alpha \approx 4$, $\beta \approx 24$ ($\Delta\varepsilon \approx 4$, $|\delta| \approx 6$). In making these estimates, the numerical values of the parameters λ_{+-}, ρ, a, ε_0, and ε_∞ given in [11] were used.

The estimates lead to the conclusion that, in calculating $\sigma_{jj}(\nu)$ for the case $\delta e_k \sim e_k$, we cannot take only the first few terms of the series (C.21), but must sum the whole series.

If, however, we are concerned only with the imaginary part of $\sigma_{jj}(\nu)$ (which is due to the substitution $\omega \to \omega + i\delta$, $\delta > 0$, $\delta \to 0$) at frequencies $\omega \ll \omega_0$, it can be shown that, when the defect has no resonance vibration frequency in that range, the Born approximation is sufficient. To perform this calculation, let us consider, for example, the equations (C.23) for point charged defects.

These are a set of linear Fredholm integral equations (see, for example, [62, 63]), in which the unknown functions $V_k\binom{-y'\nu}{j'j}$ depend on three continuous variables y'_x, y'_y, y'_z and on the variable j', which takes six values in a crystal having two ions per unit cell. The equations (C.23) can be considerably simplified by writing $V_k\binom{-y'\nu}{jj'}$ in the form

$$V_k\binom{-y'\nu}{j'j} = \delta e_k U_k\binom{-y'\nu}{j'j} + C_k\binom{y'}{j'}\cos\gamma_{j'}, \tag{C.35}$$

where $\gamma_{j'}$ is the angle between the polarization vector $\mathbf{w}\left(k\,\middle|\,\begin{smallmatrix}\nu\\ j\end{smallmatrix}\right)$ of the radiation in the crystal and the vector $\mathbf{w}\left(k\,\middle|\,\begin{smallmatrix}y'\\ j'\end{smallmatrix}\right)$ and the function $C_k\binom{y'}{j'}$ depends only on the modulus of $\mathbf{y}'$. After substituting Eq. (C.35) in Eqs. (C.23) and cancelling $\cos\gamma_{j'}$, we obtain a set of integral equations in which the integration is over the single variable $|\mathbf{y}''|$ instead of over the three variables y''_x, y''_y, y''_z.

Let us now consider the imaginary part of $\sigma_{jj}(\nu)$, where ν is the wave vector of radiation, and j denotes the transverse optical vibration branch. Substituting Eq. (C.35) in Eq. (C.22) and using the expression (C.12) for the function $g_j(\mathbf{y})$, in which the change $\omega \to \omega + i\delta$ ($\delta > 0$, $\delta \to 0$) has been made, we find

$$\begin{aligned}\operatorname{Im}\sigma_{jj}(\nu) &= \pi\frac{n_p}{N_0}\sum_k \delta e_k^2 v_a \sum_{j'}\int\left|U_k\binom{-\nu y'}{jj'}\right|^2 \delta(\omega_{j'}^2(\mathbf{y}') - \omega^2)\,d\mathbf{y}' + \\ &+ \pi\frac{n_p}{N_0}\sum_k \delta e_k v_a \sum_{j'}\int U_k\binom{-\nu y'}{jj'} \operatorname{Re} C_k\binom{y'}{j'}\delta(\omega_{j'}^2(\mathbf{y}') - \omega^2)\times \\ &\times\cos\gamma_{j'}d\mathbf{y}' + \frac{n_p}{N_0}\sum_k \delta e_k v_a \sum_{j'}\int U_k\binom{-\nu y'}{jj'}\operatorname{Im} C_k\binom{y'}{j'} P\left(\frac{1}{\omega_{j'}^2(\mathbf{y}) - \omega^2}\right)\cos\gamma_{j'}\,d\mathbf{y}'.\end{aligned} \tag{C.36}$$

The first term is the imaginary part of $\sigma_{jj}(\nu)$ as calculated in the Born approximation; the second and third terms give higher approximations.

We can estimate $\mathrm{Re}\, C_k\binom{y'}{j'}$ and $\mathrm{Im}\, C_k\binom{y'}{j'}$ by taking $U_k\binom{yy'}{jj'} = U_k^{(1)}\binom{yy'}{jj'}$. The second term in Eq. (C.36) shows that, when the inequality $\omega \ll \omega_0$ is satisfied, y' is small in the expression $\mathrm{Re}\, C_k\binom{y'}{j'}y'$, i.e., $y' \ll y_m$ (y_m is the maximum lattice wave number).

The solution of the integral equations for $\mathrm{Re}\, C_k\binom{y'}{j'}$ may therefore be here sought in the form

$$\mathrm{Re}\, C_k\binom{y'}{j'} = C_k^{(0)}(j') + C_k^{(1)}(j')\frac{y'}{y_m} + C_k^{(2)}(j')\left(\frac{y'}{y_m}\right)^2 + \cdots, \tag{C.37}$$

and algebraic equations are then obtained for the expansion coefficients. Taking only the lowest-order terms in Eq. (C.37), we find the following equations for the functions $\mathrm{Re}\, C_k\binom{y'}{j'}$:

$$\begin{aligned}
\mathrm{Re}\, C_k\binom{y'}{o\,\|} &\approx C_k^{(0)}(o\,\|) = \frac{B_1A_{22} - B_2A_{12}}{A_{11}A_{22} - A_{21}A_{12}},\\
\mathrm{Re}\, C_k\binom{y_1}{o\perp_1} &\approx C_k^{(0)}(o\perp_1) = \frac{B_2A_{11} - B_1A_{21}}{A_{11}A_{22} - A_{21}A_{12}},\\
\mathrm{Re}\, C_k\binom{y'}{a\,\|} &\approx C_k^{(2)}(a\,\|)\left(\frac{y'}{y_m}\right)^2,\\
\mathrm{Re}\, C_k\binom{y'}{a\perp_1} &\approx C_k^{(2)}(a\perp_1)\left(\frac{y'}{y_m}\right)^2,\\
\mathrm{Re}\, C_k\binom{y'}{o\perp_2} &\approx \mathrm{Re}\, C_k\binom{y'}{a\perp_2} = 0;
\end{aligned} \tag{C.38}$$

where the indices $o\,\|$, $a\,\|$ denote longitudinal optical and acoustic vibrations; $o_{\perp 1}$, $a_{\perp 1}$ denote transverse vibrations with the polarization vector in the plane of the vectors **y'** and $\mathbf{w}\left(k\,\middle|\,\begin{smallmatrix}\mathbf{v}\\ j\end{smallmatrix}\right)$ and a $o_{\perp 2}$, $a_{\perp 2}$ denote transverse vibrations with the polarization vector perpendicular to this plane.

The coefficients which appear in (C.38) are

$$\begin{aligned}
C_k^{(2)}(a\,\|) &= {}^1\!/_5\{({}^1\!/_3\alpha_{o\,\|} + \alpha_{o\perp})\alpha_{a\,\|}\omega_{a\,\|}^2 - 2C_k^{(0)}(o\,\|)\alpha_{o\,\|} + 3C_k^{(0)}(o\perp_1)\alpha_{o\perp}\},\\
C_k^{(2)}(a\perp_1) &= {}^1\!/_{10}\{({}^1\!/_2\alpha_{o\,\|} + {}^2\!/_3\alpha_{o\perp})\alpha_{a\perp}\omega_{a\perp}^2 - 3C_k^{(0)}(o\,\|)\alpha_{o\,\|} + 2C_k^{(0)}(o\perp_1)\alpha_{o\perp}\},\\
A_{11} &= 1 + {}^2\!/_3\alpha_{o\,\|} - {}^1\!/_{10}\alpha_{a\perp}\alpha_{o\,\|},\\
A_{12} &= {}^1\!/_3\alpha_{o\perp} + {}^1\!/_{15}\alpha_{a\perp}\alpha_{o\perp},\\
A_{22} &= 1 - {}^1\!/_6\alpha_{o\perp} - {}^1\!/_{10}\alpha_{a\,\|}\alpha_{o\perp},\\
A_{21} &= {}^1\!/_6\alpha_{o\,\|} + {}^1\!/_{15}\alpha_{a\,\|}\alpha_{o\,\|},\\
B_1 &= {}^1\!/_9\alpha_{o\,\|}(\alpha_{o\,\|} - \alpha_{o\perp})\omega_{o\,\|}^2 - {}^1\!/_{60}\alpha_{o\,\|}\alpha_{a\perp}\left(\alpha_{o\,\|} + \frac{4}{3}\alpha_{o\perp}\right)\omega_{o\,\|}^2,\\
B_2 &= {}^1\!/_{36}\alpha_{o\perp}(\alpha_{o\,\|} + 8\alpha_{o\perp} + 9\alpha_{a\,\|})\omega_{o\perp}^2 + {}^1\!/_{90}\alpha_{o\perp}\alpha_{a\,\|}(\alpha_{o\,\|} + 3\alpha_{o\perp})\omega_{o\perp}^2,
\end{aligned} \tag{C.39}$$

where

$$\alpha_j = \left(\frac{\delta e_k}{e_k}\right)\Delta\varepsilon_j, \quad \Delta\varepsilon_j = \frac{4\pi e^2}{v_a M\omega_j^2},$$

$$\omega_{a\perp}^2 = (2\pi)^2 v_\perp^2 y_m^2, \quad \omega_{a\,\|}^2 = (2\pi^2)v_\|^2 y_m^2.$$

The quantities $\omega_{o\perp}^2$, $\omega_{o\,\|}^2$ are, in order of magnitude, the averages, over the spectrum, of the squares of the frequencies of transverse and longitudinal optical vibrations.

The following simplifying assumptions have been made in deriving Eqs. (C.38) and (C.39). In Eq. (C.27), the function $f(|\mathbf{y}|)$ has been taken equal to unity, and the masses of the ions have been assumed equal ($m_1 = m_2$). These simplifications do not essentially affect Eqs. (C.38).

Since the parameter $\omega/\tilde{\omega}$ is small (where $\tilde{\omega}$ is of the order of the mean frequency in the lattice spectrum), we have also been able to neglect the frequency dependence of the expressions (C.38) and (C.39).

The following estimates are obtained from Eqs. (C.23), (C.32), and (C.35):

$$\operatorname{Im} C_k\begin{pmatrix} \mathbf{y}' \\ j' \end{pmatrix} \sim \left(\frac{\omega}{\tilde{\omega}}\right)^3 C_k^{(0)}(o\,\|) \text{ or } \left(\frac{\omega}{\tilde{\omega}}\right)^3 C_k^{(0)}(o\perp_1). \tag{C.40}$$

If the functions $C_k\begin{pmatrix} \mathbf{y}' \\ j' \end{pmatrix}$ are calculated with the assumption that $U_k\begin{pmatrix} \mathbf{y}\mathbf{y}' \\ jj' \end{pmatrix} = U_k^{(2)}\begin{pmatrix} \mathbf{y}\mathbf{y}' \\ jj' \end{pmatrix}$, then relations similar to Eqs. (C.38) and (C.40) are obtained for $C_k\begin{pmatrix} \mathbf{y}' \\ j' \end{pmatrix}$. The chief difference from Eqs. (C.38), (C.39), and (C.40) is that the parameter $\alpha = \widetilde{\Delta\varepsilon}\left(\frac{\delta e_k}{e_k}\right)$ is replaced by $\alpha\delta$.

Let us now estimate the order of magnitude of the various terms in Eq. (C.36). Using the relationships (C.38), (C.39), and (C.40), we find that the second and third terms are of the same order of magnitude, and the ratio between either term and the first term is, in order of magnitude,

$$\frac{C_k^{(0)}(o\,\|)}{\tilde{\omega}^2}\left(\frac{\omega}{\tilde{\omega}}\right)^2 \text{ or } \frac{C_k^{(0)}(o\perp_1)}{\tilde{\omega}^2}\left(\frac{\omega}{\tilde{\omega}}\right)^2. \tag{C.41}$$

At microwave frequencies, these ratios are small if the denominators in the expressions for $C_k^{(0)}(o\,\|)$ and $C_k^{(0)}(o\perp_1)$ are not almost zero.

If this last condition is violated, the quantities $C_k^0(o\,\|)$ and $C_k^{(0)}(o\perp_1)$ must be more accurately calculated, in such a way as to take account of their frequency dependence. This would lead to resonance behavior at certain frequencies.

In the calculation of the resonance frequencies of a real charged defect, we must take into account the quantities $U_k^{(1)}\begin{pmatrix} \mathbf{y}\mathbf{y}'' \\ jj'' \end{pmatrix}$ and $U_k^{(2)}\begin{pmatrix} \mathbf{y}\mathbf{y}' \\ jj' \end{pmatrix}$ simultaneously in the integral equation (C.23), and also the corresponding quantity due to the presence of short-range forces between the defect and the surrounding ions. Moreover, the expressions $U_k^{(1)}\begin{pmatrix} \mathbf{y}\mathbf{y}' \\ jj' \end{pmatrix}$, $U_k^{(2)}\begin{pmatrix} \mathbf{y}\mathbf{y}' \\ jj' \end{pmatrix}$, $\omega_j(\mathbf{y})$, $u_\alpha\left(k\,\middle|\,\begin{matrix} \mathbf{y} \\ j \end{matrix}\right)$, etc., for large values of $|\mathbf{y}|$ and $|\mathbf{y}'|$ must be calculated more exactly than we have done. Numerical methods would then have to be used.

Thus, the calculation of the resonance frequencies of charged defects and of the scattering by them of phonons having frequencies close to resonance is a complicated problem.

We shall not consider it here, but merely enunciate the following deduction from the foregoing calculations and estimates. The Born approximation can be used to calculate the imaginary part of $\sigma_{jj}(\nu)$ at frequencies $\omega \ll \omega_0$ if the defects have no resonance in that range.

This deduction remains valid for linear defects, since estimates analogous to (C.41) can be obtained for these.

In calculating the real part of $\sigma_{jj}(\nu)$, the Born approximation is invalid, even if the above conditions are satisfied (in the case $\delta e_k \sim e_k$). However, the correction to the real part of the permittivity is of the order of n_p/N_0 (or $N_l N_z/N_0$), and in the quantity $\tan\delta$ in which we are directly interested it leads to a correction of the order of $(n_p/N_0)^2$ (or $N_l N_z/N_0)^2$. It may therefore be neglected.

Literature Cited

1. Czerny, M., Z. Physik, 65:600 (1930).
2. Barnes, R. B. and Czerny, M., Z. Physik, 72:447 (1931).
3. Barnes, R. B., Z. Physik, 75:723 (1932).
4. Born, M. and Blackman, M., Z. Physik, 82:551 (1933).
5. Blackman, M., Z. Physik, 86:421 (1933).
6. Pauli, W., Verhandl. Deutsch. Phys. Ges., 6:10 (1925).
7. Blackman, M., Phil. Trans. Roy. Soc. London, A 236:103 (1936).
8. Neuberger, J. and Hatcher, R. D., J. Chem. Phys., 34:1733 (1961).
9. Jepsen, D. W. and Wallis, R. F., Phys. Rev., 125:1496 (1962).
10. Barnes, R. B., Brattain, R. R. and Seitz, F., Phys. Rev., 48:582 (1935).
11. Born, M. and Kun Huang, Dynamical Theory of Crystal Lattices, Clarendon Press, Oxford (1954).
12. Mitskevich, V. V., Izv. Vuzov, Fizika, No. 4, p. 6 (1960).
13. Maradudin, A. A. and Wallis, R. F., Phys. Rev., 120:442 (1960).
14. Vinogradov, V. S., Fiz. Tverd. Tela, 3:1723 (1961).
15. Kleinman, D. A., Phys. Rev., 118 (1960).
16. Szigeti, B., Proc. Roy. Soc. (London), A 258:377 (1960).
17. Vinogradov, V. S., Fiz. Tverd. Tela, 4, 712 (1962).
18. Wallis, R. F. and Maradudin, A. A., Phys. Rev., 125:1277 (1962).
19. Langer, J. S., Maradudin, A. A. and Wallis, R. F., in: R. F. Wallis (editor), Lattice Dynamics, Pergamon, Oxofrd (1965), p. 411.
20. Lax, M., J. Phys. Chem. Solids, 25:487 (1964).
21. Mitskevich, V. V., Fiz. Tverd. Tela, 3:3036 (1961).
22. Mitskevich, V. V., Fiz. Tverd. Tela, 4:3035 (1962).
23. Gurevich, L. É. and Ipatova, I. P., Zh. Eksp. Teor. Fiz., 45:231 (1963).
24. Kashcheev, V. N., Fiz. Tverd. Tela, 5:1358 (1963).
25. Kashcheev, V. N., Fiz. Tverd. Tela, 5:2339 (1963).
26. Lifshits, I. M., Zh. Eksp. Teor. Fiz., 12:117 (1942).
27. Lifshits, I. M. (Lifšic), Nuovo Cimento, 3 (Suppl.):716 (1956).
28. Montroll, E. W. and Potts, R. B., Phys. Rev., 100:525 (1955).
29. Montroll, E. W. and Potts, R. B., Phys. Rev., 102:72 (1956).
30. Maradudin, A. A., Mazur, P., Montroll, E. W. and Weiss, G. H., Rev. Mod. Phys., 30:175 (1958).
31. Progr. Theoret. Phys. (Kyoto), Vol. Suppl. (1962).
32. Lax, M. and Burstein, E., Phys. Rev., 97:39 (1955).
33. Wallis, R. F. and Maradudin, A. A., Progr. Theoret. Phys. (Kyoto), 24:1055 (1960).
34. Dawber, P. G. and Elliott, R. J., Proc. Roy. Soc. (London), A 273:222 (1963).
35. Lysenko, E. E., Zh. Tekh. Fiz., 8:1637 (1938).
36. Vinogradov, V. S., Fiz. Tverd. Tela, 2:2622 (1960).
37. Vinogradov, V. S., Fiz. Tverd. Tela, 4: 3348 (1962.
38. Vinogradov, V. S., in: R. F. Wallis (editor), Lattice Dynamics, Pergamon, Oxford (1965), p. 421.
39. Zubarev, D. N., Usp. Fiz. Nauk, 71:71 (1960).
40. Tyablikov, S. V. and Bonch-Bruevich, V. L., Perturbation Theory for Double-time Temperature-Dependent Green's Functions [in Russian], Izd. VTs AN SSSR, Moscow (1962).
41. Agranovich, V. M. and Ginzburg, V. L., Spatial Dispersion in Crystal Optics and the Theory of Excitons, Interscience, London (1966).
42. Heilmann, G., Z. Physik, 152:368 (1958).
43. Hass, M., Phys. Rev., 117:1497 (1960).
44. Klier, M., Z. Physik, 150:49 (1958).

45. Genzel, L., Happ, H. and Weber, R., Z. Physik, 154: 13 (1959).
46. Hohls, H. W., Ann. Physik, 29: 433 (1937).
47. Genzel, L. and Klier, M., Z. Physik, 144: 25 (1956).
48. McCubbin, T. K., Jr., and Sinton, W. M., J. Opt. Soc. Am., 40: 537 (1950).
49. Willmott, J. C., Proc. Phys. Soc. (London), A 63: 389 (1950).
50. Burstein, E., Oberly, J. J. and Plyler, E. K., Proc. Indian Acad. Sci., 28: 388 (1948).
51. Strong, J., Phys. Rev., 37: 1565 (1931).
52. Groth, R., Physik, 6: 328 (1960).
53. Bonch-Bruevich, V. L., Fiz. Tverd. Tela, 5: 2714 (1963).
54. Johnson, L. F., Dietz, R. E. and Guggenheim, H. J., Phys. Rev. Letters, 11: 318 (1963).
55. Kashcheev, V. N. and Krivoglaz, M. A., Fiz. Tverd. Tela, 3: 1528 (1961).
56. Heller, W. R. and Marcus, A., Phys. Rev., 84: 809 (1951).
57. Cohen, M. H. and Keffer, F., Phys. Rev., 99: 1128 (1955).
58. Rupprecht, G. and Bell, R. O., Phys. Rev., 125: 1915 (1962).
59. Breckenridge, R. G., J. Chem. Phys., 16: 959 (1948).
60. Skanavi, G. I. and Lipaeva, G. A., Zh. Eksp. Teor. Fiz., 30: 824 (1956).
61. Abrikosov, A. A., Gor'kov, L. P. and Dzyaloshinskii, I. E., Methods of Quantum Field Theory in Statistical Physics, Prentice-Hall, Englewood Cliffs (1963).
62. Smirnov, V. I., A Course of Higher Mathematics, Vol. 4, Pergamon, Oxford (1964).
63. Whittaker, E. T. and Watson, G. N., A Course of Modern Analysis, 4th ed., University Press, Cambridge (1927).

VIBRATIONAL SPECTRA OF STRONTIUM, BARIUM, AND CALCIUM TITANATES

V. N. Murzin

Transmission and reflection measurements were made on $SrTiO_3$, $BaTiO_3$, and $CaTiO_3$ crystals in a wide range of the infrared and submillimeter spectrum, using a specially designed long-wave infrared spectrometer, and the spectral dependence of the complex permittivity of these substances was calculated. Conclusions about the vibrational spectra of such crystals having a perovskite-type structure are drawn from the experimental results and the theory of symmetry. The features of the vibrational spectra are analyzed by means of the dynamical theory of crystal lattices, using Cochran's method. For barium titanate, it is found for the first time that the frequency and the oscillator strength of the lowest-frequency vibration show a considerable decrease and increase, respectively, as the phase-transition temperature is approached, in qualitative agreement with the theoretical predictions of Ginzburg, Anderson, and Cochran.

CHAPTER I

Introduction

Crystals having a perovskite-type structure, which include the titanates of strontium, barium and calcium, are distinguished by unusual properties and are of great theoretical and practical importance. There have been many investigations of these substances [1-9].

One can now identify two very important features that are typical of perovskite crystals, viz. the high permittivity and the existence of a ferroelectric state. Since these properties of the substances in question are due mainly to the nature of the crystal structure, it is obvious that the normal modes of their lattices are a very important topic. Measurements at radio and microwave frequencies have shown that these substances retain their high permittivity up to frequencies of about 10^{10} cps. Barium titanate is found not to be an exception in this respect if its properties in the single-domain state are considered [10, 11]. Thus, the processes responsible for the high permittivity of these crystals must take effect in the higher-frequency region of the spectrum, and may be reasonably assumed to be connected with the eigen-vibrations of the crystal lattices. It is also known that the static polarization in crystals having the structure concerned is a complicated quantity and is largely determined by the strong internal electric fields. We may expect this to have a considerable influence on the nature of the vibrational spectra of perovskite-type crystals.

Some crystals of this type have also been found to possess ferroelectric properties. As well as barium titanate, whose ferroelectric properties were discovered in 1946 by Vul [1], strontium titanate appears to be in this category. Various authors [12-16] have suggested that the ferroelectric phase transition in $SrTiO_3$ takes place at 30-40°K. The ferroelectric properties of these crystals lead to other interesting features of their vibrational spectra. According to the work of Ginzburg [17, 18], Anderson [19], and Cochran [20, 21] on the microscopic theory of ferroelectricity, the vibrational spectrum of crystals such as barium titanate is expected to contain a very-low-frequency vibration, which should move to lower frequencies as the temperature approaches the phase-transition point.

The present paper gives an account of the results of our research on the vibrational spectra of $SrTiO_3$, $BaTiO_3$, and $CaTiO_3$ crystals, using infrared spectroscopy. Crystals were selected which at room temperature included all three systems found with the perovskite structure (cubic, tetragonal, and orthorhombic). Transmission and reflection spectra of crystal samples were measured in the near and far infrared ($\lambda = 1-1000\ \mu$) and at submillimeter wavelengths (2.68, 4, and 8 mm); the complex permittivity was measured at radio frequencies ($f = 10^3-10^7$ cps). For the far infrared measurements, a diffraction spectrometer was designed and constructed, and was used also for measurements in the submillimeter range with a microwave radiation source. Particular attention was given to the temperature dependence of the vibrational spectrum of $SrTiO_3$ and $BaTiO_3$ crystals, because of their ferroelectric properties.

The work done hitherto on the vibrational spectra of the crystals will first be briefly reviewed.

1. Vibrational Spectrum of $SrTiO_3$-Type Crystals

The most complete investigation of the vibrational spectra of perovskite crystals is that by Last in 1957 [22]. In considering the number and symmetry of the normal modes of the crystal lattices having this structure, all the atoms in the unit cell were divided into two groups, the Sr atom and the TiO_6 octahedron. The subsequent discussion related to the latter group. From the resulting vibrational spectrum, the modes which could not be propagated translation-wise through the crystal were excluded. In particular, it was found that the spectrum of normal modes of a cubic lattice of the perovskite type must consist of three F_{1u} vibrations and one F_{2u} which is inactive in the infrared absorption spectra. This analysis easily gives the form of the normal modes. If the atoms in the unit cell are labeled as in Fig. 1, the F_{2u} vibration (denoted by ω_1) which is inactive in the infrared spectra is a deformation vibration of the form O(4)-O(5). The remaining F_{1u} vibrations can be regarded as a low-frequency vibration ω_2, Sr-TiO_3; a valence vibration ω_3, Ti−O(3); and a deformation vibration ω_4, O(4)−Ti−O(5).

The experimental part of the work was carried out by measuring the transmission and reflection spectra of single-crystal samples in the near infrared (1-35 μ). The transmission spectra of $SrTiO_3$ and $BaTiO_3$ powders showed two bands with minima at about 18 and 30 μ, which were ascribed to vibrations of the types ω_3 and ω_4, respectively. The frequency ω_2, about 200 cm^{-1} (50 μ), of the lowest-frequency vibration of the $BaTiO_3$ lattice was estimated indirectly, using the value of the specific heat at low temperatures. No significant change was observed in the transmission spectrum of tetragonal $BaTiO_3$ when the temperature was lowered beneath the phase-transition point at $\theta = 120$°C. In an orthorhombic $BaTiO_3$ crystal, the ω_3 band was split. Detailed measurements in the near infrared were subsequently made by various authors [23]. The interpretation of the experimental results was not changed.

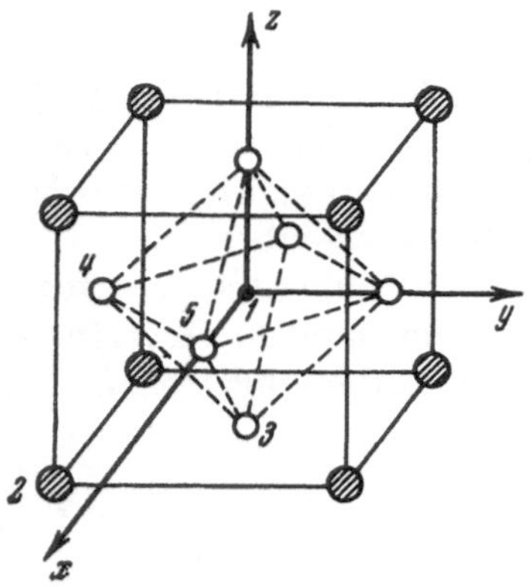

Fig. 1. Unit cell of $SrTiO_3$. 1) Ti; 2) Sr; 3)-5) O.

Raman-scattering spectra were measured [24] for $SrTiO_3$, which showed four bands having maxima at 620, 447, 335, and 90 cm^{-1} (16.2, 22.4, 30, and 111 μ).

An important finding was that the low-frequency vibration appeared at a much longer wavelength. The bands at 16.2, 30, and 111 μ were interpreted as before, being ascribed to vibrations of the types ω_3, ω_4, and ω_2, respectively. To account for the band at about 22.4 μ, a model of rotational vibration of the oxygen octahedron about the Sr–Ti axis was proposed. This vibration was incorrectly assigned to the F_{1u} class, and this explains its presence in the Raman spectrum on the same footing as the other vibrations. The observation of a Raman spectrum in $SrTiO_3$ that is forbidden by the selection rules in cubic perovskite crystals was in principle explained by distortions of the crystal structure caused by the presence of strong local electric fields. The Raman spectrum of $BaTiO_3$ single crystals was investigated in [25], but these authors observed only high-frequency vibrations. A more complete Raman spectrum of $BaTiO_3$ was obtained by Ikegami [26], using single crystals of higher purity and excitation by a line of longer wavelength. The spectrum showed lines at 1181, 823, 736, 513, 502, 416, and 359 cm^{-1}; the lines at 513 and 502 cm^{-1} were assigned to the vibration ω_3 split into two in the tetragonal phase; 359 cm^{-1} was regarded as one component of the split lowest-frequency vibration ω_2, and 416 cm^{-1} was taken to be one of the components of ω_1. The high-frequency lines at 736, 823, and 1181 cm^{-1} were regarded as second harmonics of some of the above. Perry and Hall [27] studied the temperature variation of the $BaTiO_3$ Raman spectrum, and found lines at 722, 518, 307, and 271 cm^{-1}, in fair agreement with the results of [26]. Their most interesting finding, however, was the strong temperature-dependence of the position of the line near 271 cm^{-1}, whose frequency varied from 271 cm^{-1} at T = 290°K to 230 cm^{-1} at T = 408°K. On this basis, it was interpreted as a longitudinal vibration of type ω_3.

An important contribution to the study of vibrational spectra of perovskite crystals came from investigations of these substances in the far infrared. The first measurements of transmission and reflection spectra in a wide range from 1 to 1000 μ were made in [28] for $SrTiO_3$ single crystals and in our own work [29-31] for $SrTiO_3$, $BaTiO_3$, and $CaTiO_3$ polycrystals, and also in [32] for polycrystalline $BaTiO_3$. $SrTiO_3$ and $BaTiO_3$ single crystals were thoroughly examined in the range 1-140 μ by Spitzer et al. [33, 34]. Since thin samples of these crystals proved to be opaque in the far infrared, reflection spectra were mainly used in order to study their vibrational spectra. A mathematical analysis, using the Kramers–Kronig integral relation, gave the frequency dependences $\varepsilon'(\omega)$ and $\varepsilon''(\omega)$ of the real and imaginary parts of the complex permittivity. This method has been successful in [28] and elsewhere [33, 34]. For single crystals of $SrTiO_3$ and $BaTiO_3$, the spectra of $\varepsilon(\omega)$ showed [28, 33, 34] three resonances at 18.2, 56.3, and 114 μ, and 19.6, 54.7, and 296 μ, respectively. In [28], these vibrations were interpreted in accordance with Last's model already mentioned. In [33, 34], a somewhat different interpretation of the normal modes was proposed. On the basis of a comparison of measured oscillator strengths of corresponding vibrations in $SrTiO_3$, $BaTiO_3$, and TiO_2, the low-frequency and high-frequency vibrations were identified with normal modes ω_3 and ω_4, respectively. The vibration at about 55 μ was assigned to ω_2. Similar results were obtained much later by Ballantyne [35] for $BaTiO_3$ single crystals.

In recent years, the vibrational spectra of $SrTiO_3$ and $BaTiO_3$ have also been studied by means of slow-neutron scattering spectra [36, 37]. The $SrTiO_3$ spectrum showed four vibrations having frequencies 530, 266, 170, and 97 cm^{-1} (19, 37.6, 59, and 103 μ), which are in satisfactory agreement with the infrared absorption spectra and the Raman spectra. $BaTiO_3$ showed several bands [37], those at 480, 340, and 235 cm^{-1} (20.9, 29.5, and 42.6 μ) being assigned to optical transverse vibrations. Those at 20.9 and 42.6 μ are similar to those found in $SrTiO_3$ (see above). Various mechanisms such as surface effects were proposed in order to explain the band at 29.5 μ, which was also observed in the infrared absorption spectra of powdered strontium and barium titanate and in their Raman spectra. There does not yet appear to be any definite conclusion as to the nature of this band. In [37], it is considered that the peak at 42.6 μ in the slow-neutron scattering spectra of $BaTiO_3$ is to be ascribed to an F_{2u} vibration inactive in the infrared spectra, i.e., to ω_1.

To summarize this review, we may say that, according to infrared transmission and reflection measurements [22, 23, 28, 32, 34], the normal-mode spectrum of the $SrTiO_3$ and $BaTiO_3$ lattices contains three vibrations at about 20, 55, and 100-300 μ. When these crystals are studied by slow-neutron scattering and Raman scattering, a further vibration is found at about 30-40 μ.

In referring to types of symmetry and the form of the normal modes of perovskite lattices, it should be noted that Last's treatment is only a rough approximation, and this is the reason for some variety in the interpretation of the results in [23, 24, 33]. It would, of course, be more exact to determine the symmetry types of the normal modes from the general principles of group theory. Such a treatment has been given in [38] for a cubic perovskite lattice. The work of the Czech theoreticians Dvořák and Janovec [39, 40] is extremely interesting as regards the form and spectral position of the normal modes of perovskite-type crystals. This work was likewise based on the principles of group theory, using the equivalence of the atoms O(4) and O(5), and led to a determinant which describes the normal modes of the cubic perovskite lattice of $BaTiO_3$. The model of van der Waals and long-range Coulomb forces was used. Although the absolute values are not very significant, the general assessment of the form and spectral distribution of the normal modes is very interesting, especially as these properties differ considerably in different models. The calculated form of the normal modes was found to be quite complex, though it is in general comparable with Last's model. In particular, the high-frequency vibrations appear to be due mainly to internal vibrations of the atoms of the TiO_6 octahedron, and the lowest-frequency vibration to a relative displacement of the titanium and barium atoms. It is not certain whether oxygen atoms take part in the latter vibration.

Our studies of the infrared transmission and reflection spectra of polycrystalline $SrTiO_3$, $BaTiO_3$, and $CaTiO_3$ in the range 1-1000 μ were made at the same time as the work previously cited [28, 32-34]. The samples were polycrystalline, since it was not possible to obtain single crystals of sufficient size. Since these crystals contain heavy atoms (Sr, Ba, Ca), a wide region of the spectrum was used. Most of the measurements were of transmission spectra of powers or reflection spectra of plates.

$SrTiO_3$ and $BaTiO_3$ showed vibrations at about 18, 55, and 100-140 μ. The longest-wavelength vibration in $BaTiO_3$ was detected only after a mathematical analysis of the measured reflection spectra. Since it was not visible in these spectra themselves, it was originally thought to correspond to the reflection maximum of $BaTiO_3$ at about 55 μ [29]. Here, as in [33], the spectral variation of the measured reflection coefficient of the titanates was found to show peculiarities at about 30 μ. The recent calculations by Barker [41] have shown that these features appear to be due to the influence of the phonon–phonon interaction of the three above-mentioned normal modes of these crystals. For $CaTiO_3$, measurements in a wide range of the infrared showed that the vibrational spectrum contains at least seven normal modes. The infrared studies were supplemented by measurements in the submillimeter (λ = 2.68, 4, and 8 mm) and radio ranges, for the purpose of mathematical analysis of the reflection spectra and to elucidate the contribution of the normal modes of the lattice to the general mechanism of polarization in such crystals. It was found that the polarization of $SrTiO_3$ and $CaTiO_3$ is mainly determined, over a wide frequency range, by the resonance mechanism of lattice vibrations [31]. In $BaTiO_3$, additional dispersion was found in the range $\lambda > 8$ mm.

The results were interpreted by means of theoretical ideas based on group theory. In particular, it was shown that the spectrum of normal modes in perovskite crystals in the cubic, tetragonal, and orthorhombic states consists of vibrations having symmetry types as follows: $\Gamma^0_{cub} = 3F_{1u} + 1F_{2u}$, $\Gamma^0_{tetr} = 3A_1 + 1B_1 + 4E$, and $\Gamma^0_{orth} = 4A_{1u} + 4B_{1u} + 4B_2$.

By means of symmetry coordinates, an analysis was made of the change of the infrared spectrum of such crystals between the cubic and the tetragonal and orthorhombic states.

The spectra of all three substances showed anomalous features of the lowest-frequency vibration, the oscillator strength and the degree of anharmonicity both being large. To explain these features, a calculation was made by a method similar to that of Cochran [21], giving an explicit dispersion expression for the complex permittivity of cubic perovskite crystals. The problem was solved by the use of normal coordinates. Local electric fields and the polarization of the electron shells of the ions during their vibration were taken into account. In comparing the resulting expression with the experimental data for a particular model of normal modes, estimates were made of a number of the microscopic parameters of $BaTiO_3$. In particular, it was found that the electron polarization is the main constituent of the total polarization of the crystal.

Unfortunately, the experimental results are not sufficient to determine the form of the normal modes observed. Theoretical calculations of the normal modes, based on the most suitable models of internal forces in the crystal, are therefore very important.

2. Features of the Vibrational Spectra of Perovskite Ferroelectrics

The polarization mechanism in perovskite-type crystals is known to be very complex. Its characteristics are largely determined by the strong interaction of local electric fields, due to the displacement of the ions from their equilibrium positions, with the electron shells of the ions [42]. This must obviously have a large effect on the vibrational spectrum of perovskite crystals. The presence of these local fields also appears to be responsible for the occurrence of a ferroelectric state in a number of crystals having the perovskite structure.

In recent years, there has been great interest in the temperature variation of the normal modes in $BaTiO_3$-type crystals, as a result of the work of Ginzburg [17, 18], Anderson [19], and Cochran [20, 21] on the microscopic theory of ferroelectricity, to which it is related. As early as 1949, Ginzburg applied his phenomenological theory of ferroelectric phenomena to a model of an anharmonic oscillator, and showed that the resonance frequency of a low-frequency oscillator is given, in terms of the coefficient α in the expansion of the thermodynamic potential, by $\omega_i = \sqrt{\alpha/\mu}$. Thus, since the parameter $\alpha \equiv \alpha'_\theta(T-\theta) \rightarrow 0$ as the phase-transition temperature is approached, it is clear that ω_i also must tend to zero. Rough estimates show that, even at temperatures far from the Curie point, the frequency of this vibration remains very low. According to these ideas, as later developed for first-order phase transitions, dispersion relations for the permittivity of $BaTiO_3$ can be written in terms of the phenomenological parameters as

$$\varepsilon_z(\omega, T) = \varepsilon_\infty + \frac{2\pi}{\alpha(T) - \mu\omega^2 + i\gamma\omega} \quad \text{for } T > \theta \tag{I.1}$$

and for the polarization phase

$$\varepsilon_z(\omega, \theta) = \varepsilon_\infty + \frac{2\pi}{4\alpha(\theta) - \mu\omega^2 + i\gamma\omega} \quad \text{for } T = \theta. \tag{I.2}$$

Thus, the frequency of this oscillator when $T = \theta$ is

$$\omega_{tz}(\theta) = 2\left[\frac{\alpha(\theta)}{\mu}\right]^{1/2}, \tag{I.3}$$

and its temperature variation may be conveniently written as

$$\omega_{tz}(T) = \frac{2\pi}{\mu[\varepsilon(0, T) - \varepsilon_\infty]}. \tag{I.4}$$

When the static permittivity varies according to the Curie-Weiss law, the latter expression gives

$$\omega_{tz}(T) = A\sqrt{T - T_c}. \tag{I.5}$$

According to the estimate made by Ginzburg [18] using Eq. (I.3) and $\alpha'_{\theta} = 3.7 \cdot 10^{-5}$ for the model of a Ba–TiO_3 vibration, the lowest frequency of the $BaTiO_3$ low-frequency vibration at the phase-transition temperature is $\omega_{tz}(\theta) = 12.6 \cdot 10^{11}$ cps, $\lambda_{tz}(\theta) = 1.5$ mm.

Other authors have also discussed similar questions and the mechanism of the anomalous temperature variation of the frequency of the $BaTiO_3$ low-frequency vibration near the phase-transition point [19-21]. Anderson [19], who first drew attention to the possibility of this effect, gave an almost entirely qualitative treatment, analyzing the properties of anharmonic oscillators corresponding to specified normal modes of a $BaTiO_3$-type lattice. The detailed study by Cochran [21] dealt with the problem from the standpoint of the dynamical theory of crystal lattices. By using normal coordinates and taking account of both short-range van der Waals forces and long-range Coulomb forces, Cochran showed that the ferroelectric transition in $BaTiO_3$-type crystals may be regarded as the result of instability of the crystal with respect to a certain normal mode of vibration. The calculation was based on the rigid-ion model and was made for the long-wave case. It was assumed that, at these frequencies, the ion shell electrons follow the field without inertia. Accordingly, the equations of motion of the ion cores and the corresponding electrons were written

$$m_k\omega^2 u_k = \sum_{k'\neq k}^{n} \{D_{kk'}(u_{k'} - u_k) + X_k X_{k'} C_{kk'}(u_{k'} - u_k)\} + \\ + \sum_{k'\neq k}^{n} \{F_{kk'}(v_{k'} - u_k) + X_k Y_{k'} C_{kk'}(v_{k'} - u_k)\} + K_k(u_k - v_k) + X_k Y_k C_0(v_k - u_k), \tag{I.6}$$

$$0 = \sum_{k'\neq k}^{n} \{S_{kk'}(v_{k'} - v_k) + Y_k Y_{k'} C_{kk'}(v_{k'} - v_k)\} + \\ + \sum_{k'\neq k}^{n} \{F_{kk'}(u_{k'} - v_k) + Y_k X_{k'} C_{kk'}(u_{k'} - v_k)\} + K_k(v_k - u_k) + X_k Y_k C_0(u_k - v_k). \tag{I.7}$$

Here, u_k and v_k are the displacements of the core and shell electron, respectively, of the kth ion along a triply degenerate direction, n the number of atoms per unit cell, and the other letters denote the force constants of different bonds between the atoms in the crystal.

Solving Eqs. (I.7) for v_k and substituting in Eqs. (I.6), we obtain equations which describe in matrix form the vibrations of the sublattices:

$$\omega^2 \mathbf{m}_d \mathbf{U}_c = \mathbf{M}' \mathbf{U}_c. \tag{I.8}$$

Here, $\mathbf{m}_d$ and $\mathbf{U}_c$ denote the matrices

$$\mathbf{m}_d = \begin{vmatrix} m_1 & 0 & 0 & \ldots & 0 \\ 0 & m_2 & 0 & \ldots & 0 \\ \ldots & \ldots & \ldots & \ldots & \ldots \\ 0 & 0 & 0 & \ldots & m_n \end{vmatrix}, \qquad \mathbf{U}_c = \begin{vmatrix} U_1 \\ U_2 \\ \cdot \\ \cdot \\ U_n \end{vmatrix} \tag{I.9}$$

and M' is the matrix of effective force constants, which express the difference between the short-range and the long-range (Coulomb) forces. According to Eqs. (I.8), the frequencies of the normal modes of the sublattices are given by characteristic equations of the form

$$| M'_{ik} - \omega^2 m_i \delta_{ik} | = 0. \tag{I.10}$$

Using the dielectric relations and the external electric field in Eqs. (I.8), we can find a dispersion for the permittivity, which in the work quoted was

$$\frac{\varepsilon_0 - \varepsilon_\infty}{4\pi} = \frac{e^2}{v} [\mathbf{Z}''_r] [\mathbf{M}']^{-1} [\mathbf{Z}''_c], \tag{I.11}$$

the square brackets denoting the principal minors of the matrices.

According to Born's theory, the instability of the crystal is due to vanishing of the frequency of one or more normal modes described, in particular, by equations like (I.10). Mathematically, this means that

$$\mathrm{Det}\,[\mathbf{M}'] = 0. \tag{I.12}$$

From Eq. (I.11), we can see that this must occur at the ferroelectric transition point, where the static permittivity ε_0 becomes infinite. Thus, a relationship is established between the ferroelectric transition and the instability of the crystal with respect to a certain normal mode of the lattice. The physical significance of the vanishing of Det $[\mathbf{M}']$ is given by the gradual balancing of the long-range and short-range forces in the crystal. Cochran also shows that the familiar Lyddane–Sachs–Teller equation, which is valid only for diatomic cubic crystals, can be extended to a wider class of crystals having diagonal-cubic symmetry. In this case, which includes $BaTiO_3$-type crystals, the equation becomes

$$\frac{\varepsilon_0}{\varepsilon_\infty} = \prod_{j=2}^{n} \frac{(\omega_j^2)_L}{(\omega_j^2)_T}, \tag{I.13}$$

where $(\omega_j)_L$ and $(\omega_j)_T$ are the frequencies of the longitudinal and transverse optical normal modes.

It is easily seen that, if the permittivity increases with temperature according to the Curie–Weiss law as the phase-transition point is approached, then the frequency of the corresponding normal mode responsible for the ferroelectric properties of the crystal must vary as follows:

$$\omega_i \propto \sqrt{T - T_c}, \tag{I.14}$$

which is the same as Eq. (I.5) given by Ginzburg's theory. In all the theories mentioned above [18-21], essentially the same phenomenological approach was used, based on the idea that the permittivity of ferroelectric crystals tends to very large values as the temperature approaches the phase-transition point.

The first experimental studies of this temperature variation of the vibrational spectrum of ferroelectrics were made in 1962 with $SrTiO_3$ [28]. The reflection spectrum of a single crystal was measured at 300 and 93°K. The results were analyzed by means of the Kramers–Kronig integral relation, and the spectral variation of $\varepsilon'(\omega)$ and $\varepsilon''(\omega)$ was found. It is known [15] that the static permittivity of $SrTiO_3$ varies with temperature in a manner similar to the Curie–Weiss law, the Curie point T_c being about 33°K. The measurements were therefore made at low temperatures. It was shown that, as the temperature decreases, the low-frequency vibration shifts toward still lower frequencies, as predicted by the theory. A similar result for $SrTiO_3$ was found by measurements of slow-neutron scattering [36].

The first studies of the temperature variations of the vibrational spectrum of $BaTiO_3$ were made in our work [30, 43-45], using polycrystalline samples and measuring reflection spectra in the range from 45 to 140°C, i.e., in the region of the ferroelectric phase transition of $BaTiO_3$. The change of frequency of the low-frequency vibration of $BaTiO_3$, predicted by the microscopic theory of ferroelectricity, was detected in the lowest-frequency part of the far infrared. This apparently explains why the effect was not observed previously [22, 32]. As the temperature approached the phase-transition point, the frequency of the lowest-frequency vibration was observed to decrease approximately as $\omega_i \propto \sqrt{T-T_c}$, which is in agreement with the theory. This shift of the vibration frequency was accompanied by an increase of the oscillator strength and degree of anharmonicity.

Somewhat later, Ballantyne [35] made similar measurements at 24-200°C, also using $BaTiO_3$ single crystals, in the range from 1-1000 cm^{-1}. When the temperature was above the phase-transition point, the calculated maximum corresponding to the low-frequency vibration moved from 6 cm^{-1} at 127°C to 13 cm^{-1} at 200°C. This represents a shift of the resonance frequency from about 16 to 70 cm^{-1}, in qualitative agreement with the theory given above.

CHAPTER II

Experimental Method

As has already been mentioned, investigations over the widest possible range of the spectrum are needed. Whereas there is no fundamental difficulty in the near infrared or the radio range, the far infrared and submillimeter wavelengths are very inaccessible.

The chief problem of longwave infrared spectroscopy is the lack of intense radiation sources. Attempts to construct electronic sources on the same principles as for microwaves have so far been unsuccessful, while the sources usually employed in the near and middle infrared, i.e., heated bodies, radiate only very weakly at long wavelengths. Another important obstacle is the problem of filters. Only one part in 10^8 of the power radiated by a black body at 10,000°K falls in the range from 0.2 to 0.4 mm [46]. This makes clear the importance of obtaining an effective filter system. Here, it is very annoying that the use of an initial prism monochromator, which has been successful in the near infrared, is impossible in the far infrared. Thus, a system of absorbing and reflecting filters has to be used, which greatly reduces the signal strength. Hence, we see the importance of high-quality receiving and recording equipment. In recent years, the development of long-wave infrared spectroscopy [46,47] has become possible mainly because of the construction of new fast-response receivers for infrared radiation, having an extremely low threshold sensitivity.

In order to measure the transmission and reflection spectra of crystals in the far infrared at 20-1200 μ, a vacuum diffraction spectrophotometer has been designed and constructed [48]. For measurements in the submillimeter range (λ = 2.68, 4, and 8 mm), a klystron was used as the radiation source, together with a long-wave infrared spectrometer for spatial separation of harmonics of the fundamental radiation [31]. Measurements in the near infrared were made with a standard IKS-14 spectrometer. Those in the radiofrequency range from 10^3 to 10^7 cps were made by the usual method [42].

1. Spectrophotometer for the Far-Infrared Region [48]

The Optical System. The spectrophotometer consisted of three main parts: the illuminator, the monochromator, and the receiver. Figure 2 shows the optical system.

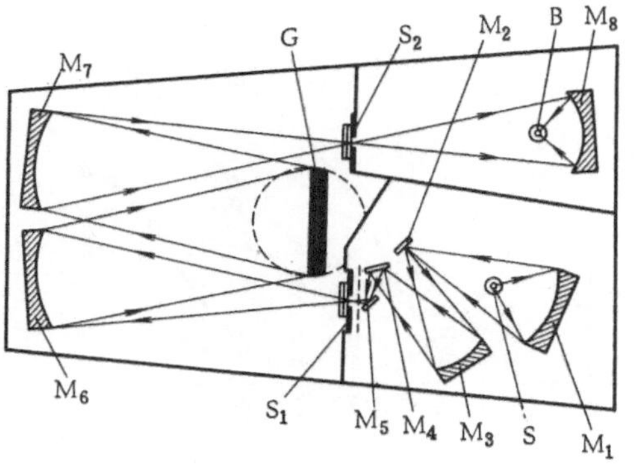

Fig. 2. Optical system of the instrument. S) Radiation source; G) echelette grating; B) bolometer detector; M_1), M_3), M_6), M_7), M_8) spherical mirrors; M_2), M_4), M_5) plane mirrors; S_1), S_2) entrance and exit slits of the monochromator.

The monochromator was of the Ebert–Fastie type [49], which can entirely eliminate coma aberration and considerably reduce the influence of aberration from astigmatism of the collimator mirrors. The latter were two spherical mirrors M_6 and M_7, with diameter 350 mm and focal length 750 mm.

The entrance and exit slits S_1 and S_2 of the monochromator were 50 mm high and could be varied from 0 to 25 mm in width. The crystal modulator for the radiation beam was placed in front of the slit S_1.

The Dispersing Element. In order to cover the whole range from 20 to 1200 μ, the dispersing element was one of five interchangeable echelette gratings* having constants $^1/_{12}$, $^1/_6$, $^1/_2$, 1.5, and 2.5 mm, and blaze angle 12.5°.

The range of applicability of a grating is conveniently determined from the expression for the half-width of the spectral distribution of intensity of radiation reflected from it, taking account of the angle of rotation:

$$\lambda_1 = 0.694 \frac{\cos\alpha}{\cos\Omega} \lambda_{bl}, \qquad \lambda_2 = 1.805 \frac{\cos\alpha}{\cos\Omega} \lambda_{bl},$$

where λ_{bl} is the wavelength in the first-order spectrum which corresponds to the blaze angle; Ω the blaze angle; α the angle of rotation of the grating; λ_1 and λ_2 the limiting wavelengths at which the intensity reflected is 50% of that reflected at the blaze angle. The two shortwave gratings had effective surfaces 200 × 200 mm, and the others 260 × 250 mm.

The holder was such that the gratings could easily be interchanged. The state was rotated by a sinusoidal mechanism which provided a spectrum scan linear in time. The rotation-rate gearing allowed the speed to be varied over a wide range, from 1° 12' to 4.5' per minute. The accuracy of the wavelength reading varied from 0.05 cm^{-1} at 100 cm^{-1} to 0.005 cm^{-1} at 20 cm^{-1}.

The Apparatus for Measuring the Sample Transmission and Reflection. The following equipment was used to measure the transmission and reflection in vacuum in the spectrophotometer: a horizontally movable holder having three cuvettes, placed behind the entrance slit of the monochromator, and a vertically movable holder having three sample-cells, placed at the position M_2 of the intermediate image of the radiation source. The angle of incidence of the radiation on the surface of the sample was less than 18°. The movement of the samples was controlled electrically.

*High-quality echelette gratings having constants $^1/_{12}$, $^1/_6$, and $^1/_2$ mm were prepared in the laboratory of F. M. Gerasimov, to whom the author is sincerely grateful.

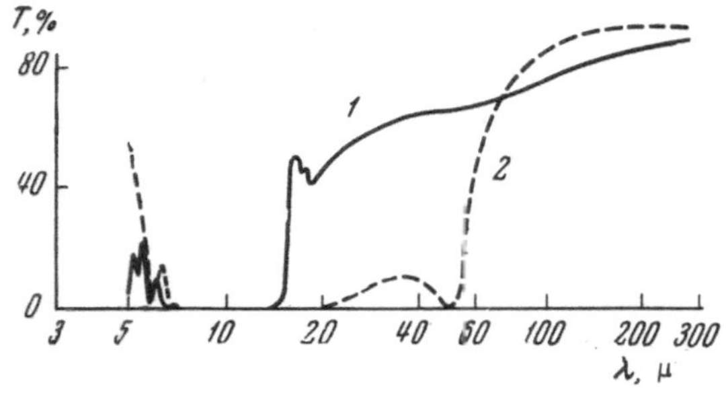

Fig. 3. Transmission in the range 5-300 μ. 1) Polyethylene (2 mm); 2) Teflon (1.5 mm).

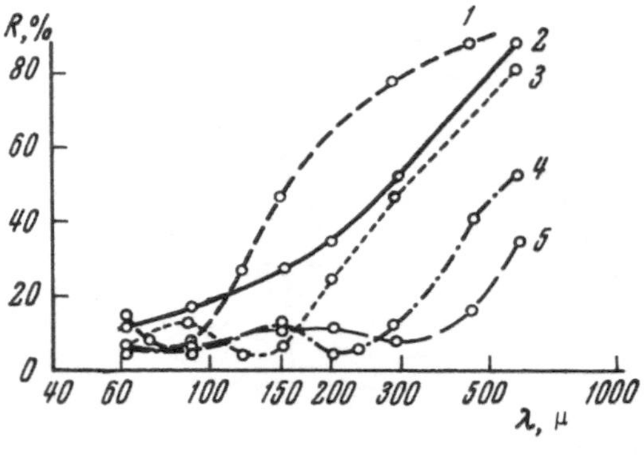

Fig. 4. Reflection from the surface of echelette filters at position of mirrors M_4 and M_5 in the range 60-1000 μ with grating constants: 1) 0.3 mm (M_4); 2) 0.5 mm (M_5); 3) 0.5 mm (M_4); 4) 0.85 mm (M_4); 5) 1.45 mm (M_4).

The Radiation Source. The chief source of radiation in the range from 50 to 1200 μ was a PRK-4 mercury–quartz lamp operating under overvoltage conditions* (current 5 A) and stabilized by a bank of 1B-10-17 barretters. In the range from 20 to 100 μ, the source was a heated silit rod.

The Radiation Detector, Amplifier, and Recorder. The far-infrared radiation detector used was a fast-response bismuth bolometer with a crystal-quartz window, constructed by M. N. Markov† on the basis of a bolometer described previously [50]. The receiving area was large, $10 \cdot 2.5$ mm, the threshold $2 \cdot 10^{-9}$ W, and the time constant 18 msec.

The modulated radiation signal was amplified by a device similar to that used in AIKS-F-4 [51, 52]. The RC circuit of the synchronous detector of the amplifier enabled the spectrum to be recorded with time constant 5, 30, 60, or 120 sec, using an ÉPP-09 electronic recorder. The equivalent noise impedance of the whole amplifier was less than 20 Ω; the noise in the spectrum recording was entirely due to the thermal noise of the bolometer strip resistances.

Filters. The lack of materials having adequate dispersion in the far infrared makes it necessary to replace the initial prism monochromator by a system of absorbing and reflecting filters.

The absorbing filters used were plates of polyethylene or Teflon covered with a dense layer of turpentine soot, paraffin, crystal quartz, and fused quartz [53, 54]. Figure 3 shows the transmission curves of polyethylene and Teflon.

The reflecting filters used were frosted plane mirrors, crystal plates of CaF_2, NaCl, KBr, CsI, and KRS-5, and echelette filters. These were in special vertically movable holders and exactly replaced the plane mirrors M_4 and M_5. The reflecting echelette filters were used at wavelengths above 200 μ. Figure 4 shows their measured reflection coefficient at the positions

*An attempt was also made to use high-pressure mercury and xenon lamps (GSVD-120, SVD-120a, SVDSh-250, SVDSh-500, SVDSH-1000), but the longwave radiation intensity in the range up to 500 μ was no greater than for the PRK-4, being subject to strong absorption in their thick quartz walls.

†The author's thanks are due for the construction and provision of trial bolometers.

TABLE 1. Filter Combinations

Spectrum range (μ)	Diffraction grating lines per mm	Filter combination	Time constant (sec)
20−40	12	2 mm polyethylene, 2 mm paraffin, two reflections from CaF_2	30
40−65	12	Polyethylene covered with soot, 2 mm paraffin, two reflections from NaCl	30
65−100	12	Polyethylene covered with soot, 2 mm paraffin, two reflections from KBr	30
90−150	6	Two Teflon covered with soot, 2 mm paraffin, frosted plane mirror, 10 mm crystal quartz	30
120−180	6	Two Teflon covered with soot, 2 mm paraffin, frosted plane mirror, reflection from CsI	60
125−200	2	Two Teflon covered with soot, 2 mm paraffin, reflection from KRS-5, 10 mm crystal quartz	60
190−350	2	Two Teflon covered with soot, 2 mm paraffin, 2 mm fused quartz, echelette filter with constant 0.5 mm (M_4)	60
350−750	1/1.5	Two Teflon covered with soot, 2 mm paraffin, 2 mm fused quartz, echelette filter with constant 0.85 mm (M_4)	120
700−1200	1/2.5	Two Teflon covered with soot, 2 mm paraffin, 3.2 mm fused quartz, echelette filters with constants 1.45 mm (M_4) and 0.85 mm (M_5).	120

of M_4 and M_5, and agrees with the calculated variation. It is seen that the filter properties become considerably less good if the grating is used at a more oblique angle to the radiation beam. The combinations of filters employed were such as to extinguish the very noticeable secondary maxima (Fig. 4).

Selective modulation was used throughout the spectrum, obtained by means of a chopper whose sectors were made of CaF_2 and NaCl crystals 10 mm thick.

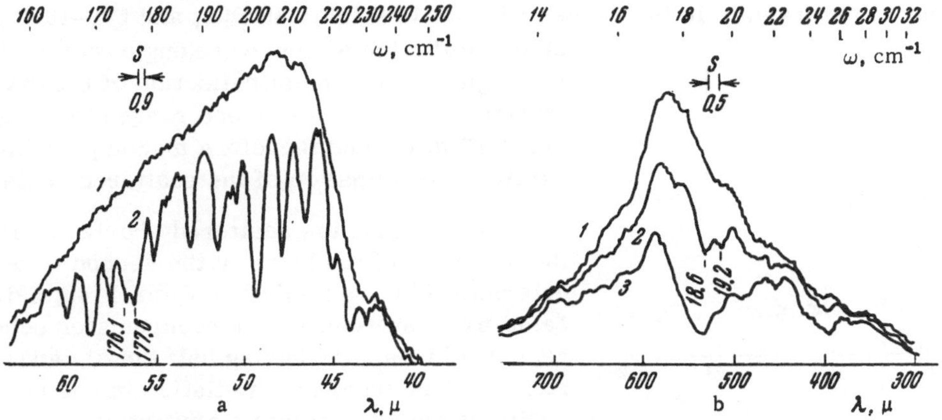

Fig. 5. Water vapor absorption spectrograms in the ranges 40-64 and 300-750 μ. a) Grating constant 1/12 mm, 2 mm polyethylene filter, covered with soot, two reflections from NaCl crystals, 1) 0.1 torr, 2) 290 torr; b) grating constant 1.5 mm, 1.8 mm Teflon covered with soot and 3 mm paraffin filters, and reflecting echelette filter with constant 0.5 mm, 1) 0.1 torr, 2) 400 torr, 3) 760 torr.

Table 1 shows the combinations of filters for which the proportion of stray radiation is less than 1-2% throughout the spectrum. For measurements which demand principally a high resolving power, the choice of filter combinations is wider (Fig. 5).

The spectrophotometer is mounted on a heavy steel plate 45 mm thick which is on a welded metal frame. The whole system rests on several layers of thick rubber.

The vacuum jacket of the spectrophotometer is 1900 × 900 × 450 mm, and is evacuated by a VN-2 backing pump to a pressure of 0.1 torr in 1 h.

Illustration of the Operation of the Spectrophotometer. To indicate the properties of the spectrophotometer, Fig. 5 shows two typical rotational absorption spectra of atmospheric water vapor, recorded with this instrument at relative air humidity 50-60% and various air pressures in the system. The resolution is seen to improve with decreasing water vapor pressure.

The resolving power was tested by the resolution of several lines, such as 176.1 and 177.6 cm^{-1} in the 40-65 μ region, 104.4 and 105.5 cm^{-1} in the 40-65 μ region, 104.4 and 105.5 cm^{-1} in the 75-100 μ region, 40.2 and 41.0 cm^{-1} in the 125-350 μ region, and 18.6 and 19.2 cm^{-1} in the 300-750 μ region. The instrument is seen to resolve bands in which the distance between peaks is less than 1.0 cm^{-1}.

In quantitative measurements in the longwave infrared, and in choosing a filter system, the quantitative determination of the scattered shortwave radiation in the range used, mainly in the second and higher orders, is particularly important. Since this topic has received little attention in the literature, it will here be discussed in some detail.

Measurements of the scattered shortwave radiation were made by the following methods, in addition to qualitative estimates from the form of the recorded absorption spectra of water vapor.

1. As far as 150 μ, the scattered shortwave radiation was as usual determined from measurements of the transmissivity of plates of CaF_2, NaCl, KBr, CsI, and KRS-5, the last four of which have a longwave transparency limit [53-55] at 25, 35, 75, and 55 μ, respectively.

2. Above 80 μ, the scattered shortwave radiation was determined by measuring the reflectivity of the same crystals in the ranges 80-150, 130-250, 240-600, and 500-1000 μ. The strong radiation whose wavelength corresponds to the region of maximum reflection of the crystals (residual rays) falls in these ranges in the second order (Fig. 6), and therefore makes possible a quantitative determination of the scattered radiation.

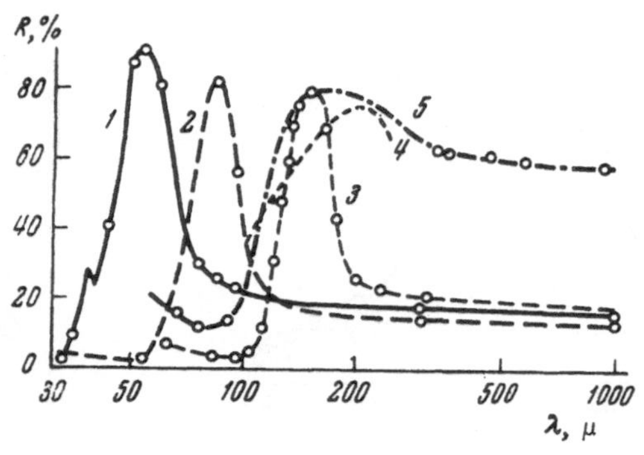

Fig. 6. Reflectivity of crystals: 1) NaCl; 2) KBr; 3) CsI; 4), 5) KRS-5. 1), 2), 4) Data of Czerny [56], Rubens [57], and McCubbin and Sinton [58]; 3), 5) author's measurements.

3. Finally, the scattered shortwave radiation in the direction of the blaze in the second order was determined by the method of Genzel [46, 59]. With the same filter system, the monochromator echelette was replaced by another having half the distance between rulings. The measured radiation intensity in the two positions gave the relative amount of radiation present in that part of the spectrum in the second order. Since the measurements were made only in the direction of the blaze, the echelettes clearly caused no important distortion of the source radiation distribution in this set of filters.

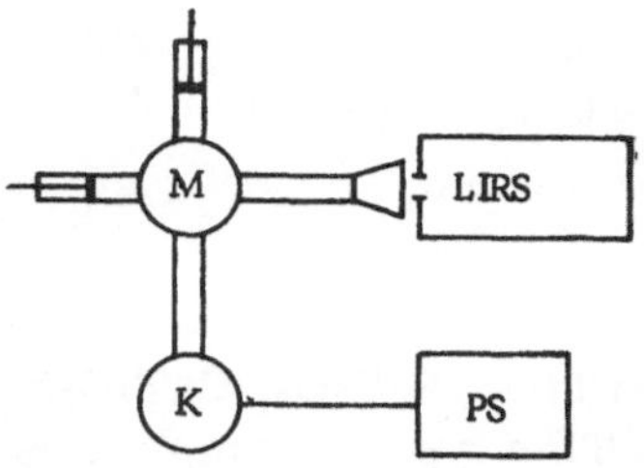

Fig. 7. Outline diagram of klystron system. K) klystron; PS) klystron power source; M) multiplying head; LIRS) far-infrared spectrometer.

Another good test of the amount of scattered radiation is to measure the absorptivity of a thin crystal plate of NaCl, which varies monotonically through the spectrum [59].

The filter combinations shown in Table 1 were derived from measurements made by the above methods.

The absorption and reflection spectra of a number of substances obtained with this spectrophotometer showed that it is suitable for quantitative measurements of both absorption and reflection spectra in the range stated.

2. Method of Measurement in the Submillimeter and Radio Ranges

The idea of using optical methods to separate the harmonics of microwave radiation is due to Genzel et al. [59] (1959). It has since been successfully applied in the USSR and elsewhere [31, 60]. The main advantages of the method are the simplicity and reliable separation of the harmonics, and the very high sensitivity and relatively simple adjustment of the receiving part of the system.

In the present work, the radiation source was a klystron, generating a fundamental mode with wavelength $\lambda = 8$ mm. (The author is most grateful to N. A. Irisova and E. M. Dianov for providing the radiofrequency equipment and for much assistance with the measurements.) This radiation was directed to a multiplying head which induced harmonics by means of a nonlinear silicon diode (Fig. 7). The harmonic antenna was a thin metal whisker making contact with the diode.

The first filter was an output waveguide with an appropriate cross section to stop the fundamental mode and the undesired low harmonics. The waveguide was terminated by a horn through which the radiation passed across the entrance slit into the spectrophotometer. The ordinary bolometer of the instrument was used to record the signal. The main difficulty was to establish effective contact between the metal whisker and the diode in the multiplying head. In order to obtain the maximum radiation, the process had to be repeated afresh for each harmonic.

This procedure gave the first five harmonics of the 8-mm fundamental mode (Fig. 8). The first, second, and third harmonics (8, 4, and 2.68 mm) were used; the signal strengths were such that the accuracy was 1-2% from five or six measurements. The fourth and fifth harmonics were recorded, but their power was insufficient for measurement. In this work, the dispersing element in the spectrometer was a diffraction grating with constant 4 or 6 mm.

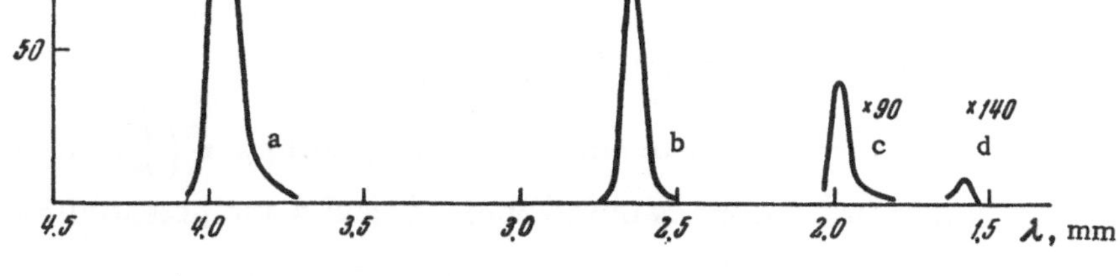

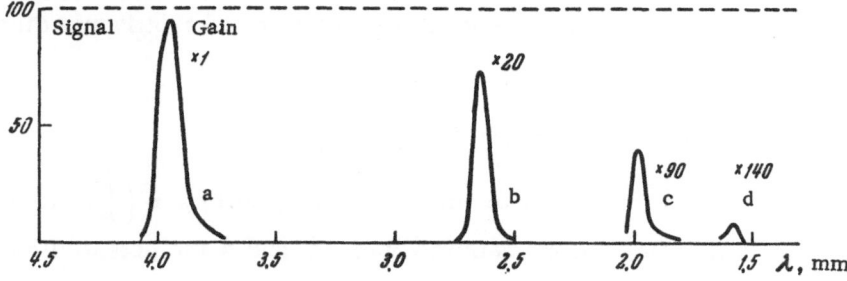

Fig. 8. Radiation in second (a), third (b), fourth (c), and fifth (d) harmonics of klystron radiation ($\lambda = 8$ mm).

Measurements of the permittivity $\varepsilon'(\omega)$ and the loss tangent δ of the crystals were made in the radio range at $f = 10^3$ cps with an MPP-300 bridge and at $f = 10^4$–10^7 cps with a 160-A Q-meter. The errors of measurement did not exceed ±3%.

3. Method of Preparation of the Samples

These investigations were done with polycrystalline samples of $SrTiO_3$, $BaTiO_3$, $CaTiO_3$, and some other substances prepared by the usual ceramic techniques. This type of sample was chosen because it is not yet possible to produce large single crystals of these materials. Their properties in the polycrystalline state have been thoroughly studied by Vul, Skanavi, and others [32]. Since the samples consisted of crystallites with a small amount of interposed vitreous phase (less than 5% by volume), their dielectric and other properties were mainly governed by the crystal lattice structure [42].

Samples used in the measurement of reflection and transmission spectra were made into disks having diameter 50-60 mm and thickness ranging from 0.1 to 10 mm. These were ground with various abrasives on both faces, and then polished. Test measurements of the reflectivity of samples having finely ground (abrasive M-10) and polished surfaces showed that, in the range $\lambda > 100\ \mu$, the manner of surface treatment has almost no effect on the results of the measurements. In the range $\lambda < 100\ \mu$, only polished samples were used in the measurements. The transmission spectra were also recorded for powers obtained by crushing the polycrystals. These samples were prepared by spraying the powder in a column of air and depositing it on a polyethylene substrate or a NaCl plate.

Since the accuracy had to be relatively high when the intensity of the far-infrared radiation is low, measurements were made point by point, using large time constants (120 sec) for signal recording. Each individual measurement then took 10-15 min. The accuracy varied from ±1.5% near 100 μ to ±2-2.5% near 800 μ.

In measurements of reflection spectra at higher temperatures, the sample-heater was placed in the reflector system M_2 of the spectrophotometer. The temperature of the reflecting layer of the sample was measured with a fine copper–constantan thermocouple attached to the illuminated surface.

CHAPTER III

The Spectrum of Normal Modes of Vibration in Crystals of Perovskite Type

According to the dynamical theory of crystal lattices developed by Born [61], the vibrations of a lattice which have a wave vector **y** which is almost zero are vibrations of appropriate sublattices, the number of these depending on the number of atoms in the unit cell. Since it will be necessary later to make use of certain consequences of this statement, they will be briefly derived here. In the general case, where the nuclei are in arbitrarily displaced positions $x\binom{l}{k} + u\binom{l}{k}$, we have the equations

$$m_k \ddot{u}_\alpha \binom{l}{k} = -\frac{\partial \Phi}{\partial u_\alpha \binom{l}{k}}. \tag{III.1}$$

Here, $x\binom{l}{k}$ is the equilibrium radius vector of nucleus k in cell l, $u\binom{l}{k}$ its displacement from the equilibrium position, m_k the mass of the nucleus, and Φ the potential energy of the crystal.

If Φ is written as a Taylor series in powers of the displacements of the nuclei, and the terms above the second order are neglected, then

$$\frac{\partial\Phi}{\partial u_\alpha\binom{l}{k}} = \Phi_\alpha(k) + \sum_{l'k'\beta} \Phi_{\alpha\beta}\binom{ll'}{kk'} u_\beta\binom{l'}{k'}, \tag{III.2}$$

and $\Phi_{\alpha\beta}$ in general denotes component α of the force on particle $\binom{l}{k}$ due to a unit displacement of particle $\binom{l'}{k'}$ in direction β. Since $\Phi_\alpha(k) = 0$ in an equilibrium configuration, substitution of Eq. (III.2) in Eq. (III.1) gives

$$m_k \ddot{u}_\alpha\binom{l}{k} = -\sum_{l'k'\beta} \Phi_{\alpha\beta}\binom{ll'}{kk'} u_\beta\binom{l'}{k'}. \tag{III.3}$$

These equations can be simplified, on account of the periodicity of the lattice. To do so, we use wave solutions having the form

$$u_\alpha\binom{l}{k} = \frac{1}{\sqrt{m_k}} W_\alpha(k) \exp\left\{2\pi i \mathbf{y}\mathbf{x}\binom{l}{k} - i\omega t\right\}, \tag{III.4}$$

where $\mathbf{y}$ is the wave vector. Substituting Eq. (III.4) in Eq. (III.3) and dividing by $\exp\left\{2\pi i \mathbf{y}\mathbf{x}\binom{l}{k} - i\omega t\right\}$, we get

$$\omega^2 W_\alpha(k) = \sum_{l'k'\beta} \frac{1}{\sqrt{m_k m_{k'}}} \Phi_{\alpha\beta}\binom{ll'}{kk'} \exp\left\{2\pi i \mathbf{y}\left[\mathbf{x}\binom{l}{k} - \mathbf{x}\binom{l'}{k'}\right]\right\}. \tag{III.5}$$

Since the periodicity of the lattice shows that $\Phi_{\alpha\beta}\binom{ll'}{kk'} = \Phi_{\alpha\beta}\binom{l-l'}{k-k'}$, i.e., it is determined by the difference $l - l'$ and not by l', the summation over $l - l'$ can be taken for any value of l; the result is then

$$\omega^2 W_\alpha(k) = \sum_{k'\beta} C_{\alpha\beta}\binom{\mathbf{y}}{kk'} W_\beta(k'), \tag{III.6}$$

where

$$C_{\alpha\beta}\binom{\mathbf{y}}{kk'} = \frac{1}{\sqrt{m_k m_{k'}}} \sum_{l'} \Phi_{\alpha\beta}\binom{l-l'}{kk'} \exp\left\{-2\pi i \mathbf{y}\left[\mathbf{x}\binom{l}{k} - \mathbf{x}\binom{l'}{k'}\right]\right\} =$$
$$= \frac{1}{\sqrt{m_k m_{k'}}} \exp\left\{-2\pi i \mathbf{y}\left[\mathbf{x}(k) - \mathbf{x}(k')\right]\right\} \sum_{l} \Phi_{\alpha\beta}\binom{l}{kk'} \exp\left[-2\pi i(\eta l)\right], \tag{III.7}$$

and $\eta\ (\eta_1\eta_2\eta_3)$ are the components of $\mathbf{y}$. Thus, we see that the equations reduce to equations involving the indices of one unit cell and the wave vector $\mathbf{y}$. In this case of long waves (compared with the size of the unit cell), $\mathbf{y} = 0$, and Eqs. (III.6) and (III.7) become

$$\omega^2 W_\alpha(k) = \sum_{k'\beta} C_{\alpha\beta}\binom{0}{kk'} W_\beta(k'), \tag{III.8}$$

$$C_{\alpha\beta}\begin{pmatrix}0\\kk'\end{pmatrix}=\frac{1}{\sqrt{m_k m_{k'}}}\sum_l \Phi_{\alpha\beta}\begin{pmatrix}l\\kk'\end{pmatrix}. \qquad \text{(III.9)}$$

It is evident that these equations are very general and describe vibrations of the corresponding sublattices containing different atoms in the unit cell. When long-range Coulomb forces and the electronic polarization of ions are considered [61], the corresponding parameters appear in the coefficients $C_{\alpha\beta}\begin{pmatrix}0\\kk'\end{pmatrix}$.

Thus, in our present case, it is easy to estimate the total number of normal modes of the lattice in substances of the $SrTiO_3$ type: 3k = 15, where k is the number of atoms in the unit cell. If acoustic vibrations due to translational motion of the whole crystal are excluded, there remain 12 optical normal modes. To determine the type of symmetry and the degree of degeneracy of these vibrations, we must take account of the symmetry of the crystals concerned. Here, we shall use Bhagavantam's method [62], which is based on the application of group theory to crystals for the case of long waves, $\mathbf{y} \approx 0$. Including translation operations, the perovskite-type lattice in the cubic, tetragonal, and orthorhombic states is described by a set of symmetry operations belonging to the symmetry groups O_h, C_{4v}, and C_{2v}, respectively. If the characters of the reducible representations are determined for all these symmetry operations, it is easy to find the number of normal modes belonging to the corresponding symmetries, using the expression

$$n_i = \frac{1}{N}\sum_R h_\rho \chi'_\rho(R)\,\chi_i(R). \qquad \text{(III.10)}$$

Here, n_i is the number of times the i-th irreducible representation is contained in a given reducible representation; $\chi'_\rho(R)$ is the character of the reducible representation for operation R; $\chi_i(R)$ is that of the i-th irreducible representation for operation R; N is the order of the symmetry group; h_ρ is the number of symmetry operations in class ρ.

Since all the other parameters are known, to solve the problem we must first find the characters of the reducible representations. Following Bhagavantam's method, we express in Cartesian coordinates the displacements of the ions in the unit cell. Then, the character of a reducible representation of the whole system is made up of the characters of the transformation of the coordinates x_i, y_i, z_i of just those ions whose position is unchanged by the symmetry operation considered. For long waves ($\mathbf{y} \approx 0$), ions of a particular sublattice have this property. Here, it is important that all the symmetry operations can be reduced to a rotation C_p or a rotary reflection S_p (the reflection $\sigma = S_0$, the inversion $i = S_2$, and the identity $E = C_1$). Then, by writing down the simple expressions for the transformation of the Cartesian coordinates of such a particle for a rotation (or a rotary reflection), we can derive the corresponding expression for the character of a reducible representation of the whole system in the crystal for the symmetry operation concerned [62].

For optical vibrations,

$$\chi'_\rho(R)_{\text{opt}} = (N_R - 1)(\pm 1 + 2\cos\varphi). \qquad \text{(III.11)}$$

For acoustic vibrations due to translation motion of the whole crystal,

$$\chi'_\rho(R)_{\text{ac}} = (\pm 1 + 2\cos\varphi). \qquad \text{(III.12)}$$

TABLE 2. Characters of Reducible Representations Calculated for All Symmetry Operations of the Groups O_h, C_{4v}, and C_{2v}, Corresponding to Cubic, Tetragonal, and Orthorhombic Perovskite Crystals, Respectively

Symmetry groups	Symmetry operations	Characters of reducible representations	Characters of reducible representations for acoustic vibrations	Characters of reducible representations for optical vibrations
O_h	E	15	3	12
	$8C_3$	0	0	0
	$6C_2$	—3	—1	—2
	$6C_4$	3	1	2
	$3C_2$	—5	—1	—4
	i	—15	—3	—12
	$6S_4$	—3	—1	—2
	$8S_6$	0	0	0
	$3\sigma_h$	5	1	4
	$6\sigma_d$	3	1	2
C_{4v}	E	15	3	12
	$2C_4$	3	1	2
	C_2	—5	—1	—4
	$2\sigma_v$	5	1	4
	$2\sigma_d$	3	1	2
C_{2v}	E	15	3	12
	$C_2(z)$	—5	—1	—4
	$\sigma(x, z)$	5	1	4
	$\sigma(y, z)$	5	1	4

Here, φ is the angle of rotation; the positive and negative signs of unity refer to rotation and rotary reflection, respectively; N_R is the number of particles whose position is unchanged by the symmetry operation R or which move to the position of another ion in the same sublattice.

Table 2 shows the characters of the reducible representations for crystal structures of the $SrTiO_3$, $BaTiO_3$, and $CaTiO_3$ type, calculated by the above method, Eqs. (III.11) and (III.12), for various symmetry operations. From these values and Eq. (III.10), we can easily determine the number of normal modes of the corresponding types of symmetry, their degree of degeneracy, and so on (Table 3). The results for the cubic, tetragonal, and orthorhombic systems may be respectively written as

$$\begin{aligned} &\Gamma^o = 3F_{1u} + 1F_{2u}, \qquad &&\Gamma^a = 1F_{1u}; \\ &\Gamma^o = 3A_1 + 1B_1 + 4E, \qquad &&\Gamma^a = 1A_1 + 1E; \\ &\Gamma^o = 4A_1 + 4B_1 + 4B_2, \qquad &&\Gamma^a = 1A_1 + 1B_1 + 1B_2, \end{aligned} \tag{III.13}$$

where Γ^0 and Γ^a are systems of optical and acoustic normal modes, respectively. The vibrations F_{1u} and F_{2u} are triply degenerate, E doubly degenerate, the remainder nondegenerate. Vibrations F_{2u} (O_h) and B_1 (C_{4v}) must be infrared-inactive by virtue of the selection rules. Accordingly F_{1u} (O_h) and F_{2u} (O_h) will not appear in the Raman scattering spectra.

In order to see how these vibrations may be transformed when the crystal changes from the cubic to the tetragonal and then to the orthorhombic state, we construct the appropriate symmetry coordinates [63], considering first the perovskite structure in the cubic and tetragonal states. It is easily shown that the Cartesian coordinates of the displacement of the

TABLE 3. Number of Normal Modes for the Various Symmetry Types, Calculated from Eq. (III.10)

Symmetry groups	Vibration symmetry types	Number of normal modes for that symmetry type	Symmetry groups	Vibration symmetry types	Number of normal modes for that symmetry type
O_h	A_{1g}	0	C_{4v}	A_1	4
	A_{1u}	0		A_2	0
	A_{2g}	0		B_1	1
	A_{2u}	0		B_2	0
	E_g	0		E	5
	E_u	0			
	F_{1g}	1	C_{2v}	A_1	5
	F_{1u}	4		A_2	0
	F_{2g}	0		B_1	5
	F_{2u}	1		B_2	5

sublattices, corresponding to the atoms of the unit cell (Fig. 1), form four groups of equivalent coordinates in the cubic crystal: I$-x_1y_1z_1$; II$-x_2y_2z_2$; III$-x_5y_4z_3$; IV$-x_3y_3x_4z_4y_5z_5$, and nine groups in the tetragonal crystal: I $- z_1$; II $- z_2$; III $- z_3$; IV $- z_4$ and z_5; V$-x_1$ and y_1: VI$-x_2$ and y_2; VII$-x_3$ and y_3; VIII$-x_4$ and y_4; IX$-x_5$ and y_5. From the well-known bases of the transformation matrices for reducible representations needed in the case of degenerate types of symmetry, we can construct the appropriate symmetry coordinates for each group:

$$S_i^\lambda = \sum_\gamma a_{i\gamma}^\lambda q_\gamma^\lambda, \tag{III.14}$$

where S_i^λ is the symmetry coordinate for symmetry type i and equivalent-coordinate group λ; q_γ^λ a Cartesian coordinate in that group; $a_{i\gamma}^\lambda$ the corresponding symmetry coefficient.

It is clear that the normal coordinate Q_i is formed only from symmetry coordinates of type i, chosen among all the equivalent-coordinate groups λ, and is a linear combination

$$Q_i = \sum_\lambda b_{i\lambda} S_i^\lambda. \tag{III.15}$$

The symmetry coordinates S_i^λ, determined as in the Appendix, may be written as follows.

For the cubic state,

$$F_{1u} \to S_{F_{1u}}^{\mathrm{I}} = z_1, \quad S_{F_{1u}}^{\mathrm{II}} = z_2, \quad S_{F_{1u}}^{\mathrm{III}} = z_3, \quad S_{F_{1u}}^{\mathrm{IV}} = \frac{1}{\sqrt{2}}(z_4 + z_5);$$
$$F_{2u} \to S_{F_{2u}}^{\mathrm{IV}} = \frac{1}{\sqrt{2}}(z_4 - z_5); \tag{III.16}$$

for the tetragonal state,

$$A_1 \to S_{A_1}^{\mathrm{I}} = z_1, \quad S_{A_1}^{\mathrm{II}} = z_2, \quad S_{A_1}^{\mathrm{III}} = z_3, \quad S_{A_1}^{\mathrm{IV}} = \frac{1}{\sqrt{2}}(z_4 + z_5);$$
$$B_1 \to S_{B_1}^{\mathrm{IV}} = \frac{1}{\sqrt{2}}(z_4 - z_5); \tag{III.17}$$
$$E \to S_E^{\mathrm{V}} = x_1, \quad S_E^{\mathrm{VI}} = x_2, \quad S_E^{\mathrm{VII}} = x_3, \quad S_E^{\mathrm{VIII}} = x_4, \quad S_E^{\mathrm{IX}} = x_5.$$

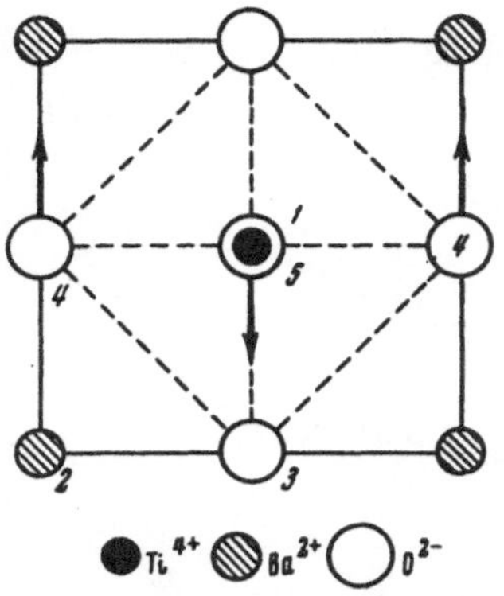

Fig. 9. Form of the F_{2u} (O_h) and $B_1(C_{4v})$ normal modes.

Since the vibrations F_{1u}, F_{2u} are triply degenerate, and E is doubly degenerate, corresponding expressions can be written also for the x and y axes for F_{1u} and F_{2u}, and for the y axis for E.

On comparing the form of the symmetry coordinates from which the normal-mode coordinates are obtained, we see that, when the crystal changes from the cubic to the tetragonal state, the normal modes are transformed as follows:

$$F_{1u}(xyz) \to 1A_1(z) + 1E(xy),$$
$$F_{2u}(xyz) \to 1B_1(z) + 1E(xy). \qquad \text{(III.18)}$$

Taking now the case of the orthorhombic perovskite lattice, we see that E (xy) normal modes must become separate vibrations along x and y.

That is to say, the infrared absorption spectrum of cubic perovskite crystals must consist of three F_{1u} (xyz) absorption bands. When the change to the tetragonal state occurs, the form of the spectrum begins to depend essentially on the polarization of the incident radiation. The absorption spectrum in polarized light with $E \parallel z$ should continue to show three A_1 (z) vibrations. When $E \perp z$, there should be not only three similar E (xy) bands but also a new E (xy) band with frequency close to that of F_{2u} (O_h). Evidently, a tetragonal perovskite crystal in natural light should show three doublet bands and a band with frequency close to ω_{2u}.

The infrared absorption spectrum of orthorhombic crystals should show a further doublet splitting of the E modes.

The Raman scattering spectrum of cubic perovskite crystals should not show any first-order lines, since they are all forbidden by the selection rules. It is interesting that the Raman spectrum of such crystals should appear when they change from the cubic to the tetragonal state, with all eight vibrations $3A_1 + 1B_1 + 4E$ present, apparently in the form of four doublet bands. The Raman spectrum of these crystals in the orthorhombic state should contain all twelve non-degenerate normal modes.

The above choice of the symmetry of normal modes of perovskite lattices may be accompanied by a remark on the form of the vibrations. Since the F_{2u} mode (III.16) and the B_1 mode (III.17) involve a relative displacement of only two sublattices, O (4) and O (5), the form of these vibrations is easily visualized (Fig. 9). In the other normal modes, several sublattices participate, and the expressions for the normal coordinates as linear combinations of the corresponding symmetry coordinates are unknown. One can attempt to write down the secular equations of motion of the ionic sublattices of the crystals by using the symmetry properties. But in the simple cubic-crystal case, for example, the resulting matrices still contain nine elastic constants [39, 44, 63], which cannot be estimated on the basis of experiment. Thus, it is not yet possible to determine from experiment the form of these normal modes. Theoretical calculations based on the most realistic models of internal forces in the crystal are therefore very important.

CHAPTER IV

Experimental Results and Their Interpretation

1. Transmission and Reflection Spectra

Since the crystals examined are opaque in the infrared even at thicknesses of about 40 μ, which is the practical limit for polycrystalline plates, we measured the transmission spectra of

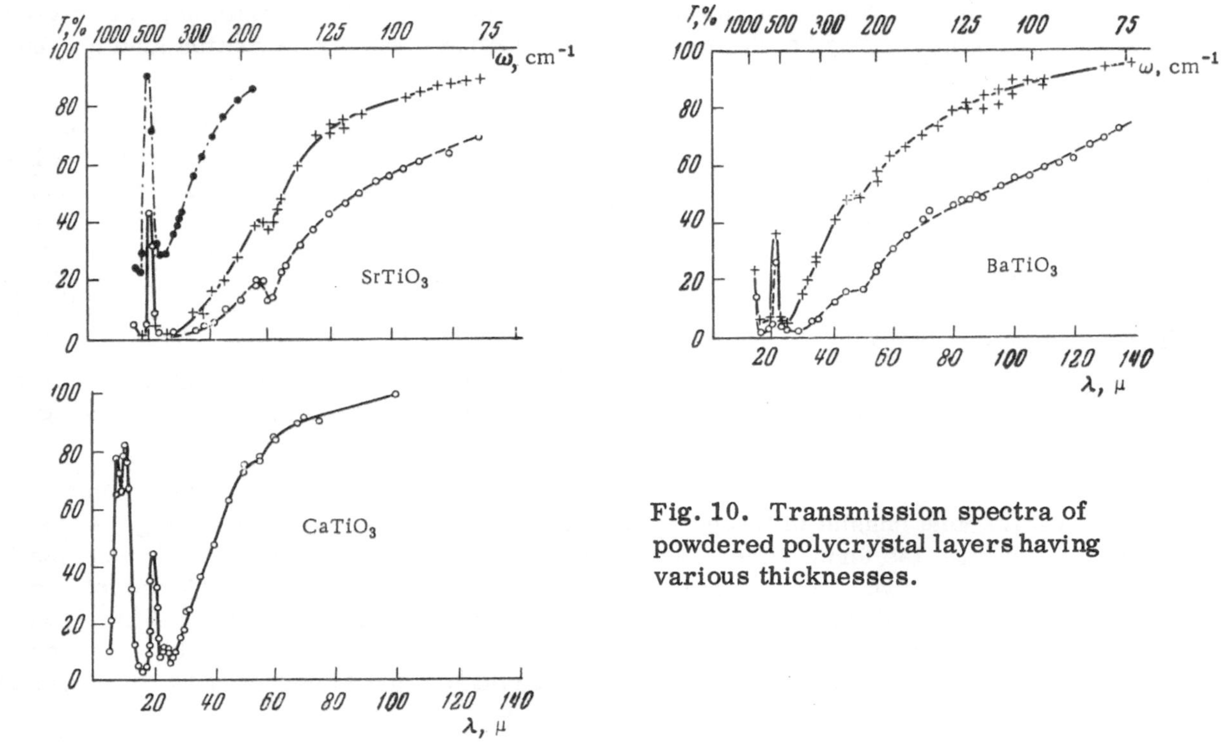

Fig. 10. Transmission spectra of powdered polycrystal layers having various thicknesses.

these substances as powders. The grain size was about 5 μ, determined by microscopy and from the near-infrared transmission spectra. The transmission spectra* of powdered $SrTiO_3$, $BaTiO_3$, and $CaTiO_3$ polycrystals measured at room temperature are shown in Fig. 10. The experimental points begin at $\lambda = 18\ \mu$ in the shortwave region, i.e., at the transmission edge of the polyethylene substrate (see Fig. 3).

The transmission spectra for all three substances show two bands with maximum absorption at about 18 and 25 μ, and a slight maximum at about 55 μ. In $CaTiO_3$, the 25-μ band is split, with two maxima at 22.3 and 26 μ. At $\lambda > 30\ \mu$, all the spectra show a steadily increasing transmission, reaching 95-98% at long wavelengths ($\lambda > 100\ \mu$) and having no further structure up to about 1000 μ. The transmission-spectrum structure found in the near infrared is in good agreement with other authors' measurements on $SrTiO_3$ and $BaTiO_3$ single crystals [22, 33]. The steady increase of transmission in the range $\lambda > 30\ \mu$ seems to be due to scattering by particles having a high permittivity [64]. This follows both from our measurements of the reflection spectra of the crystals and from measurements of the transmission of thin single crystals [28, 33].

The reflection spectra of $SrTiO_3$, $BaTiO_3$, and $CaTiO_3$, measured [30, 31, 43] at 45°C in the range from 2 to 1000 μ, are shown in Fig. 11. It is seen that, in addition to the well-known reflection band with maximum at about 18 μ, the spectra of all the substances have a wide plateau at $\lambda > 22\ \mu$. This region has a noticeable structure in $SrTiO_3$ and $BaTiO_3$, and a complicated profile with several minima in $CaTiO_3$.

Analysis of these data and comparison with the results of measuring transmission spectra of powders indicates that in either case the vibration at about 18 μ is clearly visible; that the sharp increase of the reflectivity near 22 μ to a noticeable maximum at about 30 μ corresponds in position to the minimum transmission at about 25 μ; that the $BaTiO_3$ spectrum has a maximum

*The measured transmission curves in the range 100-1000 μ showed no additional structure and are therefore omitted from the diagram.

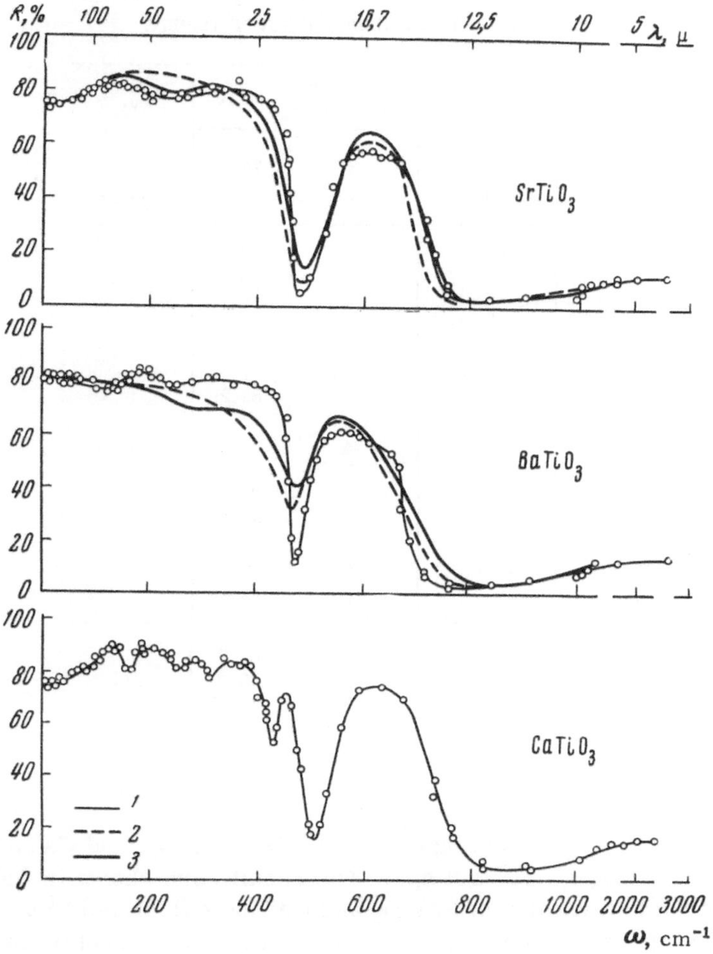

Fig. 11. Reflection spectra of polished polycrystalline $SrTiO_3$, $BaTiO_3$, and $CaTiO_3$ samples. 1) Experimental; 2) calculated from Eq. (IV.2) (see below) for a system of two oscillators, ω_2 and ω_3; 3) calculated similarly for a system of three oscillators, ω_2, ω_3, and ω_4.

reflection at about 55 μ, corresponding to a slight minimum at the same wavelength in the transmission spectrum; that the $SrTiO_3$ spectrum has a maximum reflection at about 80 μ. Finally, at long wavelengths ($\lambda > 200$ μ), the measured reflection curves for all the crystals have a linear part with R = constant.

Measurements of the reflectivity of these crystals were also made in the submillimeter range at $\lambda = 4$ and 8 mm, using the method described above. Table 4 shows the results for $SrTiO_3$ and $BaTiO_3$ polycrystals. For $CaTiO_3$, transmission of about 8% was observed in samples 2 mm thick, even in the region of $\lambda = 4$ mm. The reflectivity values for $CaTiO_3$ were found to vary considerably with the sample thickness, i.e., they were governed by interference, and they are therefore omitted from the table. It will be seen later that these measurements in the submillimeter range are especially necessary for $BaTiO_3$ to determine the spectral variation of its permittivity over a wide range of frequencies. In particular, a very important fact is that the reflectivity of $BaTiO_3$ measured in the far infrared (200–1000 μ) is almost constant up to $\lambda = 8$ mm.

TABLE 4. Reflectivities of $SrTiO_3$ and $BaTiO_3$, Measured in the Submillimeter Range

Substance	Wavelength, mm	Measurement number	R, %	R_{av}, %
$SrTiO_3$	4	1	74	
		2	74	74±1
	8	1	74	
		2	74	73.9±1
		3	73.6	
		4	74	
$BaTiO_3$	4	1	81	
		2	83.2	
		3	83	
		4	83	82.4±1
		5	82.2	
		6	81.5	
		7	82.5	
		8	82.3	
		9	82.2	
		10	83	
	8	1	82.5	
		2	83	82.9±1
		3	83	
		4	83	

Finally, the permittivity and loss tangent of $SrTiO_3$, $BaTiO_3$, and $CaTiO_3$ polycrystals were measured at radio frequencies, $f = 10^3$–10^7 cps. The results are shown in Table 5. It is immediately obvious that the reflectivity of $SrTiO_3$ and $CaTiO_3$ calculated from ε' and $\tan\delta$ measured at 10^3 and 10^6 cps is almost equal to its value for the same crystals in the far infrared at $\lambda > 200\ \mu$ (Fig. 11). Thus, for these crystals, there is no additional dispersion of the permittivity over a wide range of frequency, from 10^3 to 10^{12} cps. For $BaTiO_3$, the value R = 89.6% calculated from the measurements at 10^3 cps considerably exceeds the corresponding value of 83% measured in the far-infrared and submillimeter ranges. This indicates that in $BaTiO_3$ there is an additional dispersion in the range $\lambda > 8$ mm.

TABLE 5. Permittivity ε' and $\tan\delta$ for $SrTiO_3$, $BaTiO_3$, and $CaTiO_3$

Substance	Frequency	ε'	tg δ	R, %
$SrTiO_3$	10^3	176	0.0045	74.0
	10^6	174	0.0005	73.9
$BaTiO_3$	10^3	1210	0.01	89.2
	10^6	1170	0.011	89.1
$CaTiO_3$	10^3	150	0.0023	72.0
	10^6	150	0.00043	72.0

2. Determination of the Spectral Variation of the Complex Permittivity. The Vibrational Spectrum of $SrTiO_3$, $BaTiO_3$, and $CaTiO_3$

As already mentioned, the transmission spectra of powdered crystals are affected by the scattering of radiation in the far infrared. The vibrational spectra of these crystals were therefore studied mainly by analysis of the results of measuring reflection spectra; and, since the reflectivity is a complicated function of the optical parameters n (ω) and k (ω) or the real and imaginary parts ε'(ω) and ε"(ω) of the permittivity, their spectral variation was found by using the Kramers–Kronig integral relation, which relates these real and imaginary parts. The calculations were done on a computer, using the measurements of the reflectivity R and the equations [28]

$$re^{-i\theta} = \frac{n - ik - 1}{1 + ik + 1}, \quad n = \frac{1 - r^2}{1 + r^2 - 2r\cos\theta}, \quad k = \frac{2r\sin\theta}{1 + r^2 - 2r\cos\theta},$$
$$r^2(\omega) = R, \quad \theta(\omega) = \frac{2\omega}{\pi}\int_0^\infty \frac{\ln|r(\omega')|\,d\omega'}{\omega^2 - \omega'^2}, \tag{IV.1}$$

where r (ω) and θ (ω) are the amplitude and phase of the reflected wave.

The numerical integration was carried out with automatic selection of the step length in quadratic interpolation of the base curve R* (ω) in the frequency range from $\omega_1 = 10^{-5}$ cm^{-1} ($3 \cdot 10^5$ cps) to $\omega_2 = 3000$ cm^{-1} ($1 \cdot 10^{14}$ cps). For $SrTiO_3$ and $CaTiO_3$, the base curves of reflection R* (ω) used in these calculations were obtained from our measurements in a wide infrared and submillimeter range of the spectrum and also from measurements at radio frequencies. As already mentioned, the experimental value of R* (ω) was constant throughout the spectrum at λ > 200 μ. For $BaTiO_3$, the corresponding base curve R* (ω) was found from similar reflectivity measurements (excluding the radio frequencies), extrapolated to lower frequencies. This curve is shown in Fig. 15 below. The points represent the results of our measurements and also those of measurements [4] on similar polycrystals at microwave frequency (λ = 3.2 cm). It is clear that, in such calculations for $BaTiO_3$, we have deliberately excluded from consideration the dispersion at lower frequencies, which is situated at about 10^{10} cps and is generally recognized [4, 65, 66] to be due to relaxation of domain boundaries in $BaTiO_3$. The results of the calculations are in the form of the spectral variation of θ (ω), k (ω), n (ω), ε'(ω), and ε"(ω) in the frequency range from 1 to 1000 cm^{-1}.

Figure 12 shows the functions ε'(ω), ε"(ω), and n (ω), k (ω) for polycrystalline $SrTiO_3$, $BaTiO_3$, and $CaTiO_3$ at room temperature. The curves of ε'(ω) and ε"(ω) for these three substances show not only the clear resonance form near 18 μ but also a deep dispersion in the range from 100 to 300 μ, evidently due to the lowest-frequency lattice vibration. Below this range (λ > 400 μ for $SrTiO_3$ and $CaTiO_3$, and λ > 100 μ for $BaTiO_3$), the permittivity does not vary with wavelength, the values of ε' at λ = 1000 μ being 176 for $SrTiO_3$, 460 for $BaTiO_3$, and 150 for $CaTiO_3$. The curves of ε'(ω) and ε"(ω) also show an additional structure due to features of the corresponding reflection spectra. In particular, ε'(ω) and ε"(ω) for $BaTiO_3$ show a resonance at about 55 μ. In the spectra of both $SrTiO_3$ and $BaTiO_3$, there is also a bulge on the curve of ε"(ω) and a corresponding slight bend on that of ε'(ω) at about 30 μ. For $CaTiO_3$, the spectral variation of the complex permittivity is less simple, there being a total of seven clear resonances at 17.9, 22.8, 31.3, 38.6, 45.5, 59.0, and 87.0 μ. Table 6 gives the exact position of the maxima on the curves of ε"(ω) for these three crystals.

From these data given by the analysis of the reflection spectra of $SrTiO_3$, $BaTiO_3$, and $CaTiO_3$, and from the measurements of their transmission spectra in the powdered state, we can attempt to form a general idea of the nature of their vibrational spectra, using the results of calculating the number and types of symmetry of the normal modes of the perovskite lattices

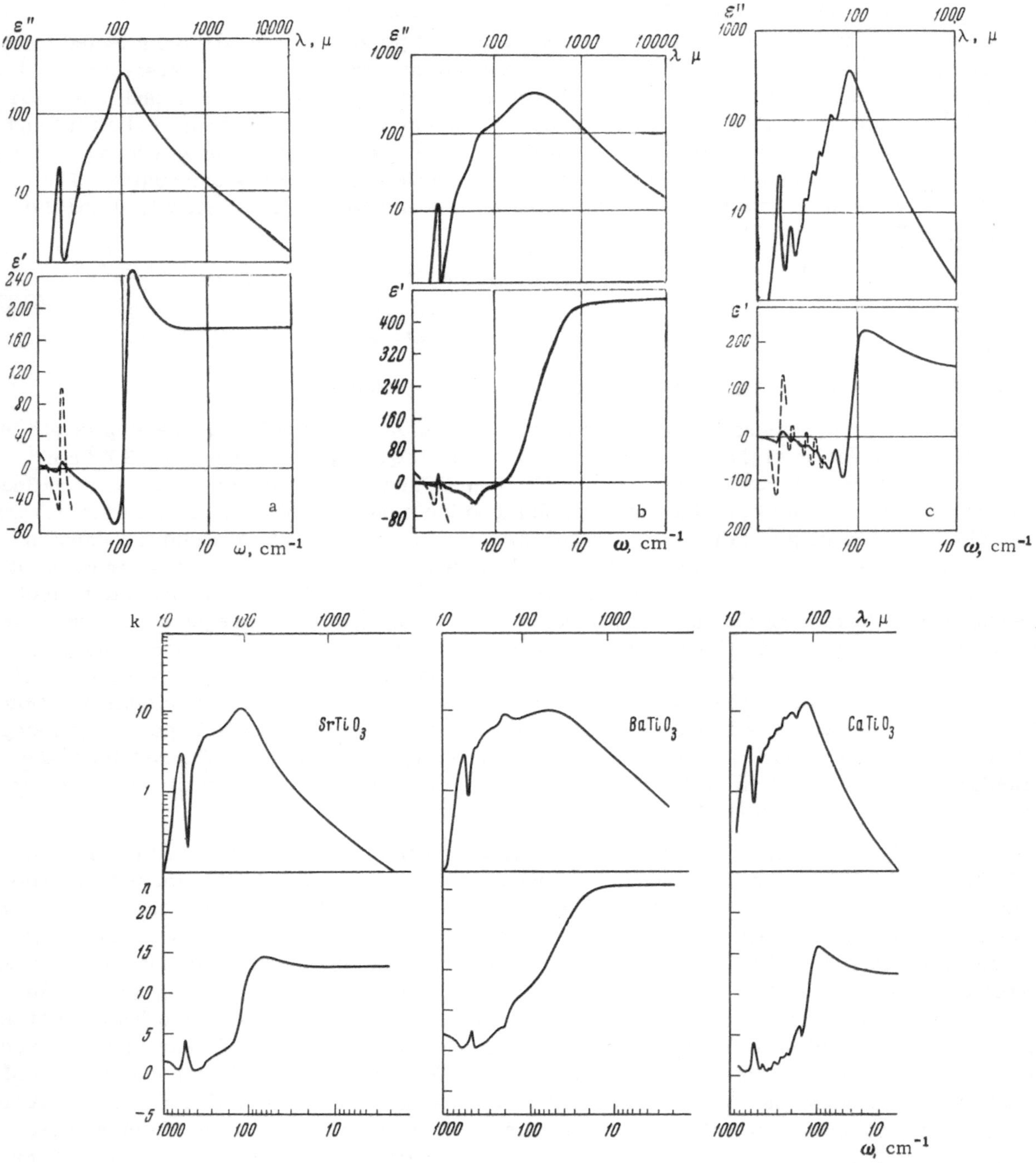

Fig. 12. Frequency and wavelength dependence of the calculated imaginary and real parts, $\varepsilon''(\omega)$ and $\varepsilon'(\omega)$, of the permittivity of (a) $SrTiO_3$, (b) $BaTiO_3$, (c) $CaTiO_3$; below, calculated spectral variation of the optical constants $k(\omega)$ and $n(\omega)$ for $SrTiO_3$, $BaTiO_3$, and $CaTiO_3$ polycrystals.

TABLE 6. Spectral Position of Maxima of Curves of $\varepsilon''(\omega)$ for $SrTiO_3$, $BaTiO_3$, and $CaTiO_3$ Crystals

$SrTiO_3$		$BaTiO_3$		$CaTiO_3$	
cm^{-1}	μ	cm^{-1}	μ	cm^{-1}	μ
550	18.2	500	20	560	17.9
—	—	—	—	440	22.8
270	37.1	300	33.4	320	31.3
—	—	—	—	260	38.6
—	—	165	60.8	220	45.5
95	105	35	290	170	59
—	—	—	—	115	87

(see Chapter III). Since the curves of $\varepsilon'(\omega)$ and $\varepsilon''(\omega)$ and the corresponding transmission spectra for $SrTiO_3$ and $BaTiO_3$ are very similar and show comparatively large differences for those of $CaTiO_3$, we shall first consider the vibrational spectrum of the former two crystals.

First of all, we note that the shortest-wavelength vibration F_{1u} in $SrTiO_3$ and $BaTiO_3$ must be taken to be that which corresponds to the resonance of $\varepsilon(\omega)$ near 18 μ. This vibration is clearly seen in the transmission spectra of the powders, and is of approximately the same type in the reflection spectra of both substances. We shall denote it by ω_3.

It is evident that the deep resonance of the complex permittivity observed at 100-140 μ in both crystals is due to the lowest-frequency vibration F_{1u} (denoted by ω_2). This vibration has important properties. The corresponding absorption band $\varepsilon''(\omega)$ is strong and broad, indicating a high degree of anharmonicity of this vibration. For $BaTiO_3$, the anharmonicity is so great that the corresponding dispersion of the permittivity is very nearly of the relaxation type. Both vibrations ω_2 and ω_3 have also been observed [28, 33] in $SrTiO_3$ and $BaTiO_3$ single crystals.

The last of the F_{1u} vibrations (denoted by ω_4) which should appear in the infrared absorption and reflection spectra of cubic perovskite crystals must apparently be taken to be that near 55 μ. This vibration, as already mentioned, was found in the transmission spectra of powdered $SrTiO_3$ and $BaTiO_3$ (see Fig. 10). In the reflection spectra of polycrystalline samples, $BaTiO_3$ showed it as a slight maximum near 55 μ (Fig. 11). It is important to note that this vibration is evident in the reflection spectra of $SrTiO_3$ and $BaTiO_3$ single crystals [33]. The reason why it was not found in the reflection spectra of polycrystalline $SrTiO_3$ is probably that in polycrystals all the marked variations of reflectivity typical of single crystals are greatly smoothed out, and also that in $SrTiO_3$ this vibration lies very close to the region of the strong low-frequency resonance.

Since the measured spectra $\varepsilon'(\omega)$ and $\varepsilon''(\omega)$ of tetragonal $BaTiO_3$ are similar to those of cubic $SrTiO_3$, it is evident that the expected splitting of the vibrations ω_2, ω_3, and ω_4 of the $BaTiO_3$ lattice in the tetragonal phase is almost absent. A similar conclusion as regards ω_3 in $BaTiO_3$ was reached elsewhere [22].

The rapid increase of the reflectivity to a maximum at about 30 μ in $SrTiO_3$ and especially in $BaTiO_3$, and the corresponding additional structure in the spectra of $\varepsilon'(\omega)$ and $\varepsilon''(\omega)$, and also the deep minimum in the transmission curves of both substances, indicate that the F_{2u} vibration ω_1 appears in this part of the spectrum. This vibration is forbidden by the selection rules in the infrared spectra of the cubic perovskite crystals. However, as has been shown in Chapter III, it should partly appear in the infrared spectra of tetragonal $BaTiO_3$. In $SrTiO_3$, it may also be detectable, since measurements of the Raman scattering spectra of $SrTiO_3$ have shown that the selection rules which forbid the appearance of this vibration in such spectra do not hold good for $SrTiO_3$ [24]. The vibration near 30 μ has also been observed both in the Raman spectra of $SrTiO_3$ and in a slow-neutron scattering study of $SrTiO_3$ and $BaTiO_3$ (see Table 8, below). An important argument against the above hypothesis, however, is that, according to [33], single crystals of $BaTiO_3$ in polarized light show the same reflection spectrum for the ordinary and extrordinary rays. This contradicts the conclusion in Chapter III that this vibration ω_1 must be polarized in exactly the z direction.

To solve the problem of the appearance of the vibration ω_1 in the infrared spectra, and to obtain a general quantitative description of the observed vibrational spectra of $SrTiO_3$ and $BaTiO_3$, we made an attempt to approximate the results by means of a set of damped dispersion-type oscillators:

$$\varepsilon(\omega) = \varepsilon_\infty + \sum_i \frac{4\pi a_i \, \omega_i^2}{\omega_i^2 - \omega^2 + i\gamma_i \, \omega_i \omega}. \qquad \text{(IV.2)}$$

The parameters ω_i, a_i, γ_i of these oscillators, determined from the spectral variation of $\varepsilon'(\omega)$ and $\varepsilon''(\omega)$ and the reflectivity curves $R(\omega)$, are given in Table 7. The spectral variation of $R(\omega)$ from Eq. (IV.2) is shown in Fig. 11, with and without allowance for the vibration ω_1, for which the optimum parameters were chosen: $\omega_1 = 250\ cm^{-1}$, $4\pi a_1 = 2.0$, $\gamma_1 = 0.2$ for $SrTiO_3$, and $\omega_1 = 310\ cm^{-1}$, $4\pi a_1 = 2.8$, $\gamma_1 = 0.4$ for $BaTiO_3$. It is seen that this type of vibration ω_1 does not lead to a satisfactory description of the reflection curve for $SrTiO_3$ or (especially) $BaTiO_3$. In this connection, reference should be made to the paper by Barker [41], who showed that these features of the spectral variation of $\varepsilon'(\omega)$ and $\varepsilon''(\omega)$ near 30 μ can be satisfactorily explained by taking account of the phonon–phonon interaction for all the other vibrations (ω_2, ω_3, and ω_4). Thus, the vibration ω_1 either does not appear at all in the infrared absorption and reflection spectra of $SrTiO_3$-type crystals, or else makes only a small contribution to $\varepsilon(\omega)$.

Table 8 collates the experimental data on the vibrational spectrum of $SrTiO_3$ and $BaTiO_3$ obtained by other authors and in our own studies of the infrared transmission and reflection spectra of polycrystalline $SrTiO_3$ and $BaTiO_3$. The notation used is as defined above. It is seen that the vibrational spectrum of the $SrTiO_3$ and $BaTiO_3$ lattices thus defined is in satisfactory agreement with the results of other authors' infrared studies and those concerning the slow-neutron scattering spectra of $SrTiO_3$ and $BaTiO_3$. The $SrTiO_3$ lattice vibration frequencies as found from the Raman spectra have somewhat higher values.

TABLE 7. Parameters of $SrTiO_3$ and $BaTiO_3$ Dispersion Oscillators

Substance	ω_i, cm^{-1} (μ)	$4\pi a_i$	γ_i
$BaTiO_3$	$\omega_2 = 72$ (139)	455	2.2
	$\omega_3 = 508$ (19.7)	1	0.095
	$\omega_4 = 182$ (55.1)	—	—
$SrTiO_3$	$\omega_2 = 98$ (102)	175	0.73
	$\omega_3 = 558$ (17.9)	1	0.090
	$\omega_4 = 167$ (60)	—	—

Note. The parameters of vibrations ω_4 cannot be determined with sufficient accuracy from the form of the curves of $\varepsilon(\omega)$.

Returning now to Table 7, we must consider in somewhat more detail the behavior of the low-frequency vibration ω_2 in $SrTiO_3$ and $BaTiO_3$, which is now put in quantitative form. It is seen that the vibration ω_2 has unusually high values of the oscillator strength $4\pi a_2$ and the damping coefficient γ_2. In particular, the oscillator strength for this vibration is in each substance some two orders of magnitude greater than for any other vibration. The vibration ω_2 is responsible for about 98% of the total low-frequency polarization caused by lattice vibrations, and is therefore the chief cause of the unusual dielectric properties of such crystals. The parameters are particularly large for $BaTiO_3$. For example, the anharmonicity γ_2 of the vibration ω_2 in $BaTiO_3$ is more than three times the corresponding quantity γ_2 for $SrTiO_3$.

TABLE 8. Comparison of Experimental Results for the Vibrational Spectra of $SrTiO_3$ and $BaTiO_3$, in cm^{-1} (μ)

Spectrum type	Form of substance	Method of measurement	ω_3	ω_1 (?)	ω_4	ω_2	Reference
			$BaTiO_3$				
T	Single crystal	IR	495 (20.3)		—	—	[22]
T	Powder	IR	545 (18.4)	400 (25.0)	—	—	
T	"	IR	578 (17.3)	410 (24.4)	—	—	[23]
R	Single crystal	IR	480 (20.9)	—	185 (54.2)	—	[32]
R	Polycrystal	IR	480 (20.9)	—	185 (54.2)	—	
R	Single crystal	IR	510 (19.6)	—	183 (54.7)	33.8 (296)	[33]
	"	NS	480 (20.9)	340 (29.4)	235 (42.6)	—	[37]
R	Polycrystal	IR	508 (19.7)	310 (32.3)	182 (55.1)	72 (139)	Own measurements
T	Powder	IR	556 (18.0)	400 (25.0)	200 (50.0)	—	
			$SrTiO_3$				
T	Powder	IR	610 (16.4)	395 (25.4)	—	—	[22]
T	"	IR	620 (16.2)	420 (23.8)	—	—	[23]
R	Single crystal	IR	550 (18.2)	—	—	—	[28]
R	"	IR	546 (18.3)	—	178 (56.3)	87.5 (114)	[33]
—	"	RS	620 (16.2)	441 (22.7)	335 (29.9)	90 (111)	[24]
—	"	NS	530 (18.9)	266 (37.6)	170 (59.0)	97 (103)	[36, 37]
R	"	IR	558 (17.9)	250 (40.0)	—	98 (102)	Own measurements
T	Powder	IR	606 (16.5)	408 (24.5)	167 (60.0)	—	

IR = infrared, RS = Raman scattering, NS = neutron scattering, R = reflection, T = transmission.

In discussing the case of $CaTiO_3$, it must first of all be mentioned that its vibrational spectrum again shows clearly the high-frequency vibration ω_3 at about 18 μ and the low-frequency vibration ω_2 at about 90 μ. These are similar in nature to ω_3 and ω_2 in $SrTiO_3$ and $BaTiO_3$. Since $CaTiO_3$ would in principle be expected to show twelve vibrational bands, whereas only seven are in fact observed, it must be admitted that the triplet splitting of the vibrations ω_2, ω_3, and ω_4 which was previously assumed does not actually appear. Possibly, $CaTiO_3$ shows the doublet splitting of the vibrational spectrum which was expected but not found in tetragonal $BaTiO_3$. It is interesting to note that the vibrational spectrum of $CaTiO_3$ shows a clear vibration at 22.7 μ, i.e., precisely in the region of the previously assumed F_{2u} vibration ω_1. Unfortunately, $CaTiO_3$ does not allow a specific interpretation of each of the observed vibrations, since the amounts of the splittings in the spectrum are comparable with the intervals between the vibrations themselves. As with $SrTiO_3$ and $BaTiO_3$, the dispersion parameters of the lowest-frequency vibration of $CaTiO_3$ may be determined: $\omega_2 = 110$ cm^{-1} (91 μ), $4\pi a_2 = 145$, $\gamma_2 = 0.47$.

Finally, let us consider the features of the measured spectral variation of $\varepsilon'(\omega)$ and $\varepsilon''(\omega)$ in $SrTiO_3$, $BaTiO_3$, and $CaTiO_3$ that are due to the polycrystalline state. As already mentioned, the samples were prepared by the ceramic technique and had only a very small vitreous content. Their principal dielectric properties were determined largely by the crystal lattice structure [42, 67]. In the present studies of $SrTiO_3$ and $BaTiO_3$, this was confirmed by the general form of the spectra of $\varepsilon'(\omega)$ and $\varepsilon''(\omega)$, which are very similar to the corresponding curves [33] measured with $SrTiO_3$ and $BaTiO_3$ single crystals. The chief difference between single crystals and polycrystals appears to be the absolute magnitude of the measured reflectivity.

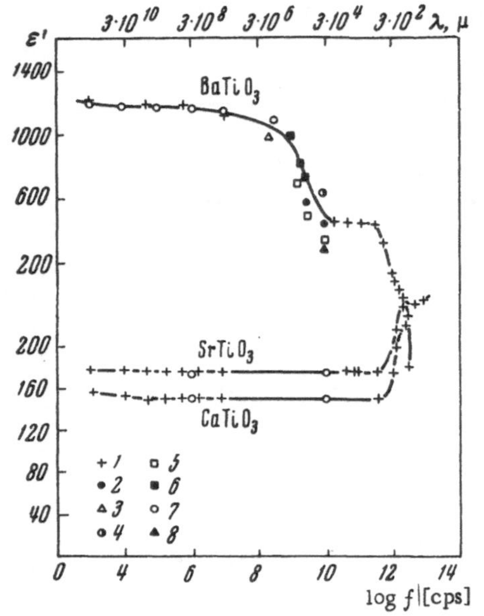

Fig. 13. Dielectric dispersion of $SrTiO_3$, $BaTiO_3$, and $CaTiO_3$ polycrystals over a wide frequency range. 1) Our measurements; 2)-8) results of [2] through [8], respectively.

This difference, however, agrees exactly with the results concerning the permittivity of such samples at radio frequencies. Thus, it may be supposed that there is some distortion of the curves of $\varepsilon'(\omega)$ and $\varepsilon''(\omega)$ when these are calculated from the measured reflection spectra of polycrystalline samples. In particular, we must expect the low-frequency normal mode ω_2 to appear at a longer wavelength than is found experimentally. As a whole, however, it has been demonstrated that infrared studies of polycrystalline $SrTiO_3$, $BaTiO_3$, and $CaTiO_3$ make possible a fairly accurate observation and analysis of the features of the vibrational spectra of these crystals.

3. Dielectric Dispersion of $SrTiO_3$, $BaTiO_3$, and $CaTiO_3$ Polycrystals over a Wide Frequency Range

In order to determine the role of crystal lattice vibrations in the general mechanism by which polarization occurs in crystals of this type, it is very important to know how their permittivity varies over a wide frequency range. Using the results of our measurements in the infrared, submillimeter, and radio ranges, and those of microwave measurements on similar polycrystals [3-9], we can plot the entire spectral variation of the permittivity of polycrystalline $SrTiO_3$, $BaTiO_3$, and $CaTiO_3$ in the range from 10^3 to 10^{14} cps (Fig. 13).

It is seen that, for $SrTiO_3$ and $CaTiO_3$, the permittivity is constant at all frequencies below about 10^{12} cps, the region of the longwave infrared resonance. Accordingly, it may be supposed that the permittivity of these substances throughout the microwave and radio range, and also in the submillimeter range, is due to the resonance vibration of their crystal lattices.

For polycrystalline $BaTiO_3$, there is not only the longwave infrared resonance at $5 \cdot 10^{11}$-10^{14} cps due to lattice vibrations and giving $\varepsilon' \approx 460$, but also a further dispersion in the range 10^9-$5 \cdot 10^{10}$ cps. This causes the additional increase of the permittivity from $\varepsilon' = 460$ at 1000 μ to 1200 at 1000 cps. In accordance with the customary view, this dispersion of the permittivity of $BaTiO_3$ would be due to relaxation of domain boundaries [65, 66]. It is important to note that, according to our measurements of the reflectivity of $BaTiO_3$ in the submillimeter range, the dispersion in question begins when $\lambda > 8$ mm. Thus, we observe a clear demarcation of the two dispersions of $BaTiO_3$ in the frequency spectrum.

Since the results of the measurements have shown that the permittivity of $SrTiO_3$ and $CaTiO_3$ crystals is governed by resonance vibration of their crystal lattices over a wide frequency range, we can estimate the contribution to the dielectric losses at microwave and radio frequencies from the corresponding wing of the infrared resonance absorption. The estimates are based on the expression

$$\tan \delta = \gamma_2 \frac{\omega/\omega_2}{1-(\omega/\omega_2)^2}, \qquad \text{(IV.3)}$$

TABLE 9. Values of Dielectric Loss Tangent, Measured and Calculated from Eq. (IV.3)

Substance	Frequency, cps	tan δ	
		Measured	Calculated
$SrTiO_3$	10^{10}	0.0025	0.0025
	10^6	0.0005	$2.4 \cdot 10^{-7}$
	10^3	0.0045	$2.4 \cdot 10^{-10}$
$BaTiO_3$	10^{10}	0.41 *	0.01
	10^6	0.011	$1 \cdot 10^{-8}$
	10^3	0.01	$1 \cdot 10^{-9}$
$CaTiO_3$	10^{10}	0.002 *	0.0016
	10^6	0.0004	$1.4 \cdot 10^{-7}$
	10^3	0.0023	$1.4 \cdot 10^{-10}$

*Results of [4].

which is derived from Eq. (IV.2). We use the previously determined dispersion parameters of the normal modes of these crystals. Since the corresponding contribution from modes ω_3 and ω_4 is extremely small in comparison with that from ω_2 (see Table 7), we shall take only the latter into account. Table 9 shows the theoretical values of tan δ calculated by this method, together with the experimental values from measurements of tan δ at frequencies $f = 10^3$, 10^6, and 10^{10} cps. It is clear that the dielectric losses in $SrTiO_3$ and $CaTiO_3$ at microwave frequencies ($f = 10^{10}$ cps) are entirely determined by the lattice-vibration mechanism. At low radio frequencies, the dielectric losses are of a different type. The large losses measured in $BaTiO_3$ at 10^3-10^{10} cps appear to be due mainly to domain-boundary relaxation.

4. Temperature Variation of the Vibrational Spectrum of $BaTiO_3$

As has been mentioned in Chapter I, the microscopic theory of ferroelectricity predicts for $BaTiO_3$ an interesting effect whereby the lowest-frequency normal mode of this lattice should vary considerably as the sample temperature approaches the phase-transition point. It was shown that the expected relation is $\omega_t \propto \sqrt{T - T_c}$. The estimate of $\omega_t(\theta)$ at the phase-transition point was $\omega_t(\theta) = 12.6 \cdot 10^{11}$ cps, i.e., $\lambda_t(\theta) = 1.5$ mm [18].

This effect was studied by measuring the reflection spectra, mainly of $BaTiO_3$, at various temperatures in the range 45-140°C.

The results showed that the reflection spectra of $SrTiO_3$ from 2 to 300 μ have no important temperature variation in this range. This is evidently due to the small variation of $\varepsilon'(\omega = 0)$ in strontium titanate at the temperatures concerned. If the variation of $\omega_t(T)$ for $SrTiO_3$ is estimated from Eq. (I.4) in accordance with the behavior of $\varepsilon'(\omega = 0, T)$ at these temperatures [16], it is found to be about 10%. Using the dispersion relation (IV.2) for a harmonic oscillator, we can determine the corresponding variation of the reflectivity near 300 μ, which is found to be less than 2%, i.e., comparable with the error of measurement. For a similar reason, the reflection spectrum of $CaTiO_3$ also should show no variation in the temperature range in question.

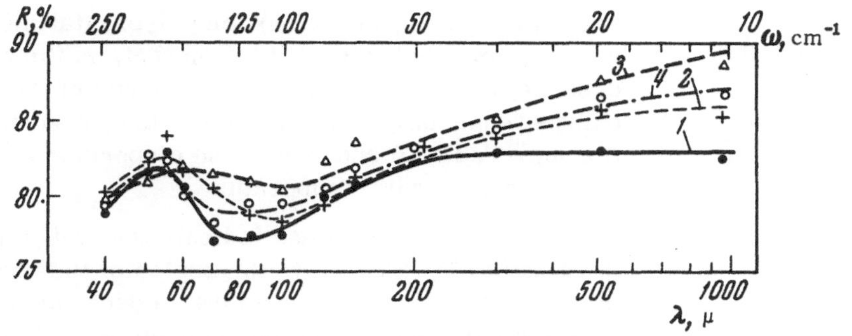

Fig. 14. Reflection spectra of $BaTiO_3$ at various temperatures: 1) 45°C; 2) 80°C; 3) 110°C; 4) 140°C.

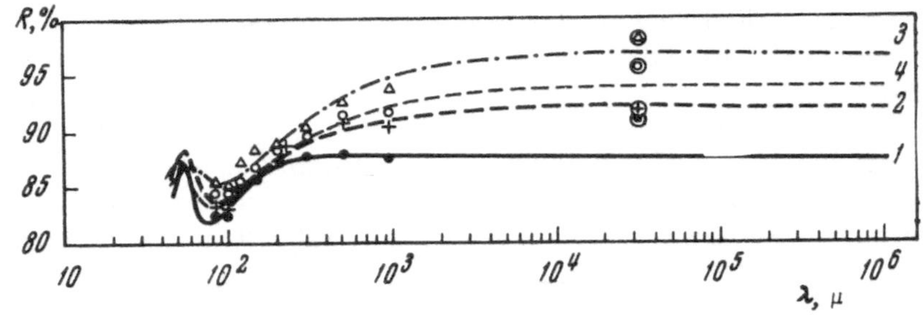

Fig. 15. Base curves of reflectivity R* (ω) of $BaTiO_3$ at various temperatures: 1) 45°C; 2) 80°C; 3) 110°C; 4) 140°C. Encircled points from [4].

Figure 14 shows the variation with temperature of the $BaTiO_3$ reflection spectrum in the far infrared between 45 to 140°C. Since the differences did not exceed 15%, each experimental point R (ω) was determined from a statistical averaging of between four and six measurements. The error of measurement was ±1-2%. It is seen from the diagram that the temperature variations of the reflectivity of $BaTiO_3$ appear mainly at the longwave end of spectrum, and to some extent between 50 and 100 μ. In this range there is a small but detectable change in the shape of the reflection maximum, which becomes flatter at 110-115°C, near the phase-transition point,* and then reappears at 51 μ. In the longwave region (200-1000 μ), as the temperature rises, the reflectivity of $BaTiO_3$ at first increases considerably, then gradually decreases above the Curie point. Figure 14 shows that these changes become larger at longer wavelengths. The observed temperature variation of the $BaTiO_3$ reflection spectrum is generally similar to the corresponding effect seen in single-crystal $SrTiO_3$ at low temperatures [28], but differs in occurring at even longer wavelengths.

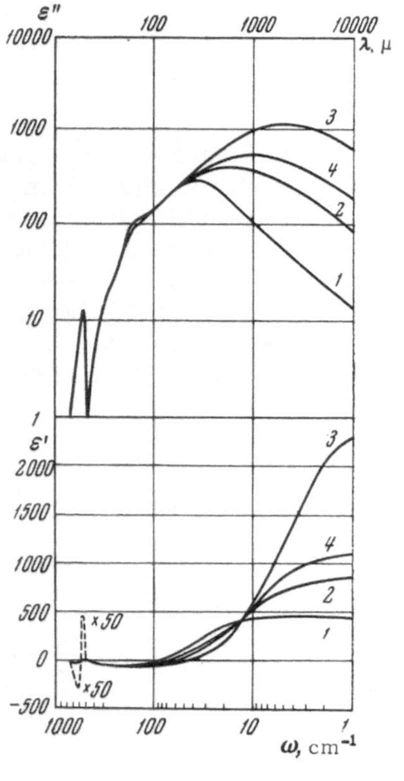

Fig. 16. Calculated spectral variation of ε"(ω) and ε'(ω) for polycrystalline $BaTiO_3$ at 1) 45°C; 2) 80°C; 3) 110°C; 4) 140°C.

These data were mathematically analyzed in the manner described previously. The base curves R* (ω) used in calculating ε (ω) were plotted from the results of our measurements in the near and far infrared and from those obtained with similar polycrystalline $BaTiO_3$ samples at microwave frequencies [4] (Fig. 15). Figure 16 shows the calculated spectral variation of ε'(ω) and ε"(ω) for polycrystalline $BaTiO_3$ at various temperatures. It is seen that, as the temperature increases, there is a considerable change of the form of the curves, especially in the region of the low-frequency vibration ω_2. Table 10 shows how the dispersion parameters of this vibration vary with temperature.

The results obtained indicate that, when the temperature is raised to 110°C, i.e., to the region of the phase transition (θ = 112°C), there is a considerable decrease of the frequency at which ε"(ω) has a maximum, from 35 cm^{-1} (λ = 0.29 mm) at 45°C to 4 cm^{-1} (λ = 2.5 mm) at 110°C. This spectral shift of the ε"(ω) band is accompanied by an increase of its maximum

*According to permittivity measurements at 10^3 cps, the $BaTiO_3$ samples used had a phase transition at θ = 112°C.

TABLE 10. Temperature Variation of Dispersion Parameters of Vibration ω_2 in $BaTiO_3$

T°, C	ω_2		$\omega_{2,\,th}$		$4\pi a_2$	γ_2
	cm^{-1}	μ	cm^{-1}	μ		
45	72	139	72	139	455	2.2
80	65	154	54	185	900	4.2
100	47	214	50	200	1500	5.2
110	33	300	31	320	2500	7.0
140	58	172	57	176	1200	5.0

value and by a considerable broadening. It is seen from Table 10 that the resonance frequency of this vibration drops from 72 to 33 cm^{-1}; the damping coefficient and, therefore, the degree of anharmonicity of the vibration increase by a factor of more than 3. The spectral variation of $\varepsilon'(\omega)$ shows that, as the temperature rises, the resonance frequency of the low-temperature vibration moves to lower frequencies, and $\varepsilon'(\omega)$ accordingly becomes much larger at low frequencies. When the temperature is further raised to 140°C, $\varepsilon'(\omega)$ and $\varepsilon''(\omega)$ show the same kind of changes, but in the opposite direction.

This new experimental observation of an anomalous temperature variation of the low-frequency vibration of the $BaTiO_3$ lattice at the phase-transition point must, of course, be explained on the basis of the microscopic theories of ferroelectricity mentioned previously [17-20]. The low-frequency vibration ω_2 acts as the vibration which in those theories is directly related to the ferroelectric properties of $BaTiO_3$. It is appropriate to mention that this vibration has a number of unusual properties in the spectra of all three crystals $SrTiO_3$, $BaTiO_3$, and $CaTiO_3$.

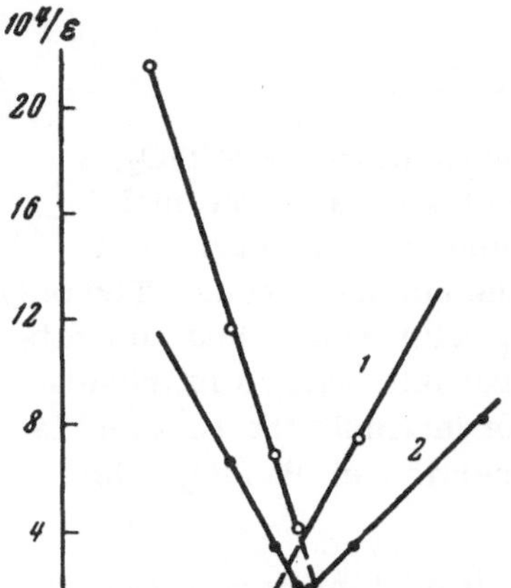

Fig. 17. Temperature dependence of reciprocal of $\varepsilon'^*(0)$. 1) From measurements in the far infrared and extrapolation to $\omega = 0$; 2) from measurements at 10^3 cps.

The experimental results may usefully be compared with theoretical estimates. To do so, we use Eqs. (I.3) and (I.5). It has been shown that, when the static permittivity varies according to the Curie–Weiss law, the expression (I.4) leads to a temperature dependence $\omega_t \propto \sqrt{T - T_c}$ for the resonance frequency of the low-frequency vibration. Accordingly, the usual graph of the reciprocal of $\varepsilon'^*(0)$ against temperature was plotted (Fig. 17). This quantity was determined by extrapolation to low frequencies of the function $\varepsilon'(\omega)$ measured over a wide range of infrared and microwave frequencies. It is clear that the influence of domain boundaries was thus also excluded. The temperature dependence of $\varepsilon'^*(0)$ is seen to be in fairly good agreement with the Curie–Weiss law, with $T_c = 90 \pm 2$°C for $T > \theta$ and $T_c = 125 \pm 2$°C for $T < \theta$. The same diagram shows results from measurement of $\varepsilon'(\omega)$ at radio frequencies. Table 10 gives the corresponding variation of the resonance frequency of the vibration ω_2, calculated from these data and Eq. (I.5). The value of A was estimated from measurements at 45°C, and was assumed constant in the temperature range concerned. The agreement found between the theoretical and experimental variation of $\omega_2(T)$ is to be regarded as entirely satisfactory.

We can also estimate the experimentally determined lowest frequency of the vibrations $\omega_2(\theta)$ at the phase-transition temperature, and compare it with the theoretical value given by Ginzburg's method, Eq. (I.3). To do so, we must know the value of

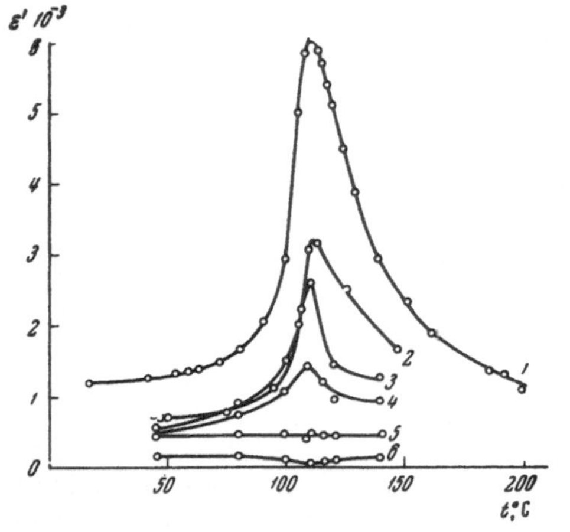

Fig. 18. Temperature variation of $\varepsilon'(\omega)$ measured in various regions of the spectrum. 1) $f = 10^3$ cps; 2) $\lambda = 3.2$ cm [4]; 3) extrapolated to $\omega = 0$; 4) $\lambda = 2.5$ mm; 5) $\lambda = 0.7$ mm; 6) $\lambda = 0.25$ mm.

α'_θ, which for polycrystalline $BaTiO_3$ differs from the value 3.7×10^{-5} valid for the single-crystal form. In accordance with the conclusion just reached, this parameter may be found from the slope of lines 1 in Fig. 17: $\alpha'_\theta = 9.2 \cdot 10^{-5}$ for $T > \theta$ and $\alpha'_\theta = 8 \times 10^{-5}$ for $T < \theta$. If we use the model of the lowest-frequency vibration of $BaTiO_3$ in the form TiO– BaO_2 and the experimental value of about 9×10^{-5}, Eq. (I.3) gives the estimate $\omega_{tz}(\theta) = 3.46 \cdot 10^{12}$ cps, $\lambda_{tz}(\theta) = 0.54$ mm. This form of the normal mode ω_2 is only a rough approximation, based on arguments to be discussed below (Chapter VI). Here it will simply be noted that this estimate of $\omega_2(\theta)$ depends essentially on the assumed relationship between the displacements of the titanium and barium atoms in this normal mode. The form of the displacements of the oxygen atoms has only a slight effect on the numerical value of $\omega_2(\theta)$. The experimental value of $\omega_2(\theta)$ is in satisfactory agreement with the theoretical value, especially if we take into account that the samples studied were polycrystalline.

The above-mentioned temperature variations of the reflection spectrum and of the permittivity of the $BaTiO_3$ polycrystal indicate that the temperature variation of $\varepsilon'(\omega)$ for $BaTiO_3$ falls into several significant sections of the spectrum (Fig. 18). The variations of $\varepsilon'(\omega)$ in this substance when $\lambda > 1$ cm must evidently be regarded as due mainly to rearrangement of domains and movement of domain boundaries. The variations when $\lambda < 1$ cm are somewhat unusual, and are due to the previously mentioned anomalous temperature variation of the low-frequency normal mode of the $BaTiO_3$ crystal.

CHAPTER V

Vibrational Spectra of Some Crystals Having Near-Perovskite Structure

Studies of the transmission and reflection spectra of the perovskite crystals $SrTiO_3$, $BaTiO_3$, and $CaTiO_3$ over a wide frequency range have shown that the observed vibrational spectra of their crystal lattices agree with those predicted by the theory of symmetry. But neither the theory nor the experimental results provide the form of the normal modes. The only exceptions are the F_{2u} (O_h) vibration ω_1 and the corresponding B_1 (C_{4v}) vibration, whose form is given by symmetry alone. The form of the normal modes for the perovskite lattice is, nevertheless, a very important problem. The knowledge of their form is of intrinsic interest and is also useful in connection with estimation of various microscopic properties of $BaTiO_3$, discussed in Chapter VI.

In order to obtain further information as to the form of the vibrational spectra of the perovskite crystals, we have studied here the spectral variation $\varepsilon(\omega)$ for certain other crystals having a perovskite or near-perovskite structure [31], in particular $MgTiO_3$ and Zn_2TiO_4 polycrystals having ilmenite and spinel structures (these, like perovskite, contain oxygen octahedra surrounding titanium atoms [67, 68]), the strontium bismuth titanates SVT-802 and SVT-227 which are based on $SrTiO_3$ and have a perovskite structure,* and the barium tetratitanate TTB, whose structure is more complicated.

*SVT-802 and SVT-227 are solid solutions based on $SrTiO_3$ with 9.6 and 19.6 mole percent bismuth, respectively. Their dielectric properties at radio frequencies have been studied in our laboratory [16].

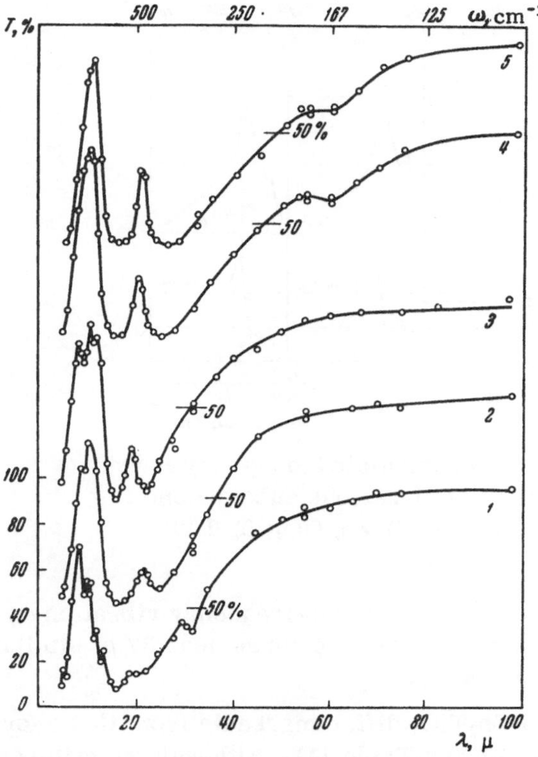

Fig. 19. Wavelength/frequency dependence of transmission of powders of various substances. 1) TTB; 2) Zn_2TiO_4; 3) $MgTiO_3$; 4) SVT-227; 5) SVT-802.

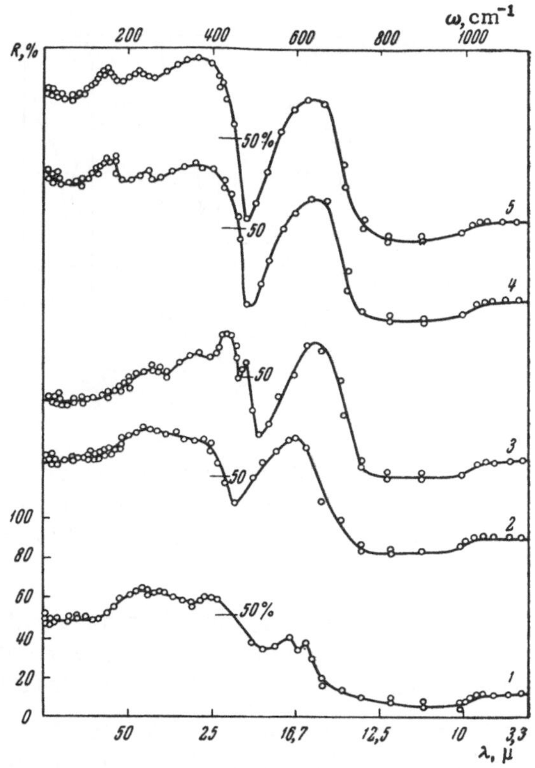

Fig. 20. Wavelength/frequency dependence of measured variation of reflectivity of various substances. 1) TTB; 2) Zn_2TiO_4; 3) $MgTiO_3$; 4) SVT-227; 5) SVT-802.

The transmission spectra of these substances in powdered form, and the corresponding reflection spectra, measured over a wide range in the infrared, are shown in Figs. 19 and 20. The reflection spectra were analyzed as described previously (see Chapter IV) to give the spectral variation $\varepsilon'(\omega)$ and $\varepsilon''(\omega)$ for these crystals (Fig. 21).

The transmission spectra for all the substances except TTB have, in the near infrared, a form very similar to that of the corresponding transmission curves for $SrTiO_3$-type crystals. All have two absorption maxima, at 18 and 25 μ. SVT-802 and SVT-227 also show a slight maximum at 50-60 μ. All the curves give increased transmission at $\lambda > 30\ \mu$ due to scattering of radiation by particles having a high refractive index.

The measured reflection spectra of these crystals, and the functions $\varepsilon'(\omega)$ and $\varepsilon''(\omega)$ derived as already described are, for SVT-802 and SVT-227, very similar to those of $SrTiO_3$. The high-frequency vibration near 18 μ and the strong low-frequency vibration near 77 μ are clearly seen, together with a vibration near 53 μ and a slight kink of $\varepsilon'(\omega)$ and $\varepsilon''(\omega)$ near 37 μ (Table 11). The features of the $\varepsilon'(\omega)$ and $\varepsilon''(\omega)$ curves for these crystals at long wavelengths ($\lambda > 300\ \mu$), and the slight shift of the low-frequency vibration to shorter wavelengths than in $SrTiO_3$ appear to be due to the very marked disturbance of their structure because of the relatively high content of bismuth impurity. Thus, measurements of SVT-802 and SVT-227 crystals have entirely confirmed the earlier picture of the vibrational spectrum of cubic perovskite crystals. The spectrum of normal modes for these crystals comprises the high-frequency

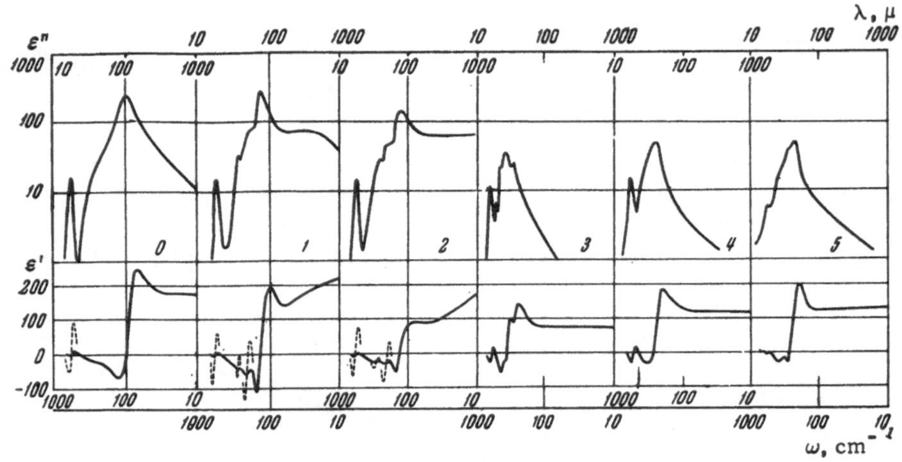

Fig. 21. Wavelength/frequency dependence of calculated imaginary and real parts $\varepsilon''(\omega)$ and $\varepsilon'(\omega)$ of the permittivity of various substances. 0) $SrTiO_3$; 1) SVT-802; 2) SVT-227; 3) $MgTiO_3$; 4) Zn_2TiO_4; 5) TTB.

vibration ω_3 at about 18 μ, the vibration ω_4 at about 55 μ, and the low-frequency vibration ω_2 near 100 μ. The spectra of $\varepsilon'(\omega)$ and $\varepsilon''(\omega)$ were also observed to have features near 37 μ, similar to those found previously at about 30 μ in $SrTiO_3$-type crystals.

The spectra of $\varepsilon'(\omega)$ and $\varepsilon''(\omega)$ for $MgTiO_3$ and Zn_2TiO_4 differ markedly from those for substances having the perovskite structure (see Fig. 21 and Table 11). Although the difference from the corresponding spectra for $SrTiO_3$-type crystals is not very great in the high-frequency range (showing clearly the vibration near 18 μ and one similar to ω_4 of $SrTiO_3$ at about 40 μ), there is a very large difference at lower frequencies. The low-frequency vibration at 80-100 μ typical of perovskite crystals does not appear in $MgTiO_3$ or Zn_2TiO_4.

For the TTB crystal, both the measured transmission spectrum and the spectral variation of $\varepsilon'(\omega)$ and $\varepsilon''(\omega)$ derived from the reflection spectrum are quite different from the corresponding curves for the perovskite crystals and for $MgTiO_3$ and Zn_2TiO_4.

The following conclusions may be drawn from these results on the vibrational spectra of the crystals in question. Firstly, the vibrational spectrum of the perovskite crystals SVT-802 and SVT-227 is, as we should expect, similar in form to that observed for $SrTiO_3$-type crystals. Secondly, the spectral variation of the complex permittivity for $MgTiO_3$ and Zn_2TiO_4 is similar

TABLE 11. Spectral Position of Maxima of Curves of $\varepsilon''(\omega)$ for SVT-802, SVT-227, $MgTiO_3$, Zn_2TiO_4, and TTB Crystals

SVT-802		SVT-227		$MgTiO_3$		Zn_2TiO_4		TTB	
ω, cm^{-1}	λ, μ	ω, cm^{-1}	λ, μ	ω, cm^{-1}	λ, μ	ω, cm^{-1}	λ, μ	ω, cm^{-1}	λ, μ
560	17.9	560	17.9	590	17.0	550	18.2	560	17.9
—	—	—	—	470	21.3	—	—	—	—
270	37.1	260	38.5	440	22.8	330	30.4	280	35.8
—	—	—	—	335	30.0	—	—	—	—
190	52.8	200	50.0	260	38.5	230	43.5	230	43.5
130	77	130	77	—	—	—	—	—	—

to that of $SrTiO_3$ at the high-frequency end, but differs greatly from it in the region of the low-frequency vibration ω_2 of $SrTiO_3$. It is therefore reasonable to suppose that the high-frequency normal modes ω_3 and ω_4 of $SrTiO_3$ are due mainly to internal vibrations of the atoms in the TiO_6 octahedron, while the low-frequency vibration ω_2 is due chiefly to the relative displacement of the titanium and strontium atoms.

Thirdly, the vibration at about 18 μ which is typical of crystal structures containing TiO_6 octahedra does not occur in the titanate TTB having a different structure.

In $MgTiO_3$, Zn_2TiO_4, and TTB crystals, the permittivity measured over a wide range up to $f \approx 5 \cdot 10^{11}$ cps is independent of the frequency, and is governed by the lattice vibrations. Estimates of tan δ similar to those made for $SrTiO_3$-type crystals in Table 9 show, as in $SrTiO_3$ and $CaTiO_3$, that the dielectric losses in the above crystals are determined in the microwave range by the wing of the far-infrared resonance due to the lattice vibrations.

CHAPTER VI

Derivation of the Dispersion Relation and Some Microscopic Properties of $BaTiO_3$

From measurements of the transmission and reflection spectra of substances of the $SrTiO_3$ type over a wide range of the infrared, followed by appropriate mathematical analysis, it has been possible to determine the frequency dependence of $\varepsilon'(\omega)$ and $\varepsilon''(\omega)$ and hence the spectrum of normal modes of perovskite crystals in the cubic, tetragonal, and orthorhombic states. These vibrations have been quantitatively described by a set of dispersion parameters. The results of the measurements and a study of crystals having a near-perovskite structure (see Chapter V) have enabled us to draw some conclusions regarding the form of the normal modes found.

Let us now consider the features found in the vibrational spectra of these perovskite crystals. It has been shown that they appear particularly in anomalous characteristics of the lowest-frequency vibration, which has considerable anharmonicity and a large oscillator strength. The latter is two orders of magnitude greater than its value for any other vibration, and causes about 98% of the total low-frequency polarization resulting from vibrations of the crystal lattice of these substances. Thus, this vibration is chiefly responsible for the unusual dielectric properties of the crystals.

It is evident that the anomalies found in the vibrational spectrum of perovskite crystals are related to features in their structure, and in particular to the presence of strong local electric fields in these crystals. It was shown some years ago that the electric polarization in such crystals is extremely large and complicated in nature [42]. The most important aspect has proved to be the interaction between the electron shells of certain atoms, especially O (3), and the internal electric fields which arise from the displacement of the ionic sublattices. Accordingly, we can suppose that the large oscillator strength of the vibration ω_2 in $SrTiO_3$-type crystals must be complicated in origin and must include a contribution from the electronic polarization of the atoms.

Using the microscopic theory of ferroelectricity evolved by Ginzburg, Cochran, and others [17-21], let us attempt to analyze some of the features of the vibrational spectra of $SrTiO_3$-type crystals. Following a method similar to that of Cochran [21], which takes account of the local electric fields and the deformation of the electron shells of the atoms, we obtain in an explicit form the dispersion expression for the complex permittivity for substances whose crystal structure is of the cubic perovskite type [69]. Comparing this with the dispersion relation (IV.2) used for the approximate representation of the experimental results, we can estimate some of the microscopic characteristics of $BaTiO_3$.

The treatment will consider the following model. Each i-th ion in the $SrTiO_3$-type crystal consists of a core having charge B^i and a peripheral electron having effective charge B'^i, so that the ion charge is $Z^i = B^i + B'^i$. Since, as shown in Chapter III, the normal modes of the cubic perovskite lattice are triply degenerate, we shall solve the one-dimensional problem for one of the degenerate directions. Let u_i and u_i' be the displacements of the core and electron of the i-th ion from their equilibrium positions in this case. According to the derivation of Eq. (III.8), the equations of motion of the ions for long wavelengths (y = 0) are those of the sublattices pertaining to the corresponding atoms in the unit cell. Here, unlike the analysis in Chapter III, we shall take into account the polarization of the electron shells of the ions, and, in accordance with Cochran's theory, we shall regard the electrons as ionic sublattices with almost zero mass. Without repeating the derivation of Eq. (III.8), we shall write the expression for the potential energy of a cubic perovskite lattice as

$$\Phi = {}^1/_2 \sum_{ik}^{n} C_{ik} u_i u_k + \sum_{ik}^{n} C'_{ik} u_i u'_k + {}^1/_2 \sum_{ik}^{n} C''_{ik} u'_i u'_k - \sum_{i} E_i \mu_i, \qquad \text{(VI.1)}$$

where the summation is over the ions and corresponding electrons in one unit cell. The term $\sum_i E_i \mu_i$, has been written separately; it takes account of the long-range Coulomb interaction between the ionic sublattices, of the local electric fields, and of the interaction between the transverse optical vibrations and the field of the external electromagnetic wave. E_i is the local electric field acting on the ionic dipole

$$\mu_i = B^i u_i + B'^i u'_i. \qquad \text{(VI.2)}$$

The remaining terms are the usual short-range interaction of the sublattices, with elasticity coefficients C_{ik}, C'_{ik}, and C''_{ik} of the same type as $C_{\alpha\beta}\left(\begin{smallmatrix} 0 \\ kk' \end{smallmatrix}\right)$ in Eq. (III.9).

In Eq. (VI.1), we expand the Coulomb interaction term by means of an expression [70, 71], which relates the local electric fields to the displacements of the ions and the electronic polarizabilities α_k of the atoms:

$$E_i = E + \sum_{k}^{n} b_{ik} (Z^k u_k + \alpha_k E_k) - u_i \sum_{k}^{n} b_{ik} Z^k = E + \sum_{k}^{n} b^*_{ik} B^k u_k + \sum_{k}^{n} b_{ik} B'^k u'_k. \qquad \text{(VI.3)}$$

Here, b_{ik} are the Lorentz structure factors,

$$b^*_{ik} = b_{ik} - \delta_{ik} \sum_{\alpha}^{n} b_{i\alpha} \frac{Z^\alpha}{B^k},$$

E is the external field strength, and δ_{ik} is the Kronecker symbol. Substituting E_i and μ_i from Eqs. (VI.3) and (VI.2) in Eq. (VI.1), we can write the equations of motion of the ion cores and peripheral electrons in forms similar to Eqs. (I.6) and (I.7):

$$m_i \ddot{u}_i + \sum_{k}^{n} [C_{ik} - 2b^*_{ik} B^i B^k]\, u_k + \sum_{k}^{n} [C'_{ik} - (b_{ik} + b^*_{ik})\, B^i B'^k]\, u'_k - EB^i = 0, \qquad \text{(VI.4)}$$

$$0 + \sum_{k}^{n} [C''_{ik} - 2b_{ik} B'^i B'^k]\, u'_k + \sum_{k}^{n} [C'_{ik} - (b_{ik} + b^*_{ik})\, B'^i B^k]\, u_k - EB'^i = 0. \qquad \text{(VI.5)}$$

In accordance with Cochran's theory, it is assumed that, at the frequencies considered, an electron follows the field without inertia, i.e., $m_i' u_i' = 0$.

The equations (VI.5) are solved for u_k' and the result is substituted in Eq. (VI.4); this leads to an equation of motion of the ion cores alone, with allowance for the local electric fields and the electronic polarization:

$$m_i \ddot{u}_i = \sum_k^n C_{ik}^0 u_k - B^{0i} E = 0. \tag{VI.6}$$

Here, C_{ik}^0 and B^{0i} are the effective elasticities and charges of the ion cores, which are somewhat complicated expressions:

$$C_{ik}^0 = \left\{ (C_{ik} - 2b_{ik}^* B^i B^k) - \frac{1}{\Delta} \sum_{\alpha\gamma} [C_{i\alpha}' - (b_{i\alpha} + b_{i\alpha}^*) B^i B'^{\alpha}] A_{\gamma\alpha} [C_{\gamma k}' - (b_{\gamma k} + b_{\gamma k}^*) B^k B^{\gamma}] \right\}, \tag{VI.7}$$

$$B^{0i} = \left\{ B^i - \frac{1}{\Delta} \sum_{k\gamma} [C_{ik}' - (b_{ik} + b_{ik}^*) B^i B'^k] A_{\gamma k} B'^{\gamma} \right\}, \tag{VI.8}$$

where Δ is a determinant having the form $\|C_{ik}' - 2b_{ik} B'^i B'^k\|$, and A_{ik} are its cofactors. Equations (VI.6) are similar to the corresponding equations (I.8) in Cochran's theory. Unlike [39], for example, they take account of the variation of the electric field when the electron shells of the ions are displaced.

Equations (VI.6) are solved by means of the normal coordinates Q:

$$u_i = \sum_{\gamma}^n a_{\gamma i} Q_{\gamma}. \tag{VI.9}$$

Substitution of this in Eqs. (VI.6) gives

$$m_i \sum_{\gamma}^n a_{\gamma i} \ddot{Q}_{\gamma} + \sum_k^n C_{ik}^0 \sum_{\gamma}^n a_{\gamma k} Q_{\gamma} - B^{0i} E = 0. \tag{VI.10}$$

If we use the orthogonality condition for this coordinate transformation [72]:

$$\sum_{ik}^n m_{ik} a_{ti} a_{\gamma k} = \delta_{t\gamma} \tilde{m}_t,$$
$$\sum_{ik}^n C_{ik}^0 a_{ti} a_{\gamma k} = \delta_{t\gamma} \tilde{C}_t^0, \tag{VI.11}$$

and multiply Eq. (VI.10) by a_{ti}, summing over i, and applying the conditions that $Q_t = Q_i, e^{i\omega t}$ when $E = E_0 e^{i\omega t}$

$$\tilde{C}_t^0 Q_t - \omega^2 \tilde{m}_t Q_t - \sum_i B^{0i} a_{ti} E = 0, \tag{VI.12}$$

where $\tilde{C}_t^0$ and $\tilde{m}_t$ are the elasticity constant and the reduced mass of normal mode t. The solution of this equation, with allowance for damping, is

$$Q_t = Q_{t0}e^{i\omega t} = \frac{(1/\widetilde{m}_t)\sum\limits_i B^{0i}a_{ti}}{\omega_t^2 - \omega^2 + i\gamma_t\omega} E, \tag{VI.13}$$

where $\omega_t = \sqrt{\widetilde{C}_t^0/\widetilde{m}_t}$ is the resonance frequency of this vibration.

To determine the complex permittivity, we use the relation

$$\varepsilon(\omega) = 1 + 4\pi\frac{P(\omega)}{E}, \tag{VI.14}$$

where

$$P(\omega) = \frac{1}{v}\sum_k^n (B^k u_k + B'^k u_k'). \tag{VI.15}$$

Substituting u_k' from Eq. (VI.5) in Eqs. (VI.14) and (VI.15), and expressing u_k in terms of the normal coordinates (VI.9), we obtain

$$P(\omega) = \frac{1}{v}\sum_\gamma \frac{(1/\widetilde{m}_\gamma)\left(\sum\limits_k B^{0k}a_{\gamma k}\right)\left(\sum\limits_i B^{0i}a_{\gamma i}\right)}{\omega_\gamma^2 - \omega^2 + i\gamma_\gamma\omega} E + \frac{1}{v}\sum_{k\gamma}\frac{1}{\Delta}A_{\gamma k}B'^k B'^\gamma E. \tag{VI.16}$$

The second term on the right represents the polarization P_∞ at high optical frequencies, as is easily verified by substituting $\omega \gg \omega_\gamma$. We can now quickly derive the final dispersion expression for the complex permittivity of cubic perovskite crystals:

$$\varepsilon(\omega) = \varepsilon_\infty + \frac{4\pi}{v}\sum_\gamma \frac{(1/\widetilde{m}_\gamma)\left(\sum\limits_k B^{0k}a_{\gamma k}\right)\left(\sum\limits_i B^{0i}a_{\gamma i}\right)}{\omega_\gamma^2 - \omega^2 + i\gamma_\gamma\omega}. \tag{VI.17}$$

For the simpler model of unpolarizable ions without allowance for the local electric fields, the expression (VI.17) becomes exactly the same as the expression for $\varepsilon(\omega)$ derived for this case in [73]. Similarly, it can be used to derive other particular forms for $\varepsilon(\omega)$ in other simplified models.

A comparison of this expression with the dispersion relation (IV.2) used to represent the experimental results shows that the quantity $A_\gamma = \sum\limits_k B^{0k}a_{\gamma k}$, which is the effective charge of normal mode γ, is given in terms of experimental data by

$$A_\gamma = \omega_\gamma\sqrt{a_\gamma v\widetilde{m}_t}. \tag{VI.18}$$

Thus it is possible to estimate some of the microscopic parameters of the cubic perovskite crystals, namely the displacement of the ions, the total and ionic polarizabilities of the unit cell, the local electric fields, and the ionic polarizabilities of the crystal atoms. The equations for these parameters are derived mainly from Eq. (VI.17).

1. The displacements u_i of all the ions are given as functions of the external field by

$$u_i = \sum_\gamma a_{\gamma i}Q_\gamma - \sum_\gamma \frac{(a_{\gamma i}/\widetilde{m}_\gamma)A_\gamma}{\omega_\gamma^2 - \omega^2 + i\gamma_\gamma\omega} E. \tag{VI.19}$$

The contributions from the various normal modes can be estimated.

2. The total polarizability of the unit cell is

$$\alpha^0_{\mathrm{uc}} = \sum_{\gamma} \frac{(1/\tilde{m}_{\gamma}) A^2_{\gamma}}{\omega^2_{\gamma} - \omega^2 + i\gamma_{\gamma}\omega}, \qquad \text{(VI.20)}$$

since

$$\varepsilon = \varepsilon_{\infty} + \frac{4\pi}{v} \alpha^0_{\mathrm{uc}} .$$

and, according to Eq. (VI.18), α^0_{uc} is entirely determined by experimental data. The form of the normal modes and the quantity $\tilde{m}_t$ are not involved in the expression for α^0_{uc}.

3. The ionic component of the total polarizability of the unit cell is

$$\alpha^{\mathrm{ion}}_{\mathrm{uc}} = \sum_{\gamma} \frac{(1/\tilde{m}_{\gamma}) A_{\gamma} \left(\sum_k Z^k a_{\gamma k} \right) \mathrm{k}}{\omega^2_{\gamma} - \omega^2 + i\gamma_{\gamma}\omega}. \qquad \text{(VI.21)}$$

This expression is derived in the same way as Eqs. (VI.19) and (VI.15), taking into account only the displacements of the ions, which is equivalent to regarding the electrons as fixed relative to the nuclei of the ions: $u_k = u'_k$. Denoting the charge on ion i by $k_i Z^i = B^i + B'^i$, where k_i is the charge coefficient, we arrive at $\alpha^{\mathrm{ion}}_{\mathrm{uc}}$ in the form (VI.21). Since the subsequent numerical estimates are very rough, we take for simplicity $k_i = k$, the same for all ions.

4. Solving Eqs. (VI.3) for E_i, we can express the latter in terms of the ion displacements, the electronic polarizabilities, and the external electric field. Then, knowing u_i in Eq. (VI.19), we can easily estimate the local electric fields E_i as functions of the applied field. Since the electron displacements are here replaced by the constant electronic polarizabilities, no account is taken of the change of the electric field when the ions move. Writing Eqs. (VI.3) in the more convenient form

$$\sum_{k}^{n} \left(b_{ik} - \delta_{ik} \frac{1}{\alpha_{ik}} \right) \alpha_k E_k = -E + \sum_{k}^{n} b_{ik} Z^k (u_i - u_k) \qquad \text{(VI.22)}$$

and solving them for E_k, we have as the final expression

$$E_k = \left[-\frac{v}{\alpha_k \Delta} \sum_i A_{ik} \right] E + \sum_i C_{ki} u_i. \qquad \text{(VI.23)}$$

Here, Δ is a determinant having the form $\left\| b_{ik} - \delta_{ik} \frac{1}{\alpha_{ik}} \right\|$; A_{ik} are its cofactors; C_{ki} are coefficients including α_k, A_{ki}, Δ, $b_{i\alpha}$, B^i, and B'^{α}.

5. Knowing the ion displacements and the local electric fields, we can find the ionic polarizability of the individual atoms:

$$\alpha^{\mathrm{ion}}_i = \frac{\mathrm{k} Z^i u_i}{E_i}. \qquad \text{(VI.24)}$$

In going on to make specific estimates, we must note that the foregoing expressions for the various parameters have been determined by using normal coordinates, and accordingly contain their coefficients $a_{\gamma i}$. Hence, in numerical estimation of these quantities, we have to know the form of the normal modes of the crystal lattice in substances of the perovskite type.

As already mentioned, the experimental data are not yet sufficient to resolve this problem. We can only make use of some general indications resulting from a comparison of the vibrational spectra of crystals having different structures. In particular, according to measurements in the near infrared, the vibrational spectra of perovskite, ilmenite, and certain other crystals are found to be similar. On the other hand, the examination of such crystals in a wider range of the infrared also showed substantial differences (see Chapter V). It was found that the low-frequency vibration at 100-150 μ, typical of perovskite crystals does not appear in crystals of the ilmenite and spinel types. These facts led to the conclusion that, in all the crystals which contain identical oxygen octahedra TiO_6, the near infrared shows normal modes due mainly to the internal motion of the titanium and oxygen atoms in the octahedra, but the lowest-frequency vibration of perovskite lattices appears to be due mainly to the relative motion of the atoms of titanium and strontium (barium, etc.). To these experimental facts should be added the very cogent argument concerning the strong bonding of the oxygen atoms in the oxygen octahedron of $BaTiO_3$ and the small deformation of the latter in the low-frequency vibration, suggested in [21] on the basis of x-ray data.

The most justifiable model of the normal modes in cubic $BaTiO_3$ at present appears to us to be that obtained from a theory [39, 40] based on taking account of van der Waals forces and long-range Coulomb forces. The form of the resulting vibrations agrees in principle with the experimental facts stated above. It is interesting that the form of the calculated normal modes was almost unaffected by the use of various methods of allowing for the motion of the electron shell of the atoms and the various values of the charge coefficient k. The only exception was the low-frequency vibration, where the direction of displacement of the O (3) atom changed with increasing k, followed by a transition of the entire crystal to a state unstable with respect to this vibration.

The numerical estimates were accordingly made only for cubic $BaTiO_3$, taking the form of the normal modes deduced in the theoretical study [39]. Since the oxygen atoms are strongly bonded in the oxygen octahedron of a $BaTiO_3$-type crystal, it was assumed that the atom O (3) moves together with O (4) and O (5) in the model of the low-frequency normal mode. The displacement of the titanium ion was determined from the condition for the center of mass of the system to be fixed. Thus, the coefficients a_{ik} of the normal coordinates Q_i for the vibrations ω_2, ω_3, and ω_4, used in the calculations, can be written as

$$u_1^{(2)} = a_2 Q_2, \quad u_2^{(2)} = -0.285 a_2 Q_2, \quad u_3^{(2)} = u_4^{(2)} = u_5^{(2)} = -0.18 a_2 Q_2;$$
$$u_1^{(3)} = 1.835 a_3 Q_3, \quad u_2^{(3)} = -0\ 035 a_3 Q_3, \quad u_3^{(3)} = -7.194 a_3 Q_3; \quad u^{(3)} = u_5^{(3)} = a_3 Q_3;$$
$$u_1^{(4)} = 0.117 a_4 Q_4, \quad u_2^{(4)} = 0.215 a_4 Q_4, \quad u_3^{(4)} = -0.198 a_4 Q_4, \quad u_4^{(4)} = u_5^{(4)} = -a_4 Q_4.$$

The extent to which these estimates depend on the chosen model of the normal modes will be discussed below.

The microscopic characteristics of $BaTiO_3$ were calculated from the values of the dispersion parameters found experimentally for polycrystalline samples in the present work (see Table 7) and for single crystals in [33]. All estimates were made for low frequencies $\omega \ll \omega_\gamma$.

Table 12 shows the calculated values of the atomic displacements u_i and the corresponding contributions from the various normal modes as functions of the external field. According to these results, a field of about 10^6 V/cm is needed in order to move the atoms of $BaTiO_3$, for example, through a distance of about 0.1 Å. Both the displacements of the atoms and the other microscopic properties are determined essentially by the dispersion parameters of the lowest-frequency vibration ω_2.

TABLE 12. Displacement of Atoms in BaTiO3

Atom	Displacements of atoms in normal modes × $10^{15}/E$, cm			Displacements of atoms due to all normal modes × $10^{15}/E$, cm	Type of sample
	ω_2	ω_3	ω_4		
u_1 [Ti]	350	1	—	3.5	Polycrystalline [43]
u_2 [Ba]	−100	−0.02	—	−1.0	
u_3 [O (3)]	−60	−4	—	−0.64	
$u_4 = u_5$ [O (4) and O (5)]	−60	0.6	—	−0.6	
u_1 [Ti]	1500	1	1	15	Single crystals [33]
u_2 [Ba]	−400	−0.02	2.5	−4	
u_3 [O (3)]	−3	−4	−2.3	−3.06	
$u_4 = u_5$ [O (4) and O (5)]	−3	0.6	−10	−3.1	

TABLE 13. Polarizability of $BaTiO_3$ Unit Cell

Type of polarizability	Contribution from the various normal modes × 10^{24}, cm^3			Polarizability from all normal modes × 10^{24}, cm^3	Type of sample
	ω_2	ω_3	ω_4		
α^0_{uc}	2300	5	—	2300	Polycrystalline [43]
α^{ion}_{uc}	800k	4k	—	800k	
α^0_{uc}	10000	5	10	10 000	Single crystal [33]
α^{ion}_{uc}	3500k	4k	20k	3500k	

The values calculated from Eqs. (VI.20) and (VI.21) for the total polarizability α^0_{uc} of the unit cell of $BaTiO_3$, and the corresponding α^{ion}_{uc} due to the ion motion only, are shown in Table 13. From these results, in both polycrystalline and single-crystal $BaTiO_3$, there follows the very interesting result that the electronic contribution to the total polarization of the crystal caused by the lattice vibration is more than 80% when k = 0.5. This confirms the arguments given earlier by Skanavi [42], Slater [74], and others to the effect that the large permittivity of perovskite titanates containing oxygen is due to a large extent to the electronic polarization of their atoms, especially O (3). The above estimate may be compared with the relative contribution of the electronic polarization to the total spontaneous polarization of $BaTiO_3$ as determined by means of x-ray data from the displacement of $BaTiO_3$ ions in the ferroelectric state. With k = 0.5, it is more than 70% [71].

TABLE 14. Local Electric Fields at Ion Sites in $BaTiO_3$

Atom	Polycrystal	Single crystal
E_1	(3+860 k) E	(3+3800 k) E
E_2	(2− 30 k) E	(2− 140 k) E
E_3	(3+700 k) E	(3+3100 k) E
E_4	(2+ 70 k) E	(2+ 310 k) E

According to the estimate of the local electric fields obtained by the above-mentioned method using Eq. (VI.23) (Table 14), these fields may be hundreds of times the applied field. These results also show that the strongest electric fields occur for the titanium and O (3) atoms. Since the electronic polarizability of the Ti atom (α^{el}_1 = $0.186 \cdot 10^{-24}$ cm^3) is small compared with that of the O (3) atom (α^{el}_3 = $2.4 \cdot 10^{-24}$ cm^3) [42], and the ionic polarizabilities

TABLE 15. Ionic Polarizabilities of Atoms in $BaTiO_3$

Type of sample	α_1(Ti) × 10^{24}, cm^3	α_2(Ba) × 10^{24}, cm^3	α_3[O (3)] × 10^{24}, cm^3	α^4[O (4) and O (5)] × 10^{24}, cm^3
Polycrystal	0.8	3	0.1	0.8
Single crystal	0.8	3	0.09	0.9

of these atoms will be shown below to be small, the total polarization of the crystal may in fact be largely determined by the electronic polarization of the O (3) atom. There is qualitative agreement between the above values and other results [71] on the spontaneous polarization of $BaTiO_3$.

Table 15 shows the values of the ionic polarizabilities of the various atoms, calculated from Eq. (VI.24). As already mentioned, the ionic polarizabilities of the Ti and O (3) atoms are in fact very small. It is interesting to note that the calculated value of α_1 (Ti) is in good agreement with $\alpha_{Ti} = 0.95 \cdot 10^{-24}$ cm^3 as estimated by Kittel [75] from more general considerations.

Despite the fact that the foregoing estimates agree quite well with existing theories and experimental results for the polarization in such crystals, it must be emphasized that these are only rough approximations. The reason is that, as already stated, the form chosen for the normal modes of perovskite-type lattices cannot be regarded as very reliable. It is therefore necessary to give further consideration to the way in which the estimates depend on the form of the normal modes of $BaTiO_3$-type crystals, especially that of the low-frequency vibration ω_2. In addition to calculations using the above model of the normal modes, similar estimates of the microscopic parameters of $BaTiO_3$ were made with other proposed forms of the normal modes. The model given in [39] was used, as were models in which the low-frequency vibration has the form Ba$-TiO_3$ or TiO (3)$-$BaO (4) O (5) or Ti$-$O (3), etc. We shall not give here the actual numerical results, but simply mention some conclusions concerning the way in which the estimates depend on the chosen form of the normal modes.

First of all, the relative displacements u_i of the atoms, the distribution of the local electric fields, and the values of the ionic polarizabilities of the atoms in the crystal are found to depend considerably on the form of the normal modes. However, the order of magnitude of the absolute displacements and strengths of the local electric fields for a given applied field E are in general the same in all the models used. Thus, the values for the ion displacements given in the table seem to be typical of $BaTiO_3$-type crystals. Since, as has been shown, the experimentally determined total polarizability of the unit cell is entirely independent of the form of the normal modes, the general quantitative conclusion as to the predominant contribution of the electronic polarization to the total polarization of $BaTiO_3$-type crystals is very significant. For the calculated theoretical model of the normal modes [39], and for the models of the vibrations ω_2 as Ba$-TiO_3$ or TiO(3)$-$BaO(4)O(5) this contribution is $\alpha_{el} / \alpha_{k=1}^0 = 80\%$, whereas, with the model of the vibration ω_2 as Ti$-$O(3), the relative magnitude of the electronic polarizability is about 50%.

The estimate of the ionic polarizability of the atoms is very sensitive to the choice of the form of the normal modes, especially the low-frequency vibration. In particular, α_1 has the extremely high value of about $25 \cdot 10^{-24}$ cm^3 for models of ω_2 as Ba$-TiO_3$ or TiO(3)$-$BaO(4)O(5). We must therefore acknowledge that these models of the vibration ω_2 are too crude. On the other hand, the values of α_1 given by the model Ti$-$O(3) are very small, about $0.24 \cdot 10^{-24}$ cm^3. It is interesting to note that an allowance [39] for a small displacement of the O(3) atom relative to the Ti ion in the low-frequency vibration changes α_1 from about $25 \cdot 10^{-24}$ to about $2 \cdot 10^{-24}$ cm^3.

Thus, these estimates lead to additional information regarding the form of the low-frequency vibration of a perovskite-type lattice, which cannot be taken in the simplified forms Ba–TiO_3 or TiO(3)–BaO(4)O(5). It has a more complicated form, but the vibration must evidently involve some relative displacement of the Ti and O(3) ions. According to the calculations in [39] and our own experimental results (see Chapter V), the vibration also cannot have the form Ti–O(3).

Since the above estimates of the microscopic parameters of $BaTiO_3$-type crystals, based on a previously chosen model of the normal modes, have proved to be entirely reasonable, we may suppose that this model is fairly realistic. The results could certainly be made more accurate by using improved models of the normal modes in these crystals.

Numerical estimates of the microscopic parameters were made for the cubic $BaTiO_3$ crystal, since only in this case was there a known theoretical model of the normal modes of the lattice. However, a comparison of the dispersion parameters of the normal modes for $BaTiO_3$ with those for $SrTiO_3$ and $CaTiO_3$ suggests that the main deductions from these estimates are valid for the latter crystals also. In particular, it is apparently true that here also there is a large electronic contribution to the total polarization.

From the results obtained, we can attempt to elucidate one of the principal features of the vibrational spectra of perovskite crystals, namely the anomalously large oscillator strength of the low-frequency normal modes. It has been shown that the electronic polarization makes a substantial contribution (65-80%) to the total polarization of the $BaTiO_3$ crystal caused by lattice vibrations. This conclusion relates primarily to the polarization caused by the low-frequency vibration ω_2, and indicates the complexity of this vibration. The calculations of Skanavi, Slater, and others [42, 74] for static polarization showed that the moving ions induce strong local electric fields at the positions of the neighboring ions. It is seen from Table 12 that the displacements of the ions at low frequencies $\omega \ll \omega_\gamma$ are determined principally by the parameters of the low-frequency normal mode. In this vibration, therefore, the general pattern of ion displacements is similar to the static one, and the strong local electric fields described above should again appear. It has also been shown above, however, that the ionic polarizabilities of the atoms such as Ti and O(3), which are subject to the strongest internal electric fields, are small in comparison with the electronic polarizabilities. Thus, we arrive at the following interpretation of the lowest-frequency vibration ω_2. The ions deviate relatively slightly from their equilibrium positions (since they make only a small contribution to the total polarization of the crystal), but in so doing they induce strong local electrical fields at the positions of the neighboring ions, and thereby entrain these ions and their electron shells into the motion. Since the ionic polarizabilities of the Ti and O(3) atoms are less than the electronic polarizabilities, the high oscillator strength of this vibration is mainly the result of the induced polarization of the electron shells of the ions. The resonance frequency can depend only on the elasticity of the bonds between ions in the crystal, and therefore lies in the infrared.

The pattern of ion displacements for other normal modes may differ considerably from the static case, and the strong local electric fields may not occur. This appears to be so for the vibrations ω_3 and ω_4, whose measured oscillator strengths are small.

Another feature of the vibrational spectra of the perovskite crystals is the anomalously high anharmonicity of the low-frequency normal mode ω_2. This was included in Eqs. (VI.12) phenomenologically (through the damping constant γ), and can be analyzed only from the form of the experimental curves $\varepsilon''(\omega)$. According to the measured values of γ (Table 7), the high anharmonicity of ω_2 occurs in all three crystals. It is larger in $SrTiO_3$ than in $CaTiO_3$, and becomes very large in $BaTiO_3$. The anharmonicity of the vibration, due to a certain loosening of the lattice, may be regarded as in some way characterizing the possible occurrence of the ferroelectric state in crystals having the perovskite structure. This supposition is confirmed by our studies of the temperature variation of the vibrational spectrum of $BaTiO_3$. The anharmonicity of the vibration ω_2 and the corresponding loosening of the $BaTiO_3$ lattice increased as the temperature approached the phase-transition point, reaching a maximum at the transition to the ferroelectric state.

On comparing the above estimates of microscopic parameters obtained from studies of $BaTiO_3$ polycrystals and single crystals, it is seen that they usually differ in absolute magnitude. For example, the ion displacements u_i, the total and ionic polarizabilities of the unit cell, and the local electric fields E_i for single crystals are 4.5 times the corresponding estimates for the same substances in the polycrystalline state, but they represent equally well the basic features of the polarization of such crystals in the field of an electric wave. In particular, the conclusion that the electronic polarization makes a relatively large contribution to the total polarization of the crystals is valid in both single-crystal and polycrystalline $BaTiO_3$. The state of the crystal also has no effect on the estimates of the ionic polarizabilities of the individual ions.

Conclusions

There has recently been a considerable increase of interest in the dynamics of crystal lattices. This can be accounted for both by new advances in the theory and by the development of methods for the study of the vibrational spectra of crystals. As well as the widely used methods for measuring the infrared absorption and reflection spectra of substances over a fairly wide range of frequencies, a very promising method is to use observations of slow-neutron scattering in order to investigate the vibrational spectra of crystals. It seems that the main attention is being given to crystals whose dielectric or other properties are unusual. There is even greater interest in the rearrangements of the crystal structure and corresponding changes in the vibrational spectra, resulting from various phase transitions.

The present work was concerned with a study of general problems of the structure and features of the vibrational spectra of a very interesting group of crystals having the perovskite structure in various states. Considerable attention was given to the most typical changes in the vibrational spectra of such crystals in the phase transition from the cubic to the tetragonal state.

The principal results obtained here are as follows.

1. From group theory and a calculation of the characters of the reducible representations, a general account is given of the number and symmetry types of normal modes of perovskite-type crystals in the cubic, tetragonal, and orthorhombic states. It is shown that the vibrational spectrum of the cubic perovskite crystals must comprise four triply degenerate vibrations $3F_{1u} + 1F_{2u}$, the F_{2u} being forbidden in the infrared spectra. In tetragonal perovskite crystals, the spectrum of normal modes must consist of eight vibrations $3A_1 + 1B_1 + 4E$, the B_1 being inactive in the infrared spectra. In the orthorhombic crystals, the degeneracy is entirely removed, and the spectrum consists of all the twelve normal modes, $4A_1 + 4B_1 + 4B_2$.

By constructing symmetry coordinates, we analyzed the effect of the change from the cubic to the tetragonal and orthorhombic states on the infrared spectrum of $BaTiO_3$-type crystals. In particular, it was shown that the spectrum of the tetragonal crystal should exhibit a doublet splitting of each of the F_{1u} vibrations together with one component of the split F_{2u}. The polarization of the normal modes was determined.

2. From measurements of the transmission and reflection spectra of polycrystalline $SrTiO_3$, $BaTiO_3$, and $CaTiO_3$ in the infrared (1-1000 μ) and submillimeter (2.68, 4, and 8 mm)

regions of the spectrum, and at radio frequencies, the spectral variation of the complex permittivity of these substances is determined over a wide frequency range. The numerical integration was done by using a computer.

3. It is shown, in accordance with the experimental results, that the vibrational spectrum of $SrTiO_3$ and $BaTiO_3$ consists of three vibrations at about 18, 55, and 100-150 μ. Thus, there is no sign in practice of the doublet splitting of the three F_{1u} modes and the appearance of the fourth normal mode E that are expected in tetragonal $BaTiO_3$. This effect does seem to occur in $CaTiO_3$, which showed seven resonances in the range 18-100 μ.

4. The measured values at frequencies f from 10^3 to 10^{14} cps show that the permittivity of $SrTiO_3$ and $CaTiO_3$ is constant over a wide range from about 10^3 to about 10^{12} cps, i.e., the region of the longwave infrared resonance, and is therefore determined entirely by the resonance mechanism of lattice vibrations.

In $BaTiO_3$, there is a further dispersion, which is shown by measurements of the reflectivity of $BaTiO_3$ in the submillimeter range to lie at $\lambda > 8$ mm, and to be due, apparently, to the relaxation of domain boundaries in polycrystalline $BaTiO_3$.

5. The vibrational spectra of these crystals exhibit the following interesting features in the lowest-frequency normal mode. The oscillator strength is extremely large, and the anharmonicity is high. These properties are especially marked in $BaTiO_3$. It is shown that this vibration is responsible for about 98% of the low-frequency polarization due to lattice vibrations in these crystals, and is therefore the principal cause of their unusual dielectric properties.

6. A method similar to Cochran's gives an explicit dispersion expression for the complex permittivity of cubic perovskite crystals. Local electric fields and the electronic polarization of the ions were taken into account. A comparison of the resulting expression with the experimental data for a particular model of the normal modes was used to obtain estimates of several microscopic parameters of $BaTiO_3$: the ion displacements, the total and ionic polarizability of the unit cell, and the local electric fields, as functions of the applied field and the ionic polarizabilities of the atoms in the crystal. It is shown that the main contribution to these parameters comes from the low-frequency vibration of the $BaTiO_3$ lattice. The estimates made lead to the very interesting result that the electronic polarization constitutes the major part (65-80%) of the total polarization of the crystals due to lattice vibrations.

The nature of the longest-wavelength normal mode of $BaTiO_3$ is elucidated, and one of its chief features, the large oscillator strength, is explained.

7. The estimates of the various microscopic parameters of $BaTiO_3$, and a separate investigation of vibrational spectra of other substances having the perovskite or a similar structure, are used to derive some conclusions as to the form of the normal modes in perovskite crystals. In particular, it is suggested that the high-frequency vibrations are due, roughly speaking, to the motion of the titanium and oxygen atoms within the TiO_6 octahedra. The low-frequency vibration appears to consist mainly of a relative displacement of the barium and titanium atoms, together with a displacement of the oxygen atom O(3) relative to the titanium atom.

8. The study of the vibrational spectrum of $BaTiO_3$ revealed for the first time the anomalous temperature variation of the lowest-frequency vibration in $BaTiO_3$. It is shown experimentally that, as the temperature approaches the phase-transition point, there is a considerable decrease of the frequency of the low-frequency vibration of $BaTiO_3$, accompanied by an increase of the oscillator strength and the anharmonicity. The results are in satisfactory agreement with the predictions of the current microscopic theory of ferroelectricity.

This work was undertaken at the suggestion of the late Professor G. I. Skanavi, and was carried out in the Semiconductor Physics Laboratory, P. N. Lebedev Physics Institute, Academy of Sciences of the USSR.

The author has to thank S. V. Bogdanov for continued interest in the work. He is much indebted to A. I. Demeshina for many discussions and valuable advice and assistance. He also thanks A. I. Balashova for preparing the samples and V. V. Buzdin for help with the measurements.

Appendix

Determination of the Symmetry Coefficients from the Known Transformation Matrices

The method of determining the symmetry coefficients $a_{i\gamma}^{\lambda}$ is derived below for the particular case of threefold degeneracy, but it is applicable also for symmetry coordinates with any degree of degeneracy [63]. Since the treatment will be given for any one group (labeled λ) of equivalent coordinates, we shall for simplicity omit the index λ. If the transformation matrices are specified for a particular symmetry type i of an irreducible representation, this means that all the coefficients α_{ml}^{i} in the transformation formula for the symmetry coordinates S_{mi} under the given symmetry operation are known:

$$\begin{aligned} S'_{1i} &= \alpha_{11}^{i} S_{1i} + \alpha_{12}^{i} S_{2i} + \alpha_{13}^{i} S_{3i}, \\ S'_{2i} &= \alpha_{21}^{i} S_{1i} + \alpha_{22}^{i} S_{2i} + \alpha_{23}^{i} S_{3i}, \\ S'_{3i} &= \alpha_{31}^{i} S_{1i} + \alpha_{32}^{i} S_{2i} + \alpha_{33}^{i} S_{3i}. \end{aligned} \tag{1}$$

Here, S_{mi} is the m-th symmetry coordinate in the i-th triply degenerate symmetry type. Each of the three coordinates S_{mi} can be represented as a linear combination of natural coordinates:

$$\begin{aligned} S_{1i} &= \sum_{\gamma} a_{1i\gamma} q_{\gamma}, \\ S_{2i} &= \sum_{\gamma} a_{2i\gamma} q_{\gamma}, \\ S_{3i} &= \sum_{\gamma} a_{3i\gamma} q_{\gamma} \end{aligned} \tag{2}$$

and conversely

$$q_{\gamma} = \sum_{im} a_{mi\gamma} S_{mi}, \tag{3}$$

where m = 1, 2, 3.

The method is as follows. We apply any symmetry operation R to S_{mi} and q_{γ}:

$$\begin{aligned} RS_{mi} &= \alpha_{m1}^{i} S_{1i} + \alpha_{m2}^{i} S_{2i} + \alpha_{m3}^{i} S_{3i}, \\ Rq_{\gamma} &= R_{\gamma j} q_j . \end{aligned} \tag{4}$$

The coefficient R shows that the effect of the symmetry operation R is to convert a natural coordinate q into q with sign $R_{\gamma j}$ (= + 1 or − 1). Then,

$$\begin{aligned} Rq_{\gamma} &= R_{\gamma j} q_j = R_{\gamma j} \sum_{im} a_{mij} S_{mi}, \\ Rq_{\gamma} &= R \sum_{im} a_{mi\gamma} S_{mi} = \sum_{im} a_{mji} RS_{mi} \end{aligned} \tag{5}$$

Substituting RS_{mi} from Eq. (4) and equating the right-hand sides of Eqs. (5), we obtain

$$R_{\gamma j}\sum_{im} a_{mij}S_{mi} = \sum_{im} a_{mi\gamma}(\alpha^{i}_{m1}S_{1i} + \alpha^{i}_{m2}S_{2i} + \alpha^{i}_{m3}S_{3i}), \quad m = 1, 2, 3. \tag{6}$$

Next, equating the coefficients of each coordinates S_{1i}, S_{2i}, and S_{3i}, we arrive at the final expressions for the symmetry coefficients $a_{mi\gamma}$:

$$\begin{aligned} R_{\gamma i}a_{1ij} &= \sum_{m}^{3} a_{mi\gamma}\alpha^{i}_{m1}, \\ R_{\gamma i}a_{2ij} &= \sum_{m}^{3} a_{mi\gamma}\alpha^{i}_{m2}, \\ R_{\gamma i}a_{3ij} &= \sum_{m}^{3} a_{mi\gamma}\, a^{i}_{m3}\,. \\ m &= 1, 2, 3 \end{aligned} \tag{7}$$

Thus, by selecting from a given set of equivalent coordinates any one q_γ and applying to the coordinates q_j the symmetry operations which convert them into q_γ with the sign $R_{\gamma j}$, we can use Eqs. (7) to express all the $a_{1i\gamma}$, $a_{2i\gamma}$, and $a_{3i\gamma}$ in terms of the three initial parameters, i.e., all the coefficients are found to within a constant factor which can be determined from the normalization conditions.

Literature Cited

1. Vul, B. M., Elektrichestvo, 13:12 (1946).
2. Bogdanov, S. V., Abstract of Thesis [in Russian], Lebedev Institute, Moscow (1950).
3. Arkhangel'skii, G. E., Zakhvatkin, G. V., and Skanavi, G. I., Prib. Tekh. Eksp., No. 3, 76 (1956).
4. Skanavi, G. I. and Lipaeva, G. A., Zh. Eksp. Teor. Fiz., 30:824 (1956); Fiz. Tverd. Tela, 2:506 (1960).
5. Yatsenko, A. F., Abstract of Thesis [in Russian], Rostov-on-Don State University (1953).
6. Petrov, V. M., Fiz. Tverd. Tela, 2:997 (1960).
7. von Hippel, A., Rev. Mod. Phys. 22:221 (1950).
8. Powles, J. G., and Jackson, W., Proc. IEE, 96 (III):383 (1949).
9. Fousek, J., Czech. J. Phys., 9:172 (1959).
10. Bond. W. L., Mason, W. P., and McSkimin, H. F., Phys. Rev., 82:442 (1951)
11. Benedict, T. S. and Durand, J. L., Phys. Rev. 109:1091 (1958).
12. Barrett, J. H., Phys. Rev., 86:118 (1952).
13. Gränicher, H., Helv. Phys. Acta, 29:210 (1956).
14. Weaver, H. E., J. Phys. Chem. Solids, 11:274 (1959).
15. Rupprecht, G., Bell, R. O., and Silverman, B. D., Phys. Rev., 123:97 (1961).
16. Kashtanova, A. M., Kurtseva, N. N., and Skanavi, G. I., Izv. AN USSR, ser. fiz., 24:114 (1960).
17. Ginzburg, V. L. Usp. Fiz. Nauk 38:490 (1949).
18. Ginzburg, V. L., Fiz. Tverd. Tela, 2:2031 (1960).
19. Anderson, P. W., Physics of Dielectrics [in Russian] (Proc. 2nd All-Union Conf. on Physics of Dielectrics), Izd. AN USSR, Moscow/Leningrad (1960), p. 290.
20. Cochran, W., Phys. Rev. Letters, 3:412 (1959).
21. Cochran, W., Adv. Phys., 9:387 (1960); 10:401 (1961).
22. Last, J. T., Phys. Rev., 105:1740 (1957).

23. Yatsenko, A. F., Zh. Tekh. Fiz., 27:2422 (1957); Izv. AN SSSR, ser. fiz., 22:1456 (1958); Physics of Dielectries [in Russian] (Proc. 2nd All-Union Conf. on Physics of Dielectries), Izd. AN SSSR, Moscow/Leningrad (1960), p. 314.
24. Narayanan, P. S. and Vedam, K., Z. Physik, 163:158 (1961).
25. Bobovich, Ya. S. and Bursian, E. V., Opt. Spektrosk., 11:131 (1961).
26. Ikegami, S., J. Phys. Soc. Japan, 19:46 (1964).
27. Perry, C. H., and Hall, D. B., Phys. Rev. Letters, 15:700 (1965).
28. Barker, A. S. Jr., and Tinkham, M., Phys. Rev., 125:1527 (1962).
29. Demeshina, A. I. and Murzin, V. N., Fiz. Tverd. Tela, 4:2980 (1962).
30. Murzin, V . N. and Demeshina, A. I., Fiz. Tverd. Tela, 5:2359 (1963).
31. Murzin, V. N., Bogdanov, S. V., and Demeshina, A. I., Fiz. Tverd. Tela, 6:3585 (1964).
32. Ikegami, S., Ueda, I., Kisaka, S., Mitsuishi, A., and Yoshinaga, H., J. Phys. Soc. Japan, 17:1210 (1962).
33. Spitzer. W. G., Miller, R. C., Kleinman, D. A., and Howarth, L. E., Phys. Rev., 126:1710 (1962).
34. Miller, R. C., Spitzer, W. G., and Kleinman, D. A., Bull. Am. Phys. Soc., 7:280 (1962).
35. Ballantyne, J. M., Phys. Rev., 136:A429 (1964).
36. Cowley, R. A., Phys., Rev. Letters, 9:159 (1962).
37. Lefkowitz, I., Proc. Phys. Soc. (London), 80:868 (1962).
38. Silverman, B. D., and Koster, G. F., Z. Physik, 165:334 (1961).
39. Dvořák, V. and Janovec, V., Czech. J. Phys., B12:461 (1962).
40. Janovec, V. and Dvořák, V., Czech. J. Phys., B13:905 (1963).
41. Barker, A. S. Jr., and Hopfield, J. J., Bull. Am. Phys. Soc., 9:215 (1964).
42. Skanavi, G. I., Physics of Dielectrics [in Russian], GITTL, Moscow (1949).
43. Murzin, V. N. and Demeshina, A. I., Fiz. Tverd. Tela, 6:182 (1964).
44. Murzin, V. N. and Demeshina, A. I., Izv. AN SSSR, ser. fiz., 28:695 (1964).
45. Murzin, V. N. and Demeshina, A. I., Bull. Am. Phys. Soc., 9:215 (1964).
46. Martin, D. H., Contemp. Phys., 4:139 (1962); 4:187 (1963).
47. Yaroslavskii, N. G., Usp. Fiz. Nauk, 62:159 (1957).
48. Murzin, V. N. and Demeshina, A. I., Opt. spektrosk., 13:826 (1962).
49. Fastie, W. G., J. Opt. Soc. Am., 42:641 , 647 (1952); 43:1174 (1953).
50. Markov, M. N., Dokl. Akad. Nauk SSSR, 108:428 (1956).
51. Markov, M. N., Zh. Tekh. Fiz., 24:1867 (1954).
52. Malyshev, V. I., Markov, M. N., and Shubin, A. A., Izv. AN SSSR, ser. fiz., 17:654 (1963).
53. Czerny, M. and Roder, H., Ergeb. Exakt. Naturw., 17:70 (1930).
54. McCubbin, T. K., Jr., and Sinton, W. M., J. Opt. Soc. Am., 40:537 (1950).
55. Plyler, E. K. and Blaine, L. R., J. Res. Natl. Bur. Std. (U.S.), 64C:55 (1960).
56. Czerny, M., Z. Physik, 65:600 (1930).
57. Rubens, H., Sitzber. kgl. preuss. Akad. Wiss. Berlin, 4 (1915).
58. McCubbin, T. K., Jr., and Sinton, W. M., J. Opt. Soc. Am., 42:113 (1952).
59. Genzel, L., Happ, H. and Weber, R., Z. Physik, 154:13 (1959).
60. Richards, P. L., J. Opt. Soc. Am., 54:1474 (1964).
61. Born, M. and Kun Huang, Dynamical Theory of Crystal Lattices, Clarendon Press, Oxford (1954).
62. Bhagavantam, S. and Venkatarayudu, T., Theory of Groups and Its Application to Physical Problems, 2nd edition, Andhra University, Waltair (1951).
63. Murzin, V. N., Lebedev Institute Report, Moscow (1963).
64. Shifrin, K., Scattering of Light in a Turbid Medium [in Russian], Gostekhteoretizdat, Moscow/Leningrad (1951).
65. Kittel, C., Phys. Rev., 83:458 (1951).
66. Sannikov, D. G., Zh. Eksp. Teor. Fiz., 41:133 (1961).

67. Skanavi, G. I., Dielectric Polarization and Losses in Glasses and Ceramics with High Permittivity [in Russian], Gosénergoizdat, Moscow/Leningrad (1952).
68. Hermann, C., Lohrmann, O., and Philipp, H., (eds.), Strukturbericht, Vol. 2, (1928-1932), Akademische Verlagsgesellschaft, Leipzig (1937), p. 484.
69. Murzin, V. N., Bogdanov, S. V., and Demeshina, A. I., Fiz. Tverd. Tela, 6:3372 (1964).
70. Kozlovskii, V. Kh., Zh. Eksp. Teor. Fiz., 20:1388 (1951).
71. Venevtsev, Yu. N., Zhdanov, G. S., Solov'ev, S. P., and Zubov, Yu. A., Kristallografiya, 3:473 (1958).
72. Vol'kenshtein, M. V., El'yashevich, M. A., and Stepanov, B. I., Molecular Vibrations [in Russian], part I, Gostekhteoretizdat, Moscow (1949), p. 191.
73. Kleinman, D. A. and Spitzer, W. G., Phys. Rev., 125:16 (1962).
74. Slater, J. S., Phys. Rev., 78:748 (1950).
75. Kittel, C., Introduction to Solid State Physics, 2nd edition, Wiley, New York (1956).